Springer Proceedings in Energy

More information about this series at http://www.springer.com/series/13370

Sachin Kumar · Samir Kumar Khanal
Y.K. Yadav
Editors

Proceedings of the First International Conference on Recent Advances in Bioenergy Research

Springer

Editors
Sachin Kumar
Sardar Swaran Singh National Institute of Bio-Energy
Kapurthala, Punjab
India

Y.K. Yadav
Sardar Swaran Singh National Institute of Bio-Energy
Kapurthala, Punjab
India

Samir Kumar Khanal
Department of Molecular Biosciences
University of Hawai'i at Mānoa (UHM)
Honolulu, HI
USA

ISSN 2352-2534 ISSN 2352-2542 (electronic)
Springer Proceedings in Energy
ISBN 978-81-322-2771-7 ISBN 978-81-322-2773-1 (eBook)
DOI 10.1007/978-81-322-2773-1

Library of Congress Control Number: 2016935576

Printed on acid-free paper

This Springer imprint is published by Springer Nature
The registered company is Springer (India) Pvt. Ltd.

Preface

The increasing prices of crude oil, the depleting nonrenewable energy resources, and the growing concern for climate change have led to a significant interest in renewable energy, including bioenergy. There is a need to shift the current energy spectrum to green energy by creating advanced technologies to utilize renewable resources (such as agro-forestry biomass) for bioenergy production to keep pace with the growing demand for energy. The research in bioenergy is receiving unprecedented attention from all over the world due to the potential of biofuels as a replacement for fossil fuels.

The First International Conference on "Recent Advances in Bio-energy Research" (ICRABR-2015) organized by the Sardar Swaran Singh National Institute of Bio-Energy as an attempt to bring together researchers to form a meaningful conversation on directions of bioenergy. This proceedings volume comprises the top 26 papers contributed by authors who participated and presented their findings at ICRABR 2015. These chapters are peer-reviewed by experts in the field.

We hope that the contents of this volume will help researchers, students, and policy makers involved in this research field.

Kapurthala, India
March 2015

Sachin Kumar
Samir Kumar Khanal
Y.K. Yadav

Acknowledgements

First, we would like to thank all the participants of ICRABR-2015, who contributed their findings to prepare these proceedings in a comprehensive nature. Apart from the authors, we would also like to acknowledge the reviewers for their constructive and valuable comments and suggestions to improve the chapters that significantly improved the quality of the proceedings:

Prof. Pranab Goswami, IIT Guwahati, India
Prof. Toshinori Kojima, Seikei University, Japan
Prof. B.S. Pathak, SPRERI Vallabh Vidyanagar, India
Dr. S. Dasappa, IISc Bangalore, India
Dr. D.K. Adhikari, CSIR-IIP Dehradun, India
Prof. B. Viswanathan, IIT Madras, India
Prof. M.M. Ghangrekar, IIT Kharagpur, India
Dr. Venkat Mohan, IICT Hyderabad, India
Dr. Dilip Ranade, ARI Pune, India
Prof. R.K. Behl, HAU Hisar, India
Prof. A.K. Jain, CUP Bathinda, India
Prof. I.M. Mishra, ISM Dhanbad, India
Dr. Rupam Kataki, Tezpur University, India
Prof. K. Mohanty, IIT Guwahati, India
Dr. Urmila Phutela, PAU Ludhiana, India
Prof. Rajesh Sani, South Dakota School of Mines and Technology, USA
Dr. D.K. Sahoo, CSIR-IMTech Chandigarh, India
Dr. Anuj K. Chandel, CTC, Brazil
Prof. Anil Kumar, Devi Ahilya University Indore, India
Dr. Pratibha Dheeran, CPUT, South Africa
Dr. R.C. Mathur, DBT-IOC Center Faridabad, India
Dr. P.V. Bhale, S.V. National Institute of Technology, Surat, India
Dr. Biplab K. Debnath, NIT Meghalaya, India
Prof. D.C. Deka, Gauhati University, India

Dr. V. Sivasubramanian, PERC Chennai, India
Dr. D.K. Bora, AEC Guwahati, India
Dr. Neeraj Atrey, CSIR-IIP Dehradun, India
Dr. Nasreen Munshi, Nirma University Ahmadabad, India
Dr. S.K. Tyagi, SSS-NIBE, Kapurthala, India
Dr. Sasikumar Elumalai, CIAB Mohali, India
Dr. T. Bhaskar, CSIR-IIP Dehradun, India
Dr. S.K. Soni, PAU Ludhiana, India
Dr. A.K. Sarma, SSS-NIBE, Kapurthala, India
Dr. V.P. Sethi, PAU Ludhiana, India
Dr. Surender Singh, IARI, New Delhi, India
Dr. Neha Mittal, University of North Carolina at Charlotte, USA
Dr. Madhulika Shukla, SSS-NIBE, Kapurthala, India

Kapurthala, India
October 2015

Sachin Kumar
Samir Kumar Khanal
Y.K. Yadav

Organizing Committees

International Advisory Committee

Dr. Robert Dekker, Lakehead University, Canada
Prof. Mohammed Farid, University of Auckland, New Zealand
Prof. S.C. Bhattacharya, International Energy Initiative, India
Prof. B.S. Pathak, SPRERI, India
Prof. Rajesh Sani, South Dakota School of Mines and Technology, USA
Prof. Toshinori Kojima, Seikei University, Tokyo, Japan
Prof. Rakesh Bajpai, University of Louisiana, USA
Prof. S.K. Sharma, PU Chandigarh, India
Dr. Samir K. Khanal, University of Hawaii, USA
Prof. Mohammad J. Taherzadeh, University of Boras, Sweden
Dr. Giuliano Grassi, European Biomass Industry Association
Prof. Giuliano C. Premier, University of South Wales, UK
Prof. I.M. Mishra, IIT Roorkee, India
Dr. Badal C. Saha, US Department of Agriculture, USA
Dr. Lalini Reddy, CPUT, South Africa

National Advisory Committee

Prof. Subhash Chand, IIT Delhi
Prof. S. Dasappa, IISc Bangalore
Prof. A.K. Jain, CUP, Bathinda
Prof. U.C. Banerjee, NIPER, Mohali
Prof. V. Sivasubramanian, Director-Tech, PERC, Chennai
Dr. D.K. Adhikari, IIP, Dehradun
Dr. Dilip Ranade, ARI, Pune
Prof. D. Das, IIT Kharagpur
Prof. Pranab Goswami, IIT Guwahati
Prof. V.K. Vijay, IIT Delhi

Prof. M.M. Ghangrekar, IIT Kharagpur
Prof. M.K. Jha, NIT Jalandhar
Dr. Ashok Pandey, NIIST, Trivandrum
Dr. D.K. Sahoo, IMTech, Chandigarh
Dr. R.K. Behl, HAU, Hisar
Dr. S.S. Sooch, Director, SES, PAU, Ludhiana
Dr. Kaustubha Mohanty, IIT Guwahati

Contents

Part I Biochemical Conversion

Comparison Between Separate Hydrolysis and Fermentation and Simultaneous Saccharification and Fermentation Using Dilute Acid Pretreated Lignocellulosic Biomass 3
Madhuri Narra, Jisha P. James and Velmurugan Balasubramanian

Liquid Hot Water Pretreatment of Paddy Straw for Enhanced Biomethanation 15
Abhinav Trivedi, Virendra Kumar Vijay and Ram Chandra

Improving Yeast Strains for Pentose Hexose Co-fermentation: Successes and Hurdles 23
Shalley Sharma, Sonia Sharma, Surender Singh, Lata and Anju Arora

Pretreatment of Paddy Straw to Improve Biogas Yield 43
Rupali Mahajan, Harmanjot Kaur, Raman Rao and Sachin Kumar

Bioethanol Production from Fermented Cashew Apple Juice by Solar Concentrator 63
Y.P. Khandetod, A.G. Mohod and H.Y. Shrirame

Potential Role of Xylose Transporters in Industrial Yeast for Bioethanol Production: A Perspective Review 81
Nilesh Kumar Sharma, Shuvashish Behera, Richa Arora and Sachin Kumar

Part II Chemical Conversion

Transesterification of Neem Oil Using Na Ion-Doped Waste Fishbone 97
Himadri Sahu and Kaustubha Mohanty

Assessment of the Seed Oils of *Persea americana* and *Melia dubua* for Their Potentialities in the Production of Biodiesel and Possible Industrial Use 111
Kariyappa S. Katagi, Ravindra S. Munnolli, Sangeeta D. Benni and Sneha S. Kulkarni

Investigation of CI Engine Fueled with Ethanol Nano Additives Blended Diesel 121
V. Karthickeyan, P. Balamurugan and R. Senthil

Designer Biodiesel: An Optimization of Fuel Quality by Blending Multiple Oils 131
Anuchaya Devi, Renuka Barman and Dhanapati Deka

Biodiesel Production from *Moringa oleifera* Oil and Its Characteristics as Fuel in a Diesel Engine 149
Mohd. Zeeshan, Mohit Vasudeva and A.K. Sarma

Growth and Lipid Production from *Scenedesmus* sp. Under Mixotrophic Condition for Bioenergy Application 159
Monika Prakash Rai and Shivani Gupta

Experimental Evaluation of Combustion Parameters of a DI Diesel Engine Operating with Biodiesel Blend at Varying Injection Timings 169
Abhishek Sharma and S. Murugan

Performance and Emission Characteristics of CI Engine Fueled with Alternative Fuel with Special Reference to Modification for Combustion: A Literature Review 179
Himansh Kumar, Y.K. Yadav and A.K. Sarma

Part III Thermochemical Conversion

Upgradation of Bio-oil Derived from Lignocellulose Biomass—A Numerical Approach 197
Anjani R.K. Gollakota, Malladi D. Subramanyam, Nanda Kishore and Sai Gu

Sustainability Assessment of Biomass Gasification Based Distributed Power Generation in India 213
Amit Kumar Singh Parihar, Virendra Sethi and Rangan Banerjee

Overview of Biogas Reforming Technologies for Hydrogen Production: Advantages and Challenges 227
Priyanshu Verma and Sujoy Kumar Samanta

Modifications in Improved Cookstove for Efficient Design 245
Amit Ranjan Verma, Rajendra Prasad, Virendra Kumar Vijay and Ratnesh Tiwari

Part IV Biomass Management

Thermochemical Characterization of Pine Needles as a Potential Source of Energy 257
Anil Kumar Varma and Prasenjit Mondal

Biomass Supply Chain Management: Perspectives and Challenges 267
Yogender Singh Yadav and Y.K. Yadav

Part V Electrochemical Processes

Enhanced Power Generation in Microbial Fuel Cell Using MnO_2-Catalyzed Cathode Treating Fish Market Wastewater 285
Md.T. Noori, M.M. Ghangrekar, A. Mitra and C.K. Mukherjee

Bioelectricity Generation from Marine Algae *Chaetoceros* Using Microbial Fuel Cell 295
P.P. Rajesh and M.M. Ghangrekar

Selective Enrichment of Electrochemically Active Bacteria in Microbial Fuel Cell By Pre-treatment of Mixed Anaerobic Sludge to Be Used as Inoculum 305
B.R. Tiwari and M.M. Ghangrekar

Performance Analysis of Separators in Dual-Chambered Microbial Fuel Cell and Treatment of Combined Industrial Effluent of South Gujarat 319
Purvi Zaveri, Tanvi Modi, Lipi Parekh, Aditi Patel, Sameer Kureshi and Nasreen Munshi

Part VI Integrated Systems

Role of Rhamnolipid: A Biosurfactant in Methane Gas Hydrate Formation Kinetics 333
Amit Arora, Swaranjit Singh Cameotra, Rajnish Kumar, Anil Kumar Singh, Pushpendra Kumar, Chandrajit Balomajumder and Sukumar Laik

Economic Analysis of Solar PV/Wind/Diesel Generator/Battery Connected Integrated Renewable Energy Systems for Residential Applications 345
Suresh Muthusamy and Meenakumari Ramachandran

Author Index 357

About the Editors

Dr. Sachin Kumar is Deputy Director in the Biochemical Conversion Division at the Sardar Swaran Singh National Institute of Bio-Energy, Kapurthala, India. Dr. Sachin Kumar obtained his Ph.D. in Chemical Engineering from Indian Institute of Technology Roorkee, India in 2011. He has more than 11 years of research experience in biochemical conversion of biomass to biofuels. His areas of research interest include biofuels including bioethanol, biogas, biobutanol, and biohydrogen, algal biomass, bioprocess engineering, enzyme technology, and metabolic engineering.

Dr. Samir Kumar Khanal is Associate Professor of Biological Engineering at the University of Hawai'i at Mānoa and Collaborating Professor at Iowa State University. Dr. Khanal obtained a Ph.D. in Civil Engineering with focus in Environmental Biotechnology from the Hong Kong University of Science and Technology (HKUST), Hong Kong in 2002. Dr. Khanal is an internationally leading researcher in the bioenergy field and has over ten active research projects in the field of bioenergy/bio-based products and environmental biotechnology.

Prof. Dr. Y.K. Yadav received his Ph.D. in Energy Studies from Indian Institute of Technology, Delhi and at present he is the Director General, Sardar Swaran Singh National Institute of Bio-Energy (Formerly Sardar Swaran Singh National Institute of Renewable Energy), Kapurthala, India. Professor Yadav has been actively engaged in research, teaching, consultancy, administration, and management in the field of renewable energy for more than two decades. He developed and taught renewable energy courses to undergraduate and postgraduate students at CCS Haryana Agricultural University, Hisar, India and established a Renewable Energy Laboratory there. He also contributed in the establishment the Department of Energy at Tezpur University, Assam, India.

About the Conference

The First International Conference on "Recent Advances in Bio-energy Research" ICRABR-2015 was organized by Sardar Swaran Singh National Institute of Bio-Energy, Kapurthala, India during March 14–17, 2015. The conference was inaugurated by Shri Upendra Tripathy, Secretary, MNRE. Dr. B.S. Dhillon, Vice-Chancellor, Punjab Agricultural University, Ludhiana, Dr. N.P. Singh, Adviser, MNRE, Prof. B.S. Pathak, Former Director, SPRERI, Vallabh Vidhyanagar, and Shri Balour Singh, Director, PEDA, Chandigarh, and Prof. Y.K. Yadav, Director General, SSS-NIBE were also present during the inaugural function. The prominent personalities among delegates from overseas were Dr. Lalini Reddy, Cape Peninsula University of Technology, South Africa, Dr. Hanna Tashyreva, Zabolotny Institute of Microbiology and Virology of National Academy of Sciences of Ukraine, and Dr. Baba Shehu Ibn Abubakar, University of Maiduguri, Nigeria.

Eminent speakers including Prof. B.S. Pathak, Dr. Lalini Reddy, Dr. Abubakar, Dr. Hanna Tashyreva, Prof. A.K. Jain, Dr. D.K. Adhikari, Mr. Vasudeo Joshi, Dr. A.S. Mathur, Dr. R.C. Ray, Dr. D.K. Sahoo, Prof. I.M. Mishra, and Mr. Parikshit Dhingra delivered the plenary speeches. Two plenary speeches were conducted through video conferencing by Prof. Lee R. Lynd, Thayer School of Engineering, Dartmouth College, USA and Prof. Ram B. Gupta, School of Engineering, Virginia Commonwealth University, USA.

Fifteen technical sessions were conducted in three parallel slots each day with different themes such as Biomass & Energy Management; Thermo-chemical Conversion; Biochemical Conversion; Chemical Conversion; Electrochemical Processes; and Integrated/Waste to Energy. About ten invited speakers delivered their research findings and review papers. More than 30 participants presented their work through oral presentation, whereas 38 participants presented their work through posters. About 218 abstracts were published in the souvenir under the different sections including plenary speakers, invited speakers, biochemical conversion, chemical conversion, biomass and energy management, thermochemical conversion, electrochemical processes, waste to energy, and integrated systems.

On the last day of event, one session was conducted for the formation of a "Bio-Energy Alliance." During the session, the representatives of different organizations including academic and research institutions, universities, industry, and international groups were present and agreed on fully supporting the Bio-Energy Alliance.

Part I
Biochemical Conversion

Comparison Between Separate Hydrolysis and Fermentation and Simultaneous Saccharification and Fermentation Using Dilute Acid Pretreated Lignocellulosic Biomass

Madhuri Narra, Jisha P. James and Velmurugan Balasubramanian

Abstract In the present study, two different processes, separate hydrolysis and fermentation (SHF), and simultaneous saccharification and fermentation (SSF) were compared. Three different lignocellulosic biomass viz. rice straw (RS), wheat straw (WS), and sugarcane bagasse (SB) were pretreated with dilute acid at two different concentrations (2 and 4 % H_2SO_4 w/v) and at two different time intervals, i.e., 30 and 60 min. RS, WS, and SB with 4 % H_2SO_4 at 121 °C for 30 min yielded maximum reducing sugars (110, 90, and 95 g l^{-1}). Delignification of the solid residues were carried out with 0.5 % NaOH, at 121 °C for 30 min. In-house cellulase produced by *Aspergillus terreus* was used for separate hydrolysis studies at 10 % solid loading and 9 FPU g^{-1} substrate enzyme loading for 0–48 h at 42 °C. Maximum yield of reducing sugars from RS, WS, and SB were 266, 242, and 254 mg g^{-1} substrate, respectively. Acid and enzymatic hydrolysates from RS, WS, and SB produced 5.1, 4.9, 5.2 g l^{-1}, and 14.0, 13.9, 12.9 g l^{-1} of ethanol with *Pichia stipitis* and *Saccharomyces cerevisiae* in 24 and 36 h, respectively. Whereas SSF at 10 % solid loading and 9 FPU g^{-1} substrate enzyme loading for different time intervals 0–72 h at 42 °C was carried out using in-house thermotolerant yeast strain *Kluyveromyces* sp. RS, WS, and SB yielded maximum ethanol of 23.23, 18.29, and 17.91 g l^{-1}, respectively. Ethanol yield was enhanced by addition of Tween 80 1 % (v/v) by 8.39, 9.26, and 8.14 % in RS, WS and SB, respectively.

Keywords Lignocellulosic biomass · Delignification · Thermotolerant yeast strain · SHF · SSF · Ethanol

---

Madhuri Narra (✉) · J.P. James · Velmurugan Balasubramanian
Sardar Patel Renewable Energy Research Institute, P. Box No. 2, Vallabh Vidyanagar 388120, Gujarat, India
e-mail: madhuri68@gmail.com

S. Kumar et al. (eds.), *Proceedings of the First International Conference on Recent Advances in Bioenergy Research*, Springer Proceedings in Energy,
DOI 10.1007/978-81-322-2773-1_1

1 Introduction

Bioethanol can be produced from any lignocellulosic biomass such as RS, WS, and SB as they are readily available renewable resources of carbohydrates for biological conversion to fuels and chemicals (Borbala et al. 2013). One of the most abundant lignocellulosic biomass in the world is RS. About 731 million tones of RS is produced annually which are distributed in Africa (20.9 million tones), Asia (667.6 million tones), Europe (3.9 million tones), America (37.2 million tones), and Oceania (1.7 million tones). Around 205 billion liters bioethanol per year can be potentially produced from this quantity of RS, which is the largest amount from a single biomass (Faveri et al. 2004). Sugarcane industries also generate huge amount of bagasse annually and some of this residue is currently used for energy cogeneration in sugar mills while the surplus being stockpiled.

Lignocellulosic biomass primarily consists of cellulose, hemicellulose, and lignin and its composition varies with different feed materials used. They are very complex materials; hence a single pretreatment method cannot be applied to all lignocellulosic biomass. Various pretreatment methods have been developed including chemical, physical, physico-chemical, and biological for lignocellulosic biomass and commonly used methods are steam, dilute acid, alkaline, and oxidative pretreatment methods. At current scenario, an up-to-date-technology like dilute acid pretreatment is used for pretreating any lignocellulosic biomass. In acid hydrolysis, removal of hemicellulosic content with small fraction of lignin takes place and the remaining part of lignin remains fixed to the cellulosic content (Kaya et al. 2000). Delignification of lignocellulosic solid biomass is essential to achieve maximum cellulosics hydrolysis due to the greater affinity of cellulase components, β-glucosidase, and endoglucanase towards lignin than to the carbohydrates, resulting in lower saccharification efficiency during enzymatic hydrolysis. Agricultural residues such as RS and herbaceous crops are very effectively pretreated using alkali agents. (Chen et al. 2007). Advantages of alkali pretreatment over other pretreatment technologies include lower temperature, pressure, and time requirement. Sodium hydroxide (alkali) has been widely studied for many years to increase the ease of access of cellulases towards cellulose and hemicelluloses by disrupting the lignin structure of the lignocellulosic biomass.

During enzymatic hydrolysis, addition of cellulases to pretreated material containing holocellulosic (cellulose and hemicelluloses) material converts into monomeric sugars and subsequent addition of yeast ferment these sugars to ethanol. SHF is a two step process where enzymatic hydrolysis is followed by fermentation. However, in SSF using single reactor both the process steps can be carried out simultaneously. SSF is more advantageous compared to SHF as it reduces the cost of reactors by performing saccharification and fermentation in single vessel and better ethanol yields by reducing the product inhibition exerted by saccharification products (Scordia et al. 2013).

The need of the hour is to develop a suitable technology for bioethanol production as a partial replacement of gasoline from lignocellulosic biomass as most of these materials are found in surplus and burnt in the open fields, thereby creating environmental pollution. In order to have an economically viable ethanol plant, the primary focus should be on low cost pretreatments, novel enzymes with higher activities, innovations on fermentation technologies for complete sugar utilization, bioreactors design as well as strains for SSF to provide higher ethanol productivity. Hence, it is essential to develop an indigenous technology with low capital cost and operational expenditure.

In this study, two different processes, SHF and SSF were compared with respect to production of ethanol from dilute acid pretreated lignocellulosic biomass. Both SHF and SSF were carried out at 10 % solid loading and 9 FPU g^{-1} substrate enzyme loading at 42 °C by the in-house cellulase produced by *Aspergillus terreus* for different time intervals 0–48 h and using thermotolerant in-house yeast strain *Kluyveromyces* sp. for different time intervals 0–72 h, respectively. Comparative performance of both the processes was reported in this paper.

2 Materials and Methods

2.1 Lignocellulosic Biomass, Media and Chemicals

Raw materials such as RS and WS were collected from local farmers, Anand, Gujarat, India. SB was procured from a local sugar factory. After collection, the raw materials were exposed to physical pretreatment, i.e., passing through 5 mm mesh in hammer mill prior to chemical pretreatment (Finex, India). The physically pretreated lignocellulosic biomass were washed thoroughly with tap water, air dried, and stored at room temperature in air tight containers. All the chemicals, reagents, and media of analytical grade were purchased from local vendors.

2.2 Acid Pretreatment

One fifty grams of pre-sized RS, WS, and SB were mixed with 750, 1500, and 3000 ml of 2 and 4 % H_2SO_4. Substrate to acid ratio (w/v) maintained were 1:5, 1:10, and 1:20, respectively. Pretreatment of RS, WS, and SB were conducted at temperature 121 °C for time period 30 and 60 min as described earlier (Narra et al. 2015). Double layered muslin cloth was used to filter the acid hydrolysate. The acid hydrolysate was detoxified with calcium hydroxide as described by Kuhad (2010) and analyzed for sugars, phenolics, and furans. The leftover solid residues after acid pretreatment were thoroughly washed with continuous flow of water till neutral pH and dried under sunlight. The dried acid pretreated solid residues were further subjected to delignification using 0.5 % NaOH at 121 °C for 30 min of 1:20 substrate to alkali ratio. The cellulosic material separated from lignin portion were filtered using double layered muslin cloth and washed thoroughly with tap water till neutral pH and sun dried. These dried solid residues were further used for SHF and SSF studies or stored at 4 °C in air tight bags.

2.3 Microorganisms and Culture Conditions

Standard cultures for SHF were viz. *Saccharomyces cerevisiae* 3570 and *Pichia stipitis* NCIM 3499 were procured from National Chemical Laboratory (NCL), Pune, India whereas, newly isolated thermotolerant yeast strain isolated from fruit waste was used for SSF. Partial sequencing of purified strain was carried out at National Collection for Industrial Microorganisms (NCIM), Pune, India. The hexose fermenting yeast strains were maintained on agar slants containing (g^{-1} l): glucose, 30.0; yeast extract, 3.0; peptone, 5.0; agar, 20.0 at pH 6.0 ± 0.2, and *Pichia stipitis* was maintained on agar slants containing (g^{-1} l) xylose, 20.0; yeast extract, 4.0; peptone, 5.0; KH_2PO_4, 1.5; $MgSO_4$, $7H_2O$, 0.5; agar, 20.0 at pH 5.0 ± 0.2 and temperature 30 °C, respectively. The cultures were stored at 4 °C.

2.4 Yeast Inoculum Preparation

The *Saccharomyces cerevisiae* inoculum was grown for 12 h at 30 ± 2 °C in a culture medium containing (g^{-1} l); glucose, 30.0; yeast extract, 3.0; peptone, 5.0; $(NH_4)_2HPO_4$, 0.25 at pH 6.0 ± 0.2 (Chen et al. 2007). *Kluyveromyces* inoculum was prepared at 42 ± 2 °C for 12 h. After incubation, the flask contents were aseptically collected, centrifuged, and used for SHF and SSF studies. An optical density of 0.6–0.8 at 620 nm was used for cell culture. Inoculum of *Pichia stipitis* was prepared as described by Nigam (2001) using (g^{-1} l); xylose, 50.0; yeast extract, 3.0; malt extract, 3.0; peptone, 5.0 at pH ± 0.2, and temperature 30 °C. A 12 h seed culture of *Pichia stipitis* with 1 % v/v were inoculated in separate flasks.

2.5 SHF of Cellulosic Residue

2.5.1 Cellulase Preparation

Crude cellulase used for separate hydrolysis and SSF was indigenously produced by *Aspergillus terreus* under solid state fermentation as described earlier (Narra et al. 2012). RS was used as a substrate for cellulase production and the crude enzyme contained FP activity, β-glucosidase, and endoglucanase, of 0.98 ± 0.13, 5.2 ± 0.30, and 14.2 ± 0.40 U ml^{-1}, respectively.

2.5.2 Enzymatic Hydrolysis of Delignified Cellulosic Residue

Enzymatic hydrolysis was carried out with cellulosic solid residues of RS, WS, and SB at 42 °C in 50 ml capacity oak ridge wide mouth bottles at 16 rpm for 4–40 h with an enzyme load of 9 FPU g^{-1} substrate. Total volume of the system was 20 ml (0.05 M citrate buffer, pH 4.8). Other conditions were kept constant as described earlier (Narra et al. 2012). At regular intervals, the supernatant samples were analyzed for total reducing sugars by DNSA method (Miller 1959) after centrifugation at 10,000 g for 15 min. Saccharification efficiency was calculated as mentioned previously (Narra et al. 2012).

2.5.3 Ethanol Fermentation

The enzymatic and acid hydrolysate were fermented with *Saccharomyces cerevisiae* and *Pichia stipitis,* respectively, in a 100 ml stoppered flask at 30 ± 2 °C for 40 h. Nutrients containing NH_4Cl, 0.5, KH_2PO_4, 0.15, yeast extract, 3.0 were added to the acid hydrolyse (20 ml) containing (20.0 g l^{-1}) sugars and the pH was adjusted to 5.5 ± 0.2. While the cellulosic hydrolysate having 40.0 g l^{-1} sugar was supplemented with yeast extract 3.0 g l^{-1} and $(NH_4)_2HPO_4$, 0.25 g l^{-1}. The flasks were inoculated with 10 % (v/v) inoculum and incubated at 30 °C for 60 h at 150 rpm.

2.6 SSF

SSF was performed with cellulosic solid residues of RS, WS, and SB in 50 ml capacity oak ridge wide mouth bottles at 42 °C. The total volume of the system maintained was 20 ml (0.05 M citrate buffer, pH 4.8). Crude cellulases were used to hydrolyze the cellulosic substrates at 10 % solid loading and 9 FPU g^{-1} substrate enzyme loads. After 6 h of hydrolysis at 42 °C, an in-house yeast strain *Kluyveromyces* sp. was added under sterile conditions for better conversion of cellulosic material. The experimental flasks placed in a rotating assembly and were rotated at 16 rpm for 60 h as described earlier (Narra et al. 2015).

2.7 Analytical Methods

Endoglucanase activity was assayed using 2 % carboxymethyl cellulose (CMC, Sigma Chemical Co.) in 0.05 M sodium acetate buffer, pH 4.8 as substrate. The release of reducing sugars in 20 min at 50 °C was determined by DNSA (Miller, 1959). β-glucosidase assay was carried out using *p*-nitrophenyl—β-D-glucopyranoside (PNPG, Sigma Chemical Co.) as substrate at 50 °C for 30 min. The reaction was terminated by addition of 4 ml NaOH-glycine buffer (0.2 M, pH 10.6).

FP activity was measured as describer earlier (Narra et al. 2012). Lignin, cellulose, and hemicellulose contents of the untreated and pretreated RS, WS, and SB were analyzed according to Goering and Vansoest (1975).

The samples from SHF and SSF were withdrawn at regular intervals, centrifuged at 10,000 × g for 15 min, and the supernatant was analyzed for residual sugars by DNSA method as described earlier. Ethanol was estimated using high performance liquid chromatography (Shimadzu, Japan) equipped with a refractive index detector (RID) and packed with an Aminex-HPX-87 column (Biorad, Hercules, USA, CA) with dimension of 300 mm × 7.8 mm. Samples were eluted using 5 mM H_2SO_4 with the flow rate of 0.6 ml min^{-1}. Column temperature was maintained at 65 °C. The saccharification efficiency and the theoretical yield of ethanol were calculated as described by Narra et al. (2012, 2015).

3 Results and Discussion

3.1 Biomass Composition Analysis

RS, WS, and SB contained cellulose content of 41.02 ± 1.45 %, 38.50 ± 1.07 %, 39.00 ± 1.83 %; hemicellulose content of 28.47 ± 1.91 %, 27.00 ± 1.36 %, 25.00 ± 1.44 %; lignin content of 9.20 ± 1.12 %, 12.82 ± 1.27 %, 14.21 ± 1.62 % and moisture content of 7.04 ± 1.21 %, 6.93 ± 1.17 %, 8.14 ± 1.15 %, respectively. After pretreatment, the cellulose, hemicellulose, and lignin contents of RS, WS, and SB were (80.00 ± 1.58 %; 67.00 ± 1.60 %; 70.00 ± 1.32 %); (3.00 ± 1.91 %; 6.54 ± 1.42 %; 4.36 ± 1.44 %); (2.01 ± 1.16 %; 4.98 ± 1.97 %; 5.13 ± 1.48 %), respectively (Narra et al. 2015).

3.2 Dilute Acid Hydrolysis of Lignocellulosic Biomass

RS yielded maximum amount of reducing sugars (110 g l^{-1}) followed by WS and SB (90 and 95 g l^{-1}) when treated at optimum conditions, i.e., 4 % H_2SO_4 at 121 °C for 30 min. The substrate to acid ratio maintained was 1:5 (w/v). As the treatment time increased from 30 min to 60 min, there was no substantial difference in the reducing sugar yield (102, 85, and 91 g l^{-1}). The acid hydrolysate obtained at optimum conditions of hydrolysis contained major amount of xylose (90.23 ± 2.34 g l^{-1}), glucose (5.01 ± 1.29 g l^{-1}), arabinose (7.48 ± 2.69 g l^{-1}), phenolics, furfural, and HMF. The major degradation products of pentose, hexose sugars, and lignin are furfural, HMF, and phenolics, respectively (Kuhad et al. 2010). These toxins have the ability to decrease the activities of several yeast enzymes, e.g., alcohol dehydrogenase, aldehyde dehydrogenase, and pyruvate dehydrogenase during fermentation (Modig et al. 2002). In order to reduce the inhibitor concentrations in acid hydrolysates, sequential addition of overliming and activated charcoal was used. Overliming followed by activated charcoal treatment resulted in reduction of furfural (96.31, 95.15, and 96.20 %), HMF (93.36, 92.18, and 92.19 %), and phenolic (93.36, 92.18, and 92.19 %) in RS, WS, and SB, respectively. As the acid hydrolysate was rich in xylose it was used to produce ethanol using pentose utilizing yeast strain *Pichia stipitis* NCIM 3499.

3.3 Delignification of Lignocellulosic Biomass

Delignification of acid pretreated solid biomass was carried out with 0.5 % NaOH at 121 °C for 30 min to maximize reducing sugar yield. The cellulose content in the pretreated biomass increased substantially with simultaneous reduction in lignin content after alkaline pretreatment (Narra et al. 2015). The cellulose content was found to be increased by 95.02, 74.02, and 79.48 %, while the lignin removals were 78.16, 61.15, and 63.90 % in RS, WS, and SB, respectively. It was also observed that during the treatment most of the pentosan was solubilized. The increase in cellulose content during pretreatment might be due to removal of lignin which might have increased the enzyme effectiveness by eliminating nonproductive adsorption sites and increasing access to cellulose and hemicelluloses (Lu et al. 2002). Kumar et al. (2009) reported that solubilization of other components in the aqueous alkali solution is also responsible for increase in the cellulose content. These findings were consistent with earlier published data on effects of delignification of different lignocellulosic biomass RS, WS, and rapeseed straw by alkaline pretreatment (Narra et al. 2012; Nopparat et al. 2013).

3.4 Enzymatic Saccharification of Delignified Cellulosic Material

When delignified cellulosic substrates were subjected to enzymatic saccharification, an increase in reducing sugar concentration was observed till 40 h time period and thereafter it remained almost constant. Maximum yield of saccharification from RS, WS, and SB was 266, 242, and 254 mg g^{-1} substrate, respectively (Fig. 1). Tween 80 1 % (v/v) addition also found to increase the saccharification yield by 12.1, 11.4, and 10.6 %, respectively (Fig. 2). According to Kaar and Holtzapple (1998), all through the enzymatic saccharification process, thermal deactivation of enzymes was prevented by the addition of Tween. This outcome may be due to the surface activity of Tween which resulted in shorter enzyme contact with air-liquid interface. The decrease in surface tension of the solution not only permits the saccharifying exoglucanase more active sites to cellulose, but it also prevents the nonproductive part of the exoglucanase to the lignin surface which yielded in increased sugar level (Hematinejad et al. 2002).

Fig. 1 Enzymatic hydrolysis of RS, WS, and SB at different time intervals. Temperature 42 °C, substrate load 10 % (w/v), enzyme load 9 FPU g^{-1} substrate, pH 4.8, rpm 16

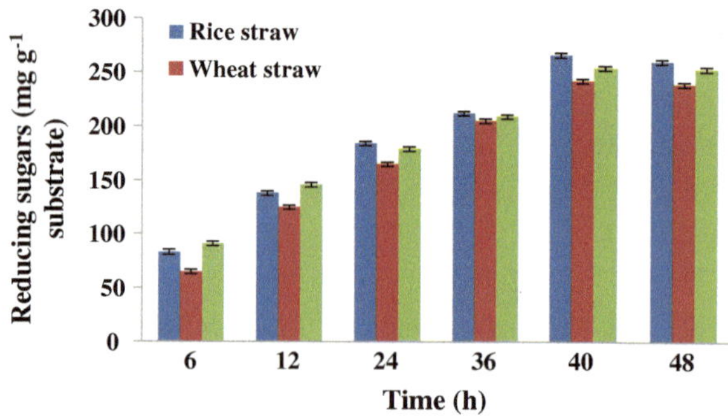

Fig. 2 Influence of Tween 80 on enzymatic hydrolysis of RS, WS, and SB. Substrate load 10 % (w/v), enzyme load 9 FPU g^{-1} substrate, pH 4.8, time 48 h, rpm 16, Tween 80 1 % (w/v)

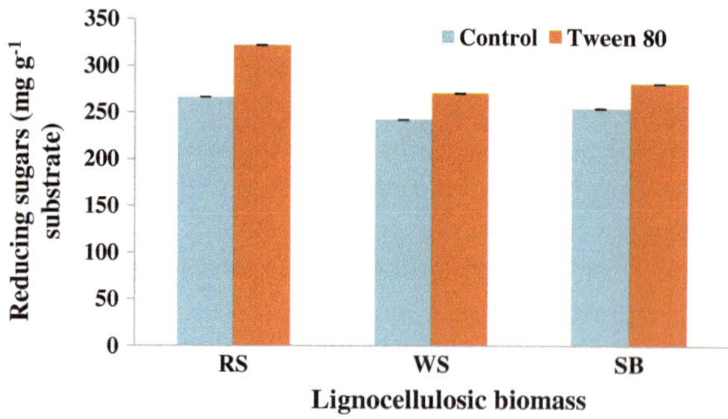

3.5 *Fermentation of Hemicellulosic and Enzymatic Hydrolysate*

Maximum amount of ethanol production from RS, WS, and SB was obtained up to 5.1, 4.9, 5.2 g l^{-1}, respectively. Table 1 shows the profile for ethanol fermentation by *Pichia stipitis* 3499 NCIM. It was observed that there was increase in ethanol

Table 1 Fermentation profile of hemicellulosic hydrolysate of lignocellulosic biomass using *Pichia stipitis*

Time (h)	Ethanol (g l^{-1})			Sugar (g l^{-1})			Ethanol yield (g g^{-1})		
	RS	WS	SB	RS	WS	SB	RS	WS	SB
0	0.04	0.05	0.05	20.00	19.80	20.00	0.00	0.00	0.00
6	1.62	2.25	1.97	19.12	18.45	19.26	0.08	0.11	0.99
12	3.11	3.78	3.95	12.58	13.54	12.95	0.16	0.19	0.19
24	5.10	4.90	5.20	6.21	6.92	6.09	0.26	0.25	0.26
36	3.27	4.17	4.19	5.98	5.21	5.04	0.16	0.21	0.21
48	2.61	3.09	3.35	5.26	4.97	4.70	0.13	0.16	0.17
60	2.21	2.85	3.01	4.18	4.26	4.13	0.11	0.14	0.15

Table 2 Fermentation profile of enzymatic hydrolysate of lignocellulosic biomass using *Saccharomyces cerevisiae*

Time (h)	Ethanol (g l^{-1})			Sugar (g l^{-1})			Ethanol yield (g g^{-1})		
	RS	WS	SB	RS	WS	SB	RS	WS	SB
0	0.08	0.07	0.08	40.00	40.00	40.00	0.00	0.00	0.00
6	10.18	6.95	6.29	11.89	13.28	14.09	0.25	0.17	0.16
12	12.84	9.01	7.58	10.37	12.04	12.54	0.32	0.22	0.19
24	13.53	11.26	9.26	8.43	9.30	9.01	0.34	0.28	0.23
36	14.00	13.90	12.90	7.27	6.95	8.15	0.35	0.35	0.32
48	13.84	11.78	11.27	5.31	5.29	6.26	0.34	0.29	0.28
60	12.63	11.05	10.98	4.98	4.42	5.01	0.32	0.32	0.27

and increase in the growth of fermenting yeast but the consumption of reducing sugars was relatively poor. The incomplete utilization of reducing sugars by fermenting yeast may be due to some kind of inhibitors present in the acid hydrolysate which might have not been removed during the detoxification process. Similar observation was also made by Kuhad et al. (2010) that the yeast could not tolerate the higher amount of sugar concentration. The ethanol yield obtained from cellulosic hydrolysate (RS, WS, and SB) using *Saccharomyces cerevisiae* was 0.35, 0.35, and 0.32 g g^{-1}, respectively (Table 2).

3.6 SSF

Maximum ethanol yield achieved at 60 h with 10 % solid load from RS, WS, and SB (23.23, 18.29, and 17.91 g l^{-1}) which was equivalent to 51.29, 48.22, and 45.19 % of maximum theoretical yield (Fig. 3). The earlier reports have shown that increase in substrate load beyond certain extent could have caused decrease in hydrolysis rate due to the product inhibition. The extent of inhibition usually

Fig. 3 SSF of RS, WS, and SB at different time intervals. Temperature 42 °C, substrate load 10 % (w/v), enzyme load 9 FPU g^{-1} substrate, pH 4.8, rpm 16

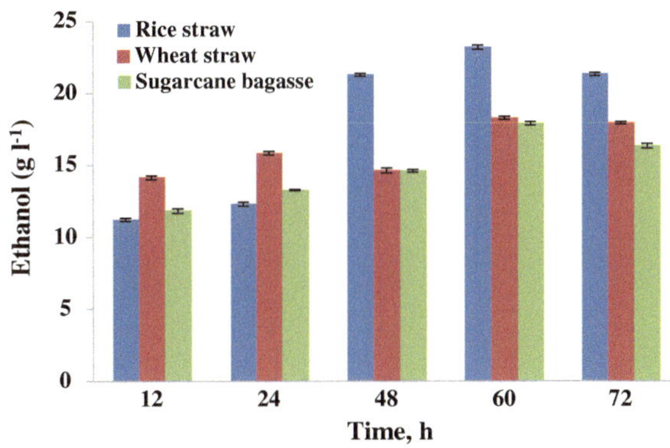

depends on the ratios of substrate to enzyme load (Wang et al. 2011; Xin et al. 2010). The present results also revealed that higher ethanol yields at lower solid loads may be due to lower glucose accumulation compared to higher solid loads. The current results were in accordance with Naveen et al. (2011) that at 8 % solid loading, higher ethanol yields were observed compared to 12 % solid loading when pretreated switchgrass was used for SSF studies by *Kluyveromyces marxianus* IMB3.

Enhanced ethanol yields were achieved by the addition of 1 % Tween 80 to the reaction mixture in comparison to the control (Fig. 4). The yields were increased by 8.39, 9.26, and 8.14 % with 10 % solid loading at 60 h from RS, WS, and SB, respectively. Maximum ethanol yield from RS, SB, and WS were found to be 25.18, 19.57, and 18.19 g l^{-1}, respectively. Kaar and Holtzapple (1998), Zhu et al. (2014) similarly have found that addition of surfactant enhanced the rate and extent of hydrolysis as the surfactants could have changed the nature of the substrate either by increasing the available cellulose surface area or by eliminating inhibitory lignin.

The present results have shown that 1 kg untreated RS, WS, and SB contained 410, 370, 380 g, and 252, 205, 207 g cellulose and hemicelluloses, respectively. This amount can be theoretically enough for production of 232, 217, 215 g and 129, 105, 106 g ethanol from cellulose and hemicellulose, respectively. Based on the best yields attained in the current work, the ethanol production from RS, WS, and SB through SSF was found to be 201, 162, 156 g kg^{-1} raw biomass, respectively. Whereas, the ethanol production from RS, WS, and SB through SHF was found to be 95, 75, 82 g kg^{-1} raw biomass, respectively (Table 3). The present results revealed that SSF could prove enhanced ethanol production from delignified biomass compared to SHF. The reason could be that the optimum condition for separate enzymatic hydrolysis might be higher than 42 °C, and secondly the product inhibition exerted by saccharification might have occurred during the process. Whereas in case of SSF, pre-hydrolysis was carried out at 42 °C for 6 h followed by yeast addition to ease the problems caused by production inhibition.

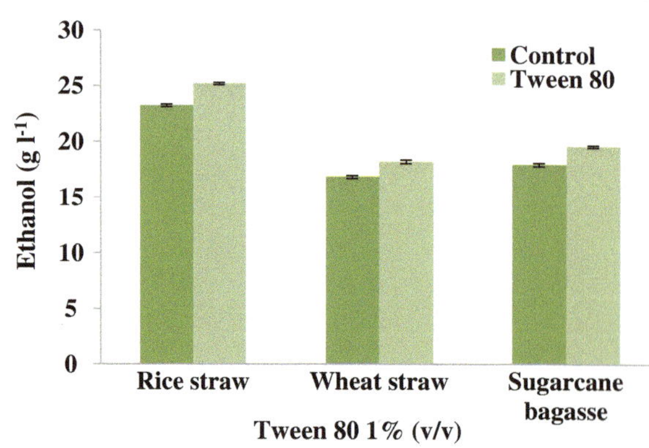

Fig. 4 Influence of Tween 80 on SSF of RS, WS, and SB. Substrate load 10 % (w/v), enzyme load 9 FPU g^{-1} substrate, pH 4.8, time 60 h, rpm 16

Table 3 Comparison of ethanol yields from SSF and SHF of different lignocellulosic biomass

Process	SSF			SHF		
Lignocellulosic substrates	RS	WS	SB	RS	WS	SB
Optimum conditions	SSF: Temperature: 42 °C, substrate load: 10 % (w/v), enzyme load: 9 FPU g^{-1} substrate, duration: 48 h, Tween 80: 1 % (v/v)			SH: Temperature: 42 °C, substrate load: 10 % (w/v), enzyme load: 9 FPU g^{-1} substrate, duration: 48 h, Tween 80: 1 % (v/v) Fermentation: Temperature: 30 °C, duration: 24 h, sugar concentration: 2 %		
Theoretical ethanol yield, g kg^{-1} substrate						
Cellulosic substrate	232	217	215	232	217	215
Hemicellulosic substrate	129	105	106	129	105	106
Ethanol yield, g kg^{-1} substrate						
Cellulosic substrate	129	108	95	23	21	21
Hemicellulosic substrate	72	54	61	72	54	61

4 Conclusion

Cellulosic ethanol is considered to be a globally accepted alternative fuel. The current results demonstrated that SSF could ascertain improved ethanol production from delignified lignocellulosic biomass compared to SHF in terms of total ethanol production time and ethanol yield. RS yielded higher amount of ethanol followed by WS and SB using in-house cellulases produced by *Aspergillus terreus* and in-house thermotolerant yeast strain *Kluyveromyces* sp.

Acknowledgements The authors are thankful to the Director, Sardar Patel Renewable Energy Research Institute (SPRERI), Vallabh Vidyanagar, Gujarat for allowing us to carry out research at SPRERI. The financial support from Department of Biotechnology (DBT), Government of India is highly acknowledged.

References

Borbala E, Mats G, Guido Z (2013) Simultaneous saccharification and co-fermentation of whole wheat in integrated ethanol production. Biomass Bioenergy 56:506–514

Chen HZ, Xu J, Li ZH (2007) Temperature cycling to improve the ethanol production with solid state SSF. Appl Biochem Microbiol 43:57–60

Faveri DD, Torre P, Perego P, Convarti A (2004) Statistical investigation on the effects of starting xylose concentration and oxygen mass flow rate on xylose production from RS hydrolysate by response surface methodology. J Food Eng 65:383–389

Goering K, Van Soest PJ (1975) Forage fiber analysis-agriculture research series. Handbook 379

Hemantinejad N, Vahabzadeh F, Koredestani SS (2002) Effect of surfactant on enzymatic hydrolysis of cellulosic fabric. Iranian Poly J 11:333–338

Kaar WE, Holtzapple MT (1998) Benefits from tween during enzymatic hydrolysis of corn stover. Biotechnol Bioeng 419–427

Kaya F, Heitmann JA, Thomas WJ (2000) Influence of lignin and its degradation products on enzymatic hydrolysis of xylan. J Biotechnol 80:241–247

Kuhad RC, Gupta R, Khasa YP, Singh A (2010) Bioethanol production from Lanata Camara (red sage): pretreatment, saccharification and fermentation. Bioresour Technol 641–654

Kumar R, Mago G, Balan V, Wyman CE (2009) Physical and chemical characterization of corn stover and poplar solids resulting from leading pretreatment technologies. Bioresour Technol 100:3948–3962

Lu Y, Yang B, Gregg D, Saddler J, Mansfield S (2002) Cellulase adsoption and an evaluation of enzyme recycling during hydrolysis of steam-exploded soft wood residues. Appl Biochem Biotechnol 98–100

Miller GL (1959) Use of dinitrosalicylic acid reagent for determination of reducing sugars. Anal Chem 31:426–428

Modig T, Liden G, Taherzadehm MJ (2002) Inhibition effects of furfural on alcohol dehydrogenase, aldehyde dehydrogenase and pyruvate dehydrogenase. Biochem J 363:769–776

Narra M, Dixit G, Madamwar D, Shah AR (2012) Production of cellulases by solid state fermentation with *Aspergillus terreus* and enzymatic hydrolysis of mild alkali-treated RS. Bioresour Technol 121:355–361

Narra M, James J, Balasubramanian V (2015) SSF of delignified lignocellulosic biomass at high solid loadings by a newly isolated thermotolerant *Kluyveromyces* sp. for ethanol production. Bioresour Technol 179:331–338

Naveen KP, Hasan KA, Mark RW, Danielle DB, Ibrahim MB (2011) SSF of Kanlow switchgrass by thermotolerant *Kluyveromyces marxianus* IMB3: The effect of enzyme loading, temperature and higher solids. Bioresour Technol 102:10618–10624

Nigam JN (2001) Ethanol production from WS hemicellulose hydrolysate by *Pichia stipitis*. J Biotechnol 87:17–27

Nopparat S, Khatiya W, Navadol L, Verawat C (2013) Optimize simultaneous saccharification and co-fermentation of RS for ethanol production by *Saccharomyces cerevisiae* and *Scheffersomyces stipitis* co-culture using design of experiments. Bioresour Technol 142:171–178

Scordia D, Cosentino SL, Jeffries TW (2013) Enzymatic hydrolysis, simultaneous saccharification and ethanol production of oxalic acid pretreated giant reed (*Arundodonax* L.). Ind Crops Prod 49:392–399

Wang W, Kang L, Wei H, Arora R, Lee Y (2011) Study on the decreased sugar yield in enzymatic hydrolysis of cellulosic substrate at high solid loading. Appl Biochem Biotechnol 164:1139–1149

Xin F, Geng A, Chen ML, Gum MJM (2010) Enzymatic hydrolysis of sodium dodecyl sulphate (SDS) pre-treated newspaper for cellulosic ethanol production by *Saccharomyces cerevisiae* and *Pichia stipitis*. Appl Biochem Biotechnol 162:1052–1064

Zhu JQ, Lie Q, Li BZ, Yuan YJ (2014) Simultaneous saccharification and co-fermentation of aqueous ammonia pretreated corn stover with an engineered *Saccharomyces cerevisiae* SyBE005. Bioresour Technol 169:9–18

Liquid Hot Water Pretreatment of Paddy Straw for Enhanced Biomethanation

Abhinav Trivedi, Virendra Kumar Vijay and Ram Chandra

Abstract Indian agriculture produces nearly 150–160 million tons of paddy straw per annum. 70–80 % of total produced rice straw is burned in the open field for cultivation of the next crop. Proper disposal of paddy straw in an economical and environment friendly manner is a serious concern in India, where rice is the major agricultural crop in some states. The work presented in the paper is focused on pretreatment of paddy straw using liquid hot water treatment followed by biomethanation. Results observed during batch process are encouraging and work is being carried out for bigger models. Under batch condition with ~5–7 kg of paddy straw can yield 1.0 m^3 of biogas with average methane content of 60.0 and 25–30 % carbon dioxide content during retention period of 40 days.

Keywords Agricultural residue · Lignocellulosic biomass · Biomass pretreatment · Liquid hot water pretreatment · Biogas from paddy straw · Biomethane production

1 Introduction

Paddy straw is among ubiquitous lignocellulosic agricultural waste biomass which has been rarely utilized as a source for biofuel production. For the year 2013, annual production of rice was reported as 159.200 million metric tons in India (FAOSTAT 2013). With such colossal production of rice, around 238.50 million metric tons of paddy straw is generated per annum with an average of 1.0–1.5 kg of paddy straw per kilogram of the rice grain harvested (Maiorella 1985) of which two–third is burnt in the field itself (Phutela 2011). Some amount of unburnt residue is used as a fuel for modern biomass-based power plant, while some amount is used

Abhinav Trivedi (✉) · V.K. Vijay · Ram Chandra
Biogas Production and Enrichment Laboratory, Centre for Rural Development and Technology, Indian Institute of Technology Delhi, Hauz Khas, New Delhi 110016, India
e-mail: abhinavtrivedi22@gmail.com

S. Kumar et al. (eds.), *Proceedings of the First International Conference on Recent Advances in Bioenergy Research*, Springer Proceedings in Energy,
DOI 10.1007/978-81-322-2773-1_2

to fuel biomass cookstoves in rural area for cooking and heating applications (Gadde et al. 2009).

The portion of the paddy straw residue that remains in the field as uncollected is subsequently plowed back into the soil (residue incorporation into soil), which serve as biofertilizer for upcoming crops. This is widely practised across the country and it provides a nutrient source for upcoming crops but also conductive to crop diseases (Hrynchuk 1998). This activity stimulates methane emissions due to decomposing paddy straw in the field (Yuan 2013).

The research paper published in recent years has been perceived that open field burning of paddy straw residues contributes toward emissions of polycyclic aromatic hydrocarbons, i.e., PAHs (Korenaga et al. 2001), as well as polychlorinated dibenzo-p-dioxins (PCDDs), and polychlorinated dibenzofurans (PCDFs), referred to as dioxins (Gullett and Touati 2003). It has also been reported that open burning of one ton of paddy straw emits 3 kg particulate matter, 60 kg CO, 1460 kg CO_2, 199 kg ash, and 2 kg SO_2 (Jenkins and Bhatnagar, 2003a, b). These compounds are toxic and carcinogenic for humans and have a long term impact on human health (Meesubkwang 2007).

Anaerobic digestion of rice straw for methane production is considered as the best option among all thermochemical and biochemical processes for providing energy in form of methane, thus reducing air pollution up to considerable limits and at the same time producing digestate solids, which is a potential biofertilizer for land application (Nand 1999). The anaerobic digestion technology is a most efficient way in term of energy output/input ratio for handling of biomass resources to produce energy and biofertilizer. Furthermore, raw biogas after enrichment is equivalent to natural gas and can be alternatively used in mobile and static internal combustion engines.

The present research focused on evaluation of liquid hot water (LHW) pretreatment method for delignification of paddy straw and subsequently anaerobic digestion of the substrates. In this method of pretreatment the destruction of lignocellulosic structure occurs by super-heated water at temperature 200 and 240 °C. The work was carried out using nonstirred reactor having volume of 2.0 liter, which can be operated to maximum working temperature of 300 °C. The degraded paddy straw samples were evaluated for its biomethane potential under laboratory conditions.

2 Materials and Methods

2.1 Characterisation of Paddy Straw

The characterization of paddy straw was carried out using proximate and ultimate analyses. Initially paddy straw was dried at 105 °C using hot air oven and ground to nearly 1.0 mm size using grinder. Percentage of moisture content, total solids, volatile solids, and ash contents (proximate analysis) in paddy straw was estimated using standard methods as described by APHA while basic elements such as carbon, hydrogen, and nitrogen (ultimate analysis) were determined using 'Vario EL' elemental analyzer (Perkin Elmer, USA Made). 20 mg of homogenized samples were packed in tin foil and placed into the carousel of the automatic sample feeder of the analyzer and analyzed as per procedure described in the instruction manual of the instrument supplied by the manufacturer of CHN elemental analyzer.

2.2 Pretreatment of Paddy Straw

Fresh paddy straw samples were pretreated in a hydrothermal reactor at reactor operating temperatures of 200 and 240 °C for pretreatment time duration of 15–20 min. During pretreatment the water-to-paddy straw ratio was maintained as 8:1, this ratio is maintained in order to get nearly 10 % of total solids concentration in the digesting substrate. Appropriate amount of NaOH was added in the pretreated paddy straw substrate to maintain pH around 7.5–7.8. To maintain the homogeneity of the substrates in reactor bottles, the bottles were shook daily manually. The prepared substrates were kept at 37 °C temperature in an incubator for stabilization of pH before being fed to anaerobic digester (Chandra et al. 2012).

2.3 Experimental Setup

The biogas production potential of pretreated paddy straw samples was estimated using 1.0 L glass bottles (Schott Duran). Each reactor consisted 30 g of hydrothermally pretreated paddy straw and 240 mL of process water along with 480 mL of fresh slurry taken from cow dung-based biogas plant. The C/N ratio of substrate was adjusted in a range of 20–25 by addition of 0.75 g of urea in the substrate. Biogas production was monitored for 40 days after which biogas production had been found ceased. The reactor bottles contained the desired substrates along with inoculum were placed in an incubator and maintained constantly at 37 °C operating temperature. The daily biogas volume was measured by using inverted water displacement system. The biogas produced from various anaerobic reactors was analyzed for methane as well as carbon dioxide contents using a gas chromatograph (Agilent 7890) equipped with Porapak Q column and thermal conductivity detector (TCD).

3 Results and Discussion

3.1 Properties of Paddy Straw

The results of proximate and ultimate analysis of untreated paddy straw are shown in Table 1. The proximate analysis revealed that untreated paddy straw contains up

Table 1 Proximate and ultimate analysis of nontreated paddy straw

Sl. No.	Properties parameter	Observed value (%)
Proximate properties		
1	Moisture content (wet basis)	08.00
2	Total solids (wet basis)	92.00
3	Volatile solids (dry basis)	81.50
4	Ash content (dry basis)	18.50
Ultimate properties		
5	Carbon	35.20
6	Hydrogen	5.596
7	Nitrogen	0.745
8	*C*/*N* ratio	47.248

to 8.0 % moisture and 92.0 % total solids on wet weight basis, while 81.50 and 18.50 % volatile solid matter and ash matter, respectively, on dry weight basis. The ultimate analysis resulted into 35.2 % carbon, 5.5 % hydrogen, and 0.745 % nitrogen contents on dry weight basis. Upon elemental analysis, it has been found that the amount of nitrogen content present in rice straw biomass is very low (*C*/*N* ratio = 47.2) to maintain an appropriate range of *C*/*N* ratio (20–30) for effective methane fermentation.

3.2 *Effect of Liquid Hot Water Pretreatment on Structure of Paddy Straw*

Figs. 1, 2, 3, and 4 show the scanning electron microscopic images of nontreated paddy straw and various hydrothermal pretreated paddy straw such as 200 and

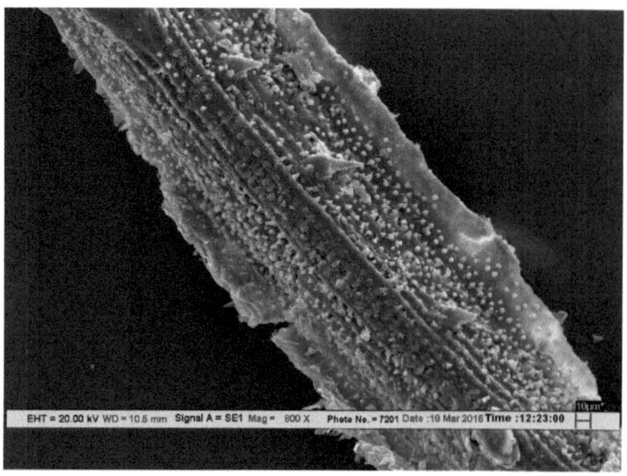

Fig. 1 SEM image of nontreated paddy straw

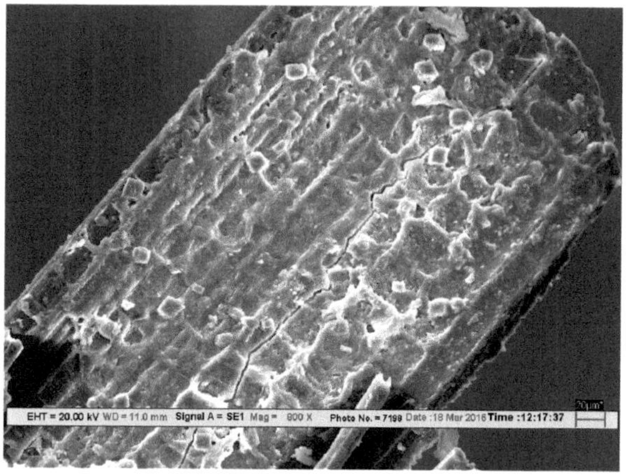

Fig. 2 SEM image of pretreated paddy straw (200 °C)

Fig. 3 SEM image of pretreated paddy straw (240 °C)

240 °C, respectively. It is clearly evident from Fig. 1 that the microfibrils are very tightly packed in case of nontreated paddy straw. However, in case of hydrothermal pretreated paddy straw the original structure of paddy straw has been visibly disrupted depending upon the level of pretreatment temperature. The microfibrils are loosed and open for microbial degradation. The brightness in the various images is due to charging effect during SEM imaging of volatile fatty acids released during the pretreatment process.

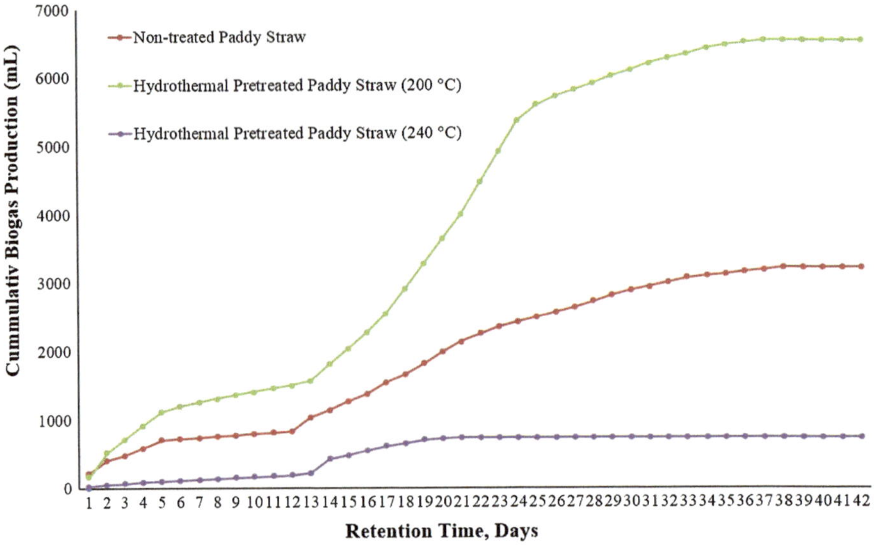

Fig. 4 Cumulative biogas production for various substrates of paddy straw

Table 2 Cumulative biogas production and average methane content in the biogas

Sl. No.	Treatment	Cumulative biogas production (mL)	Biogas production (m^3/kg TS)	Methane production (m^3/kg TS)
1	Nontreated paddy straw	3215	0.107	0.058
2	Hydrothermal pretreated paddy straw (200 °C)	6540	0.218	0.144
3	Hydrothermal pretreated paddy straw (240 °C)	740	0.024	Not measured

3.3 *Biomethane Potential Analysis*

The cumulative biogas and methane production yields observed during the course of investigation are shown in Table 2. The input feed material in the substrates remained constant as 30.0 g which were fed to each anaerobic reactor. Nontreated paddy straw substrate resulted to produce 3215 mL of biogas over a retention period of 39 days. After that biogas production was completely ceased. However, in case of hydrothermal pretreated paddy straw (200 °C) substrate the biogas production was observed to be 6540 mL over a retention period of 37 days and afterward there was no occurrence of biogas production. In case of hydrothermal pretreated paddy straw (240 °C) substrate the biogas production was observed to be 740 mL over a retention period of 21 days. After that biogas production was completely ceased.

The pattern of biogas production with respect to retention time is shown in Fig. 4. It is clearly evident that in case of non-pretreated substrate of paddy straw the biogas production rate was slow and lasted for longer duration of retention time in comparison to the hydrothermal pretreated paddy straw (200 °C) substrate. The biogas production from hydrothermal pretreated paddy straw (240 °C) substrate was found very low in comparison to hydrothermal pretreated paddy straw (200 °C) substrate as well as non-pretreated substrate of paddy straw. This might be due to lower pH level of the digesting substrate during the anaerobic digestion process.

The average concentration of methane in the produced biogas was found as 54.0 % in case of non-pretreated substrate of paddy straw. However, the average methane content in the biogas produced from hydrothermal pretreated paddy straw (200 °C) substrate was observed to be 66.0 %. The specific biogas and methane production from nontreated paddy straw substrate was found to be 0.107 and 0.058 m^3/kg TS, respectively. However, the specific biogas and methane production from hydrothermal pretreated paddy straw (200 °C) substrate was found to be 0.218 and 0.144 m^3/kg TS, respectively.

4 Conclusions

The results of experiments so far conducted on batch scale biogas production assays revealed that liquid hot water pretreated paddy straw substrate results in the increase in amount of biogas and methane production yield. The biogas and methane production yields after liquid hot water pretreated paddy straw at 200 °C was found more than 2.0 times of untreated paddy straw substrate. There are many other processes for pretreatment of lignocellulosic waste but most of these processes are in the laboratory–bench or pilot–plant stage and at the present time and the economic feasibility of these available processes are not certain in Indian context. Further, the work has been extended in the same laboratory to explore the potential of liquid hot water pretreatment at lower temperatures and toward designing farm scale biogas digestion for paddy straw.

Acknowledgments Authors gratefully acknowledge the financial support of the Indian Institute of Technology Delhi for conducting this research. Authors also thanks to biogas research group for their support during the course of investigation.

References

Chandra R, Takeuchi H, Hasegawa T (2012) Hydrothermal pretreatment of rice straw biomass: a potential and promising method for enhanced methane production. Appl Energy 94:129–140

Gadde B, Bonnet S, Menke C, Garivait S (2009) Air pollution emissions from rice straw open field burning in India, Thaiand the Philippines. Environ Pollut 157:1554–1558

Gullett B, Touati A (2003) PCDD/F emissions from burning wheat and rice field residue. Atmos Environ 37:4893–4899

Hrynchuk L (1998) Rice straw diversion plan. In: McGuire, Terry (ed) California air resources board. California. p 23

Jenkins BM, Bhatnagar AP (2003a) Electric power potential from paddy straw in Punjab and the optimal size of the power generation station. Bioresour Technol 37:35–41

Jenkins BM, Bhatnagar AP (2003b) On electric power potential from paddy straw in Punjab and the optimal size of power generation station. Bio-res Tech 37:35–41

Korenaga T, Liu X, Huang Z (2001) The influence of moisture content on polycyclic aromatic hydrocarbons emissions during rice straw burning. Chemosphere-Glob Change Sci 3:117–122

Maiorella BL (1985) Ethanol fermentation in comprehensive biotechnology. Young M (ed) Pergamon Press, Oxford. 3:861–914

Meesubkwang S (2007) Chiang Mai's polluted air: public health office issues warnings. Chiang Mai Mail, vol VI, no. 3. http://www.chiangmaimail.com/212/news.shtml

Nand K (1999) Biotechnology from agro-industrial and food processing wastes. In: Joshi VK, Pandey A (eds) Biotechnology, vol II. Educational publishers and distributors, Kerela, India, pp 1349–1372

Phutela Urmila G, Sahni N, Sooch SS (2011) Fungal degradation of paddy straw for enhanced biogas production. Indian J Sci Tech 4(6):660–665

Yuan Q, Pump J, Conrad R (2013) Straw application in paddy soil enhances methane production also from other carbon sources. Biogeosci Discuss 10:14169–14193

Web Refrence, FAO STAT (2013) Division Production quantities by country in food and agriculture organization of the united nations statistics. http://faostat3.fao.org/browse/Q/QC/E. Accessed 07 Oct 2015

Improving Yeast Strains for Pentose Hexose Co-fermentation: Successes and Hurdles

Shalley Sharma, Sonia Sharma, Surender Singh, Lata and Anju Arora

Abstract Bioethanol, a second generation biofuel, is considered to be one of the best alternatives to conventional petroleum-based liquid fuels. In the present scenario, it is being majorly produced by fermentation of hexoses coming from the cellulosic fraction of the lignocellulosic biomass. Biomass also comprises of up to 33 % hemicellulose, therefore, its fermentation would lead to enhanced bioethanol fermentation productivities. Bioethanol can be produced through biochemical as well as thermochemical processes. A biochemical process, which is environmentally favorable involves the use of microbes, e.g., yeast and bacteria. Bioethanol production from yeast, i.e., *Saccharomyces cerevisiae* has already been commercialized. However, demerit of this commercially viable strain is that it utilizes only hexoses, while pentoses are left unused. This paper discusses different strategies for improving the potential of yeast strains for mixed sugar fermentation to ethanol. There are ways to genetically improve yeast strains to enable them to ferment mixture of hexoses and pentoses. However, there are several physiological hurdles which can limit the success of conventional genetic approaches like cofactor imbalance, excessive by product formation, glucose repression, etc., which need to be tackled, in order to obtain enhanced yield. Metabolic engineering of the yeast strains is a way for enhancing bioethanol fermentation efficiency.

Keywords Bioethanol · Hexose · Pentose · Fermentation · Metabolic engineering

1 Introduction

In recent years, research over the bioethanol production has increased, particularly focusing on the bioconversion of forestry, agricultural, and municipal residues into ethanol. With increasing demand and limited resources of the fuels, bioethanol

Shalley Sharma · Sonia Sharma · Surender Singh · Lata · Anju Arora (✉)
Division of Microbiology, ICAR-Indian Agricultural Research Institute,
New Delhi 110012, India
e-mail: anjudev@yahoo.com

S. Kumar et al. (eds.), *Proceedings of the First International Conference on Recent Advances in Bioenergy Research*, Springer Proceedings in Energy,
DOI 10.1007/978-81-322-2773-1_3

production has become a necessity, but most of the production is based on the agricultural crops (such as corn, wheat, sugarcane, rice, etc.) which may raise debate over food versus. fuel. Solution to the problem is utilization of lignocellulosic biomass (it consists of agricultural wastes, i.e., the leftover biomass after crop processing) as it is rich in cellulose and hemicellulose and hence is a good source of sugars to be fermented for the production of bioethanol. Also, lignocellulosic biomass is more abundant than crops, which is more economical and also does not hamper the stability of environment (Stephanopoulos 2007). Hydrolysis of lignocellulosic biomass yields a mixture of sugars, i.e., hexoses (glucose, mannose, and galactose) and pentoses (xylose and arabinose), of which glucose is the most abundant sugar followed by xylose.

Although lignocellulosic biomass is rich in hexose as well as pentose sugars, its efficient conversion into ethanol is not an easy task as pentose fermentation is not an efficient process because pentoses are not readily utilized by yeasts and bacteria (Chu et al. 2007) as in the case of hexoses. Even *S. cerevisiae*, well known fermentative yeast is unable to utilize xylose, when grown on media containing xylose as the sole carbon source (Kuhn et al. 1995). However, there are reports suggesting the presence of xylose transport and xylose metabolizing genes such as xylose reductase (XR), xylulokinase, etc., in this yeast. Earlier it was thought to be the reason that xylose could not be fermented by yeasts and the strategies employed for its fermentation involved: conversion of xylose to xylulose in vitro using glucose isomerase and then fermentation of this xylulose to ethanol (Chandel et al. 2011). Nevertheless, this scenario changed with the discovery of pentose fermenting yeasts during 1980s (Hahn-Hägerdal et al. 2007; Chandel et al. 2008a). As mentioned above, commercial strain of *S. cerevisiae* is unable to utilize and ferment pentose sugars. This has led the researchers to study the metabolic pathway of this yeast. Studies suggest that *S. cerevisiae* can uptake xylose via glucose transporters and hence the rate of transport becomes slow (van Vleet et al. 2009). Metabolic studies of *S. cerevisiae* found out that this organism has the transport system for xylose and all the enzymes required for the xylose metabolic pathway (Gárdonyi et al. 2003a) but is still unable to utilize it for biomass conversion. This is due to the fact that xylose fermentation requires oxygenic conditions which result in the oxidation of ethanol by the end of the process, thus lowering the fermentation efficiency of the strain.

Several research groups have been trying to genetically modify the strain so that it can utilize xylose for the production of ethanol. Efforts going on in this area of research have been discussed in this chapter.

2 Xylose Transport and Metabolism

In order to understand the co-fermentation of glucose and xylose, strong understanding of the xylose transport and its metabolic pathway is required. Pentose fermenting yeast species include *Pachysolen tannophilus*, *Scheffersomyces* (*Candida*) *shehatae*, and *Scheffersomyces* (*Pichia*) *stipitis*. Although, these strains are xylose fermenting, but the rate of fermentation is much lower than that for glucose fermentation (Jeffries 1985). This is due to the fact that these are intolerant to high ethanol concentration (Maleszka and Schneider 1982) as well as by-product formation, such as xylitol.

Xylose transport plays an important role in its metabolism which can occur only after the uptake of the sugar. There are different transport systems in different microorganisms for the uptake of xylose. These include—facilitated diffusion and active mechanism. In active transport system, metabolic energy is required and uptake occurs against a concentration gradient. Active transport systems can be categorized depending upon the type of metabolic energy used, such as chemo-osmotic, direct energization, or group translocation. While facilitated diffusion does not require any metabolic energy for the uptake of sugars. As both the systems need carrier proteins for sugar uptake making these systems specific to the substrate type and the maximum rate of uptake and transport is dependent upon the expression and activity level of the carrier protein and other enzymes involved in the transport system (e.g., transmembrane diffusion gradient in the case of facilitated diffusion). Studies also suggest that these systems can be subjected to competitive inhibition by other sugars or their analogs, due to substrate specificity. Since there are many chemo-osmotic proton symport-based transport systems, so they are strictly pH-dependent (McMillan 1993). Previous studies on xylose-utilizing yeast strains clearly indicate the presence of two import systems in xylose transport: a high capacity, low-affinity facilitated diffusion system for both, xylose and glucose; and a high-affinity proton symport system specific for xylose (Does and Bisson 1989, 1990; Kilian and Vanuden 1988; Lucas and Vanuden 1986).

After being transported into the cells, xylose metabolism starts through a two-step oxidoreductive isomerization and is converted to xylulose (Webb and Lee 1990). Xylulose is being converted to xylulose-5-phosphate through the action of enzymes, NADH- or NADPH-dependent xylose reductase (EC 1.1.1.21; XR), and NAD-dependent xylitol dehydrogenase (EC 1.1.1.14; XDH). Whole pathway has been presented in the Fig. 1. There is an alternate pathway for the conversion of

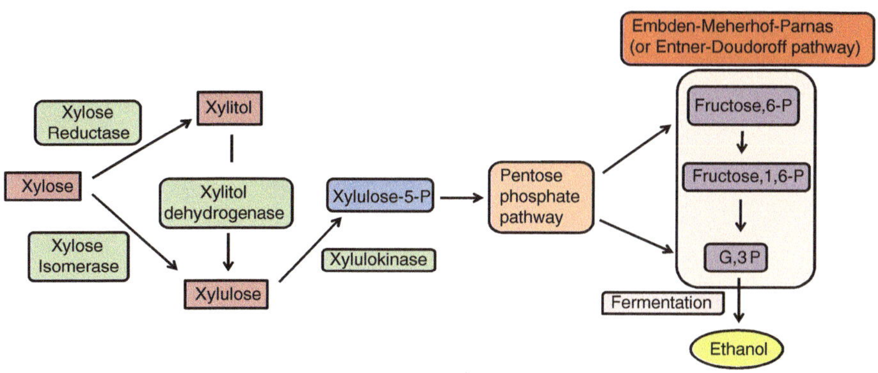

Fig. 1 Xylose metabolism to produce bioethanol

xylose to xylulose which is catalyzed by xylose isomerase, but this pathway is generally adopted by bacteria (Mishra and Singh 1993) and some anaerobic fungi (Mishra and Singh 1993) but not by yeast strains. Xylose metabolism has been extensively studied in *Pichia stipitis* and the xylose is metabolized in the strain as illustrated above.

2.1 Xylose Metabolism in S. Cerevisiae

Xylose uptake in yeasts can occur by both facilitated diffusion and active transport processes. In *S. cerevisiae*, a yeast generally recognized not to metabolize xylose, xylose is transported by facilitated diffusion (Jeffries 1983; Lucas and van Uden 1986). Although *S. cerevisiae* possess facilitated diffusion system for sugar uptake, but inefficiency to metabolize xylose is due to the fact that this system exhibits more affinity toward glucose than xylose. However, it has also been studied that *S. cerevisiae* possesses homologs for genes involved in xylose metabolism, but lacks the functional metabolic pathway. These homologs are GRE3, YPR1, YJR096 W (xylose reductase), YLR070C, SOR1, YDL246C (xylitol dehydrogenase), and YKS1 (xylulokinase).

As stated earlier, *Saccharomyces* cannot utilize xylose but it can grow on xylulose, suggesting that this yeast is unable to convert xylose to xylulose, therefore is unable to metabolize xylose for ethanol fermentation (Toivari et al. 2004). While in most of the yeast strains other than *Saccharomyces*, xylose uptake occurs after the depletion of glucose, its metabolism does not start immediately. Metabolism starts after a lag period of productive processes, especially for the formation of xylose metabolizing enzymes (Gong et al. 1981).

In a study by Toivari et al. (2004), it was found that when these endogenous genes of *Saccharomyces* for xylose metabolism were overexpressed, xylitol formation was increased, suggesting the existence of efficient xylose metabolic pathway. As for example, overexpression of *GRE3* and *ScXYL2* enabled the yeast to grow on xylose, even in the presence of glucose under aerobic conditions. While the accumulation of xylitol could be due to low XDH activity and the specificity of Gre3p toward NADPH because XRs cannot supply NAD^+ required for the activity of XDH. So, this reaction cannot occur under anaerobic conditions due to inability of NAD^+ regeneration, resulting in non-utilization of xylose. Due to the NADPH specificity of Gre3p, redox imbalance is created which results in the accumulation of xylitol and hence xylose metabolism cannot occur anaerobically in yeast.

2.2 Native Pentose Fermenting Yeasts

There have been several reports of pentose fermenting yeast strains which can not only utilize xylose but can efficiently, ferment it for ethanol production. These include *Pichia, Pachysolen, Brettanomyces, Candida, Clavispora, Kluyveromyces, Schizosaccharophora,* etc. According to the literature, *Candida shehatae, Pachysolen tannophilus,* and *Pichia stipitis* have been reported to be the best native yeast species capable of efficient xylose fermentation with high yield (Mcmillan 1993). However, it has always been debated as to which of these three strains should be used for commercial processes, but the research has been focused over xylose fermentation through these three yeast species.

Native strains of *S. cerevisiae* exhibit similar characteristic as that exhibited by the commercial strain by not utilizing xylose as the carbon source. While "torula yeast" aka *Candida utilis* was grown over xylose but it was unable to produce ethanol as this yeast is strictly aerobic. With the discovery of pentose fermenting yeast species, mentioned above, intensive screening efforts have been made and it has been observed that some yeast species can convert xylose to ethanol directly under oxygen-limited conditions (Skoog and Hahn-Hägerdal 1990; Fonseca et al. 2007).

Considering the presence of pentose fermenting yeast species in nature, many strategies can be adopted for improving the co-fermentation efficiencies of the yeast species. These may include genetic modification through recombinant DNA technology or through evolutionary approaches, so that simultaneous utilization and fermentation of glucose and xylose can be achieved in the improved strain.

3 Scope for Improvement

Since many pentose fermenting yeasts are known, their metabolic pathways need to be studied for improving the fermentation efficiencies of the native as well as commercial strains. Various strategies (Fig. 2) have been employed for the improvement of the

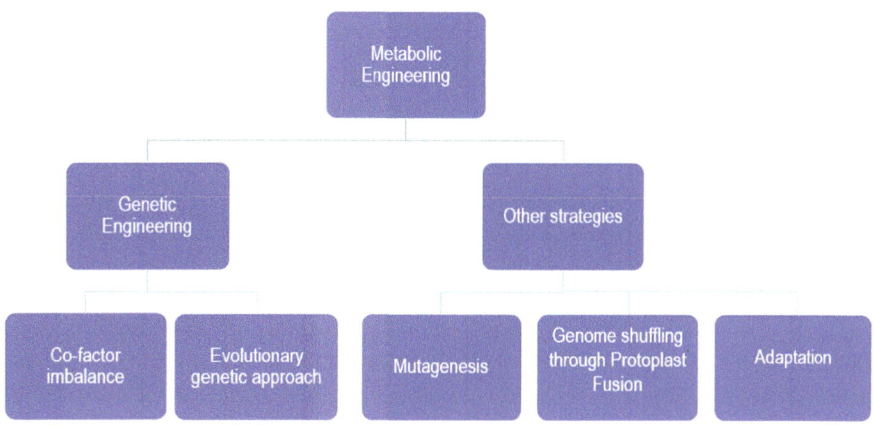

Fig. 2 Possible strategies for metabolic engineering of yeast

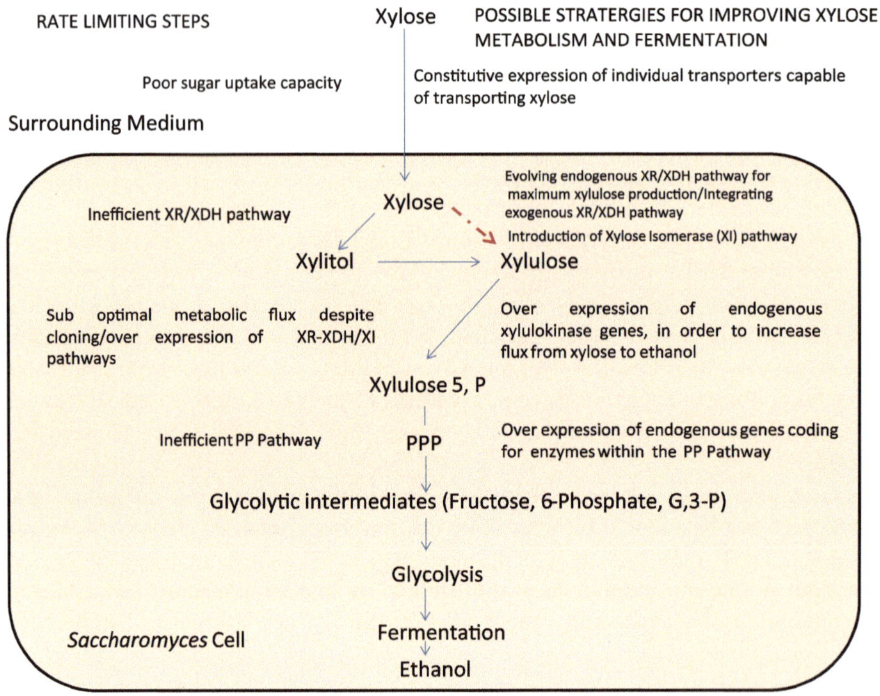

Fig. 3 The rate limiting steps for xylose metabolism and fermentation and possible remedial measures

strains. Possible strategies include metabolic engineering (Fig. 3) and evolutionary engineering. These have been discussed in detail below.

3.1 Metabolic Engineering Aspects

Study of xylose metabolism in yeast suggests that the strains can be engineered for the efficient utilization and fermentation of xylose sugar. Strains capable of pentose fermentation can be engineered and screened for their co-fermentation potential. This can be done through metabolic engineering to incorporate the genes encoding the enzymes required for xylose metabolism. Metabolic engineering is defined as the strategy to improve the product formation through the modification of a specific biochemical reaction or the introduction of a new one through recombinant DNA technology (Bailey 1991; Stephanopoulos et al. 1998). *S. cerevisiae* has been engineered to express these genes and it shows a theoretical yield of 0.51 g g^{-1} from xylose fermentation, but with low maximal productivities (Ostergaard et al. 2000).

Characteristics of an ideal microorganism for the efficient production of bioethanol/biofuel include: specific productivity, broad substrate range, and high ethanol tolerance as well as tolerance to the inhibitors present in lignocellulosic hydrolysates (Aristidou and Penttila 2000).

3.1.1 Genetic Engineering Approach

Since most of the xylose metabolism has been studied in *P. stipitis*, efforts have been made to genetically engineer the commercial strain of the yeast, i.e., *S. cerevisiae* by expressing the genes from *P. stipitis*, involved in xylose metabolism. In one such effort, a research group has expressed XYL1 and XYL2 genes (encoding XR and XDH from *P. stipitis*) in *S. cerevisiae*, which led to enhanced bioethanol production utilizing xylose. When an endogenous gene, XKS1 which encodes for XK was overexpressed in the recombinant strain of *S. cerevisiae* (already harboring XYL1 and XYL2 gene from *P. stipitis*), ethanol yield from xylose was enhanced but it was observed that the rate of fermentation is still low and a huge amount of xylitol was accumulated during the process, making this process unsuitable for commercial applications. So, it was concluded that various metabolic strategies need to be configured out for the construction of a genetically improved strain for bioethanol production. These strategies are for optimization of the ratio of the genes XR/XDH/XK (Eliasson et al. 2001; Matsushika and Sawayama 2008), heterologous expression of transporters involved in xylose uptake and the use of enzymes with decreased/altered coenzyme affinity (Matsushika et al. 2008a, b), these enzymes include XDH and XR which show affinity toward NAD+ and NADH, respectively. All these strategies can be applied over *S. cerevisiae* strain, expressing XR, XDH, and XK.

First report of successful construction of a recombinant strain of *S. cerevisiae* came from the works of Ho and coworkers (Ho et al. 1998). This strain could efficiently ferment xylose alone as well as exhibit co-fermentation of xylose and glucose (Ho et al. 1998). In this study XR and XDH genes from *P. stipitis* and XKS gene from *S. cerevisiae* were used for the transformation of *S. cerevisiae* and co-fermentation of glucose and xylose was observed in the resultant recombinant strain. While in another report (Rodriguez et al. 1998) wherein overexpression of the endogenous gene XKS from *S. cerevisiae* was carried out, strain could not grow over xylulose. Toivari and coworkers overexpressed the endogenous genes XKS 1, PsXR, and XDH and the recombinant strain successfully fermented xylose to ethanol (Toivari et al. 2001). Many such reports have come ever since Ho et al. made an effort to genetically improve the yeast strains (Matsushika et al. 2009, Matsushika et al. 2008; Karhumaa et al. 2005; van Vleet et al. 2009). As we analyze these reports, it can be inferred that the controlled expression of XKS gene in *S. cerevisiae* can improve the xylose utilization ability of the strain as well as xylose fermentation properties. Jeffries and coworkers have been very active in this field to genetically modify the *S. cerevisiae* strain through recombinant DNA technology and they have generated some good reports. One of these reports suggests that the recombinant strain of *S. cerevisiae* which was subjected to spontaneous or chemical mutations, could overcome growth inhibition caused by the overexpression of ScXKS and PsXKS gene (Ni et al. 2007).

Since, aerobic conditions are required for xylose fermentation, it was found in one of the study carried out by (Shi et al. 2002, 1999) that respiratory deficient strains can accumulate more ethanol than the normal ones. They created a strain, whose cytochrome pathway was inhibited by way of heme protein disruption in cytochrome *c*. Results were quite interesting as this mutant produced more ethanol (0.036 g g^{-1}) than the wild type (0.027 g g^{-1}).

Classic strain improvement approaches result in random changes in the genome, which may be undesirable and may have adverse effect on the strains and the overall productivity of the yeast strains for ethanol production. However, genetic engineering approaches target the specific genes in the entire genome, involved in the metabolic pathway. Only limitation of this approach is the little knowledge of the metabolic pathways and the target points needed to generate a desired strain with higher mixed sugar fermentation efficiencies.

Cofactor Imbalance

Although, there have been several reports of genetically engineered strains harboring genes from other pentose fermenting yeasts, most of these genetically modified strains have some flaws. These flaws are due to the fact that cofactor imbalance occurs as XR and XDH are specific to NADPH and NAD, respectively, and regeneration of NADH requires aerobic conditions. To address this issue of cofactor imbalance, researchers have been trying to construct strains with altered affinities toward these cofactors. These modifications can be of three types: increasing the affinity of XRs toward NADH than NADPH, increasing the affinity of XDHs toward NADPH than NAD, and modification of both XRs and XDHs.

It is worthwhile to mention that there are two yeast species of whom XRs have more affinity toward NADH than NADPH, these are *Candida parapsilosis* and *Spathospora passalidarum*. Among the two yeast species, *S. passalidarum* exhibits xylose fermentation efficiently under anaerobic conditions (Hou et al. 2009). Specificity of XRs and XDHs of various strains has been altered by different groups. Some of these efforts have been listed in Table 1:

Basis of these mutations is the fact that XRs possess an IPKS (Ile-Pro-Lys-Ser)-conserved motif and that Lys residue interacts with the 2’-phosphate group of NADPH (Lee 2008).

Another important factor to be kept in mind while developing a potential strain for xylose utilization and henceforth its efficient fermentation is the choice of the host. Generally, *S. cerevisiae* is used as the host, but its strains also show varied efficiencies during fermentation. Therefore, it should possess the ability to ferment xylulose and be tolerant to inhibitors present in the lignocellulosic biomass (Eliasson et al. 2000; Garay-Arroyo et al. 2004; Nilsson et al. 2005; Sonderegger et al. 2004; Yu et al. 1995).

Table 1 Mutations caused in the variants to alter the affinities of XRs and XDHs

Yeast	Mutation	Purpose	Reference
XR mutants *C. tenuis*	K274M	NADPH to NADH preference	Petschacher et al. (2005)
	K274R	NADPH to NADH preference	
	S275A	NADPH to NADH preference	
	N276D	NADPH to NADH preference	
	R280H	NADPH to NADH preference	
S. stipitis	K341R/N343D	NADPH to NADH preference	Dmytruk et al. (2008)
	E223A	Strict NADPH dependence	Khattab et al. (2011)
	E223D	Strict NADPH dependence	
	S271A	Strict NADPH dependence	
	K270 M	NADPH to NADH preference	Kostrzynska et al. (1998)
XDH mutants *S. stipitis*	1208R	NAD to NADP preference	Watanabe et al. (2005)
	F209S	NAD to NADP preference	
	N211R	NAD to NADP preference	
	D270G	NADP preference	Metzger and Hollenberg (1995)
	D270G/D210G	NAD to NADP preference	

Evolutionary Genetic Approach

Apart from the genetic engineering approach, evolutionary genetic approaches may be employed as well for the construction of an improved strain. Here a genetically modified strain is further subjected to improvement strategy through evolutionary engineering. This approach includes exposing a strain to varying stress conditions and then screening the best isolate among others.

Mutagenesis is the most common approach used under this strategy.

In this approach, firstly, a recombinant strain is chosen with desired traits, which is further mutagenized and then subjected to stress conditions to select the best strain of industrial relevance. An example of this approach is the construction of a vector carrying the expression cassettes for xylose utilization (Demeke et al. 2013). In this strategy, a diploid recombinant industrial strain is mutagenized and those mutants which are able to grow on xylose are selected. Their genome is shuffled and the best D-xylose utilizers are selected in D-xylose pretreated spruce hydrolysate. The cultures are then subjected to evolutionary adaptations and the best clones are selected at different stages and the best strain from the last stage is assumed to be the best performer for xylose utilization and fermentation.

An advantage of this study is that no preliminary knowledge of the metabolic networks or metabolic pathways is required as in the case of genetic engineering approach, where a gene is being targeted. Multigenic traits can as well be addressed with this approach. Only limitation associated with this issue is the identification of genetic modifications of the improved strain as the genetic changes are random due to the random mutagenesis of the genetically modified strain.

3.1.2 Other Strategies

Efforts have been made to improve pentose fermentation efficiency of yeast strains by employing strategies other than metabolic engineering. These may include targeting the native PPP. As an explanation to this, enzymes xylulokinase, transaldolase, and transketolase have been supplemented by heterologous native genes, which have been imported from *P. stipitis*; for example, PsTAL1 may be expressed to supplement ScTAL1 (Jin et al. 2003). It was evident from the enzyme kinetic studies that the transaldolase from the *Pichia* was more efficient than from the *Saccharomyces*. This lead to the conclusion that there might be the existence of more efficient PPP enzymes involved in xylose consumption and hence bioprospecting might lead to the identification of such enzymes. Apart from targeting PPP, other strategies for strain improvement include mutagenesis, adaptation, genome shuffling through protoplast fusion, etc. (Fig. 2).

Mutagenesis

Mutations can improve the strains and these are of two types, either random mutations or site-directed mutagenesis. After a mutation has been produced, efficient screening process is required to select the desired mutant. Recombinant strains have been mutagenized showing enhanced ethanol production than the wild type strain. In one such report, a recombinant strain of *S. cerevisiae*, TJ1 was mutated with the chemical mutagen, ethyl methane sulfonate (EMS). It exhibited lower XR and higher XD and XKS activities than the parental strain leading to 1.6-fold increase in ethanol production (Tantirungkij et al. 1994). Another example is a mutant of *S. cerevisiae* TMB 3001 which can utilize xylose even under anaerobic conditions. It was developed by sequential EMS mutagenesis and then adaptation under microaerophilic and anaerobic conditions (Sonderegger et al. 2003).

A similar strategy was employed over the mutagenized strains of *S. cerevisiae* (3399 and 3400) which showed improved growth on glucose and xylose (Wahlbom et al. 2003). As is evident from the different experiments carried out by the researchers that most of the xylose utilizing and fermenting yeast species do not work best under anaerobic conditions, which is required for the co-fermentation of glucose and xylose, natural selection and random mutagenesis can be employed in such cases (Matsushika et al. 2009). Apart from EMS, other mutagens used for mutagenesis include 2-deoxyglucose, UV, etc. 2-deoxyglucose (2-DOG) has been used to produce a mutated strain, which displayed substantial improvement in xylose utilization (Sreenath et al. 2000). According to another report, a UV mutagenized *P. stipitis* NRRL Y-7124 exhibited higher ethanol production than the wild type strain (Bajwa et al. 2009).

Another strategy is site-directed mutagenesis to obtain mutants with improved xylose utilization and fermentation efficiencies. Site-directed mutagenesis was used to target NAD^+ dependent XDH from *P. stipitis*, so that a structural zinc atom could be introduced to completely reverse the specificity of the enzyme (Watanabe et al. 2005). After the screening process, the selected mutants exhibited thermo stability and increased catalytic activity with $NADP^+$. Same experiment was carried out to reverse the specificity of PsXDH toward $NADP^+$. To add to this date for mutagenesis, one of the *S. cerevisiae* mutant (MA-R5), showed high ethanol yield and low xylitol yield from xylose, under the influence of a strong promoter. This strain also exhibited fermentation with a mixture of sugars, with sugars being, glucose and xylose (Watanabe et al. 2007; Matushika et al. 2008; Matsushika et al. 2009).

Genome Shuffling Through Protoplast Fusion

Genomes of the modified and improved strains can be mixed or shuffled to produce a new improved strain of fermentative importance. Protoplast fusion is a means of genome shuffling between different yeast species. It has advantages such as promotion of high frequency of genetic information. Protoplast fusion provides characteristic advantage such as promotion of high frequencies of genetic information between organisms for which poor or no genetic exchange has been demonstrated or which are genetically uncharacterized (Heluane et al. 1993). For efficient protoplast fusion, a fusogenic reagent is required, such as PEG (polyethylene glycol). In the presence of PEG, protoplasts fuse and form diploids or hybrids. This technique has been used for fusing the pentose and hexose fermenting yeasts and the resultant hybrids displayed higher ethanol formation and yielded more biomass. An example of this work is the transfer of xylose-utilizing genes from *P. tannophilus* to *S. cerevisiae* (Heluane et al. 1993). Hybrids produced were morphologically similar to *S. cerevisiae* but were able to utilize pentose like *P. tannophilus* did. Similar work was done by Lin and coworkers who fused *Schizosaccharomyces pombe* and *Lantinula edodes* which resulted in a fusant capable of utilizing xylan as the sole carbon source (Lin et al. 2005). A thermo tolerant *S. cerevisiae* and mesophilic xylose-utilizing *C. shehatae* were electro fused leading to the formation of a fusant exhibiting 90 % fermentation efficiency and higher temperature tolerance up to 40°C (Pasha et al. 1997). A fusant F6 was obtained by fusing *C. shehatae* and *S. cerevisiae*, displaying increased ethanol production of 28 % than its parent strain (Li et al. 2008). Li and coworkers used the combinatorial strategy of firstly adapting the *C. shehatae* strain for ethanol tolerance and then mutagenizing it by UV irradiation and then screened a mutant RD-5 which was deficient in respiration. Then, the protoplasts of this mutant strain were fused with that of *S. cerevisiae* and the fusant was obtained which showed 28 % higher ethanol production than the parent *C. shehatae* strain. Pure genome shuffling experiments include the several rounds of shuffling and then finally screening strains for better ethanol yield, lesser xylitol production and enhanced tolerance to inhibitors. In one such experiment, genomes of six UV mutagenized *P. stipitis* strains were used for shuffling and 3rd and 4th round of genome shuffling, improved mutant colonies were pooled, regrown, and spread on hardwood spent sulphite liquor (HWSSL) gradient plate again (Bajwa et al. 2010). Two mutants from the 4th round were able to grow in 80 % (v/v) HWSSL while the other two mutants from the 3rd round could grow in 85 %/85 % (v/v) HWSSL. The conclusion of this study was that the mutants after protoplast fusion exhibited more inhibitor tolerance toward HWSSL than the parent strains.

Adaptation

Adaptation approach is an alternative approach for the improvement of yeast strains for xylose utilization and fermentation as it is time consuming and may not produce the desired results. Yet, there have been reports of successful enhancement of ethanol production form yeast strains using this strategy. For example, modified strains of *P. stipitis* and *C. shehatae* have been obtained for the fermentation of detoxified and partially detoxified hydrolysates (Zhu et al. 2009). Another example includes ethanologenic yeast adapted to inhibitors due to repeated subculturing over the medium with furfural and HMF up to 10–20 mM concentration. This yeast strain could grow more efficiently in the presence of inhibitors than its parent strain (Liu et al. 2004). Using combinatorial strategy, a nonrecombinant strain of *S. cerevisiae* was developed which could grow upon xylose, efficiently than its parent strain. This combinatorial strategy includes natural selection and breeding. This strategy was employed by Attfield and Bell to produce a nonrecombinant strain of *S. cerevisiae* (MBG-2303), which could grow aerobically on xylose and exhibited considerable increase in biomass yield as compared to the parent strain (Attfield et al. 2006).

Evolutionary adaptation approaches have been adopted for recombinant strains to enhance their fermentation efficiencies. Earlier, these efforts were applied on the genetically engineered strains but a new strategy involves genetic engineering, followed by mutagenesis with EMS and then two-step evolutionary adaptation (under sequential aerobic and oxygen limited conditions) (Wisselink et al. 2009). Improved strain demonstrates a fourfold increase in the specific growth compared to the parent strain. It was also found during the study, that crucial enzymes of xylose metabolism (XR, XDH, and XKS) remain unchanged, indicating that this strategy created new genetic traits letting the mutants to metabolize xylose (Liu and Hu 2010). These traits still need to be studied.

Although no prior information is required for the modification of the strains, using the above-mentioned strategies, there are some limitations associated with these strategies. Major drawback of the above-discussed strategies is the screening as it becomes tedious to screen a single strain with the desired traits amongst others. Therefore, a high-throughput technology is required for the screening processes. Also, there are a very less no. of selection criteria for the improved strains through evolutionary engineering.

Since researchers have developed a number of strategies for improvement of yeast strains for enhancing lignocellulosic ethanol production, various efforts have been made to improve strains using these strategies. A very few of them have been listed in Table 2.

Of all the methods described here, genetic engineering is the most robust technique in which all the underlying processes are known. But this strategy has not generated any potent strain of industrial relevance as there is a no. of constraints associated with this method. Therefore, other strategies are used in conjunction with genetic engineering approach.

4 Constraints

Although pentose fermenting yeasts have been discovered, but the problem of co-fermentation of glucose and xylose still remains unsolved. This is due to the fact that glucose fermentation precedes that of xylose but by the time glucose is completely fermented, yeast strains being ethanol intolerant are unable to utilize xylose efficiently. Along with ethanol intolerance, cofactor imbalance is another problem. This problem arises as a result of affinity of XR and XDH for NADPH and NAD, respectively, which leads to the accumulation of NADP and NADH causing cofactor imbalance. This issue has already been discussed in the previous sections of this review. To regenerate these cofactors, oxygen is required; hence, aerobic conditions become mandatory for xylose metabolism. Another disadvantage of aerobic fermentation is the concurrent utilization of ethanol even when xylose is not completely utilized. All these problems can be addressed by genetic manipulations or by evolutionary engineering techniques allowing the organism to grow in the presence of mixed sugars and hence fermentation to occur. One more problem of concern is the presence of inhibitors after pretreatment which adversely affect microbial growth and fermentation. This issue can be resolved by letting the strains grow in the presence of inhibitors and then screening out for the best strain capable of efficient fermentation. Although all these issues have been addressed by various research teams and several improved strains have been developed, but a strain of commercial importance has not been produced till date.

Table 2 Strain improvement strategies and ethanol yield obtained

Yeast strain	Yeast improvement strategy	Method of cultivation	Sugar concentration	Ethanol yield (gg^{-1})	Ethanol productivity ($gL^{-1}h^{-1}$)	Reference
S. cerevisiae ScF2	Genome shuffling product	Fermentative	50 gL^{-1} glucose + 50 gL^{-1} xylose	0.40	NA	Zhang et al. (2012)
S. cerevisiae A	Industrial strain	Anaerobic batch fermenstation	50 gL^{-1} glucose + 50 gL^{-1} xylose	0.42	NA	Martinez et al. (2001)
S. cerevisiae and P. stipitis	Co-culture	Fermentative	30 gL^{-1} glucose + 30 gL^{-1} xylose	0.41	NA	Rouhollah et al. (2007)
P. stipitis and K. marxianus	Co-culture	Fermentative	30 gL^{-1} glucose + 30 gL^{-1} xylose	0.36	NA	Rouhollah et al. (2007)
S. cerevisiae ITV01 *and P. stipitis Y*-7124	Co-culture	Aerobic batch fermentation	75 gL^{-1} glucose + 30 gL^{-1}xylose	0.40	NA	Gutierrez-Rivera et al. (2012)
S. cerevisiae Fusant 1	Protoplast fusant	Fermentative	30 gL^{-1} glucose + 20 gL^{-1}xylose	0.19	NA	Yan et al. (2009)
S. cerevisiae YD43-4	Protoplast fusant	Fermentative	75 gL gL^{-1} glucose + 30 gL^{-1}xylose	0.348	NA	Chmielewska et al. (2003)
S. stipitis BCRC21777	Adaptation	Adapted to increasing concentration	Rice straw hydrolysate	NA	0.44	Huang et al. (2009)
S. cerevisiae TMB 3008	Recombinant	NA	50 gL^{-1} xylose	76	0.27	Jeppsson et al. (2002)
S. cerevisiae BH42	Recombinant	NA	50 gL^{-1} glucose + 50 gL^{-1} xylose	56	NA	Sonderegger et al. (2004)
P. stipitis FPL-UC7	Recombinant	Fermentative	Xylose 80 g/L in complex medium	0.38	0.41	Shi et al. (1999)
P. stipitis FPL-Shi21	Recombinant	Fermentative	Xylose 80 g/L in complex medium	0.46	0.43	Shi et al. (1999)
S. cerevisiae (δX-27)	Recombinant	Fermentative	NA	0.35	NA	Kato et al. (2013)

5 Future Prospects

As utilization of pentoses, particularly, xylose has become a prerequisite for cost effective bioethanol production from lignocellulosic biomass, most of the research has centered on developing strategies for metabolic engineering of the yeast and bacterial strains to develop a novel strain with the desired characters. Although, various groups have been doing research and have developed new strains through genetic engineering approaches, but no successful lead has been achieved as fermentation efficiency of xylose fermentation has been observed to be insignificant for commercial purposes. Hence, strategies should be emphasized over developing strains with higher substrate tolerance, tolerance toward hydrolysate inhibitors and ethanol tolerance. The approach should be focused on efficient utilization of pentoses along with hexose for enhancement in overall fermentation efficiency of the lignocellulosic substrate. Efforts are already under process to develop a modified strain of commercial importance using high-throughput screening techniques and better expression systems for efficient production of membrane proteins. Earlier efforts included integration of genes into the plasmids and then transformation of the yeasts strains with these recombinant plasmids which lead to the development of inefficient strains, but, recently chromosomal integration of these genes has been successfully reported in yeast strains for commercial production of bioethanol from xylose (Matsushika et al. 2009). So, deep understanding of the metabolic pathways of sugar utilization in yeast is required to combine biochemical engineering with metabolic engineering strategies for the betterment of the microbial (yeast and bacterial) strains with respect to fermentation efficiency improvement.

6 Conclusion

A lot of work has been done for the improvement of yeast strains with better xylose utilization and fermentation abilities. This work has generated a lot of data for reference. Of the various strategies employed for the betterment of the yeast strains, genetic engineering is the most precise one, as it targets the specific genes involved in the pentose metabolism. Though studies suggest that none of the strategies can alone be used for the improvement of the strains. So, a combinatorial strategy is required. These combinatorial studies include strategies as listed in the previous sections, such as genetic engineering, evolutionary engineering like mutagenesis, adaptation, mutations, and genome shuffling through protoplast fusion. Some studies also suggest that there are some hidden traits of the xylose metabolism which are still uncovered and a complete understanding of these traits can help in constructing a strain of commercial reputation.

References

Aristidou A, Penttila M (2000) Metabolic engineering applications to renewable resource utilization. Curr Opin Biotechnol 11:187–198

Attfield PV, Bell PJ (2006) Use of population genetics to derive non recombinant *Saccharomyces cerevisiae* strains that grow using xylose as a sole carbon source. FEMS Yeast Res 6:862–868

Bailey JE (1991) Toward a science of metabolic engineering. Science 252(5013):1668–1675

Bajwa PK, Shireen T, D'Aoust F, Pinel D, Martin VJJ, Trevors JT et al (2009) Mutants of the pentose-fermenting yeast*Pichia stipitis* with improved tolerance to inhibitors in spent sulfite liquor. Biotechnol Bioeng 104:892–900

Bajwa PK, Pinel D, Martin VJJ, Trevors JT, Lee H (2010) Strain improvement of the pentose-fermenting yeast *Pichia stipitis* by genome shuffling. J Microbiol Methods 81:179–186

Chandel AK, Chandrasekhar G, Radhika K, Ravinder R, Ravindra P (2011) Bioconversion of pentose sugars into ethanol: a review and future directions. Biotechnol Mol Biol Rev 6(1):008–020

Chandel AK, Narasu ML, Ravinder R, Edula JR, Pasha C, Ravindra P, Rao LV (2008a). Forecasting bioethanol production from agro crop residues in Andhra Pradesh state: a case study. Technol Spectrum 1(2):12–27

Chmielewska J (2003) Selected biotechnological features of hybrids of *Saccharomyces cerevisiae* and *Yamadazyma stipitis*. Elec J Po Agri Univ Biotechnol 6:1–13

Chu Byron CH, Lee Hung (2007) Genetic improvement of *Saccharomyces cerevisiae* for xylose fermentation. Biotechnol Adv 25:425–441

Demeke Mekonnen M, Dietz Heiko, Li Yingying, Foulique-Moreno Maria R, Mutturi Sarma, Deprez Sylvie, Den Abt Tom, Bonini Beatriz, Liden Gunnar, Dumortier Francoise, Verplaeste Alex, Boles Eckhard, Thevelein Johan M (2013) Development of a D-xylose fermenting and inhibitor tolerant industrial *Saccharomyces cerevisiae* strain with high performance in lignocellulose hydrolysates using metabolic and evolutionary engineering. Biotechnol for biofuels. 6:89

Dmytruk OV, Dmytruk KV, Abbas CA, Voronovsky AY, Sibirny AA (2008) Engineering of xylose reductase and overexpression of xylitol dehydrogenase and xylulokinase improves xylose alcoholic fermentation in the thermotolerant yeast *Hansenula polymorpha*. Microb Cell Fact 7:21–28

Does AL, Bisson LF (1989) Characterization of xylose uptake in the yeasts *Pichia heedii* and *Pichia stipitis*. Appl Environ Microbiol 55:159–164

Does AL, Bisson LF (1990) Isolation and characterization of *Pichia heedii* mutants defective in Xylose uptake. Appl Environ Microbiol 56(11):3321–3328

Eliasson A, Boles E, Johansson B, Österberg M, Thevelein JM, Spencer-Martins I, Juhnke H, Hahn-Hägerdal B (2000) Xylulose fermentation by mutant and wild-type strains of *Zygosaccharomyces* and *Saccharomyces cerevisiae*. Appl Microbiol Biotechnol 53(4):376–382

Eliasson A, Hofmeyr J-HS, Pedler S, Hahn-Hägerdal B (2001) The xylose reductase/xylitol dehydrogenase/xylulokinase ratio affects product formation in recombinant xylose-utilising *Saccharomyces cerevisiae*. Enzyme Microb Technol 29:288–297

Fonseca C, Spencer-Martins I, Hahn-Hägerdal B (2007) L-arabinose metabolism in *Candida arabinofermentans* PYCC 5603^T and *Pichia guilliermondii* PYCC 3012: influence of sugar and oxygen on product formation. Appl Microbiol Biotechnol. doi:10.1007/s00253-066-0830-7

Garay-Arroyo A, Covarrubias AA, Clark I, Niño I, Gosset G, Martinez A (2004) Response to different environmental stress conditions of industrial and laboratory *Saccharomyces cerevisiae* strains. Appl Microbiol Biotechnol 63(6):734–741

Gárdonyi M, Jeppsson M, Liden G, Gorwa-Grauslund MF, HahnHägerdal B (2003) Control of xylose consumption by xylose transport in recombinant Saccharomyces cerevisiae. Biotechnol Bioeng 82(7):818–824

Gong C-S, Chen L-F, Flickinger MC, Chiang L-C, Tsao GT (1981) Production of ethanol from D-xylose by using D-xylose isomerase and yeasts. Appl Environ Microbiol 41:430–436

Gutierrez-Rivera B, Waliszewski-Kubiak K, Carvajal-Zarrabal O, Aguilar-Uscanga MG (2012) Conversion efficiency of glucose/xylose mixtures for ethanol production using *Saccharomyces cerevisiae ITV01* and *Pichia stipitis NRRL Y-7124*. J Chem Technol Biotechnol 87:263–270

Hahn-Hägerdal B, Karhumaa K, Fonseca C, Spencer-Martins I, Gorwa Grauslund MF (2007) Towards industrial pentose-fermenting yeast strains. Appl Microbiol Biotechnol 74:937–953

Heluane H, Spencer JFT, Spencer D, de Figueroa L, Callieri DAS (1993) Characterization of hybrids obtained by protoplast fusion between *Pachysolen tannophilus* and *Saccharomyces cerevisiae*. Appl Microbiol Biotechnol 40:98–100

Ho NW, Chen Z, Brainard AP (1998) Genetically engineered *Saccharomyces* yeast capable of effective cofermentation of glucose and xylose. Appl Environ Microbiol 64:1852–1859

Hou J, Vemuri GN, Bao X, Olsson L (2009) Impact of overexpressing NADH kinase on glucose and xylose metabolism in recombinant xylose-utilizing *Saccharomyces cerevisiae*. Appl Microbiol Biotechnol 82:909–919

Huang CF, Lin TH, Guo GL, Hwang WS (2009) Enhanced ethanol production by fermentation of rice straw hydrolysate without detoxifiation using a newly adapted strain of *Pichia stipitis*. Bioresour Technol 100:3914–3920

Jeffries T (1983) Utilization of xylose by bacteria yeasts and fungi. Adv Biochem Eng Biotechnol 27:1–32

Jeffries TW (1985) Emerging technology for fermenting d-xylose. Trends Biotechnol 3:208–212

Jeppsson M, Johansson B, Hahn-Hägerdal B, Gorwa-Grauslund MF (2002) Reduced oxidative pentose phosphate pathway flux in recombinant xylose-utilizing *Saccharomyces cerevisiae* strains improves the ethanol yield from xylose. Appl Environ Microbiol 68:1604–1609

Jin Y-S, Ni H, Laplaza JM, Jeffries TW (2003) Optimal growth and ethanol production from xylose by recombinant *Saccharomyces cerevisiae* require moderate D-xylulokinase activity. Appl Environ Microbiol 69:495–503

Karhumaa K, Hahn-Hägerdal B, Gorwa-Grauslund MF (2005) Investigation of limiting metabolic steps in the utilization of xylose by recombinant *Saccharomyces cerevisiae* using metabolic engineering. Yeast 22:359–368

Kato H, Matsuda F, Yamada R, Nagata K, Shirai T, Hasunuma T et al (2013) Cocktail δ-integration of xylose assimilation genes for efficient ethanol production from xylose in *Saccharomyces cerevisiae*. J Biosci Bioeng 116:333–336. doi:10.1016/j.jbiosc.2013.03.020

Khattab SMR, Watanabe S, Saimura M, Kodaki T (2011) A novel strictly NADPH-dependent *Pichia stipitis* xylose reductase constructed by site-directed mutagenesis. Biochem Biophys Res Commun 404:634–637

Kilian SG, Vanuden N (1988) Transport of xylose and glucose in the xylose fermenting yeast *Pichia stipitis*. Appl Microbiol Biotechnol 27:545–548

Kostrzynska M, Sopher CR, Lee H (1998) Mutational analysis of the role of the conserved lysine-270 in the *Pichia stipitis* xylose reductase. FEMS Microbiol Lett 159:107–112

Kuhn A, van Zyl C, van Tonder A, Prior BA (1995) Purification and partial characterization of an aldo-keto reductase from *Saccharomyces cerevisiae*. Appl Environ Microbiol 61:1580–1585

Lee SK, Chou H, Ham TS, Lee TS, Keasling JD (2008) Metabolic engineering of microorganisms for biofuels production: from bugs to synthetic biology to fuels. Curr Opin Biotechnol 19 (6):556–563

Li J, Zhao X, Liu C, Li F, Ren J, Bai F (2008) Construction of yeast strains for efficient ethanol fermentation from xylose by protoplast fusion. Abstracts V4-P-043. J Biotechnol 136:402–459

Lin C, Hsieh P, Mau J, Teng D (2005) Construction of an intergeneric fusion from *Schizosaccharomyces pombe* and *Lentinula edodes* for xylan degradation and polyol production. Enzyme Microb Technol 36:107–117

Liu E, Hu Y (2010) Construction of a xylose-fermenting *Saccharomyces cerevisiae* strain by combined approaches of genetic engineering, chemical mutagenesis and evolutionary adaptation. Biochem Eng J 48:204–210

Liu ZL, Slininger PJ, Dien BS, Berhow MA, Kurtzman CP, Gorsich SW (2004) Adaptive response of yeasts to furfural and 5-hydroxymethylfurfural and new chemical evidence for HMF conversion to 2,5-bis-hydroxymethylfuran. J Ind Microbiol Biotechnol 31:345–352

Lucas C, Vanuden N (1986) Transport of hemicellulose monomers in the xylose-fermenting yeast *Candida shehatae*. Appl Microbiol Biotechnol 23:491–495

Maleszka R, Schneider H (1982) Concurrent production and consumption of ethanol by cultures of *Pachysolen tannophilus* growing on d-xylose. Appl Environ Microbiol 44:909–912

Martinez A, Rodriguez ME, Wells ML, York SW, Preston JF, Ingram LO (2001) Detoxification of dilute acid hydrolysates of lignocellulose with lime. Biotechnol Prog 17(2):287–293

Matsushika A, Sawayama S (2008) Efficient bioethanol production from xylose by recombinant *Saccharomyces cerevisiae* requires high activity of xylose reductase and moderate xylulokinase activity. J Biosci Bioeng 106(3):306–309

Matsushika A, Watanabe S, Kodaki T, Makino K, Sawayama S (2008a) Bioethanol production from xylose by recombinant *Saccharomyces cerevisiae* expressing xylose reductase, NADP+-dependent xylitol dehydrogenase, and xylulokinase. J Biosci Bioeng 105(3):296–299

Matsushika A, Watanabe S, Kodaki T, Makino K, Inoue H, Murakami K, Takimura O, Sawayama S (2008b) Expression of protein engineered NADP+-dependent xylitol dehydrogenase increase ethanol production from xylose in recombinant *Saccharomyces cerevisiae*. Appl Microbiol Biotechnol 81(2):243–255

Matsushika A, Inoue H, Murakami K, Takimura O, Sawayama S (2009a) Bioethanol production performance of five recombinant strains of laboratory and industrial xylose-fermenting *Saccharomyces cerevisiae*. Bioresour Technol 100:2392–2398

Matsushika A, Inoue H, Watanabe S, Kodaki T, Makino K, Sawayama S (2009) Efficient bioethanol production by recombinant flocculent *Saccharomyces cerevisiae* with genome-integrated NADP+-dependent xylitol dehydrogenase gene. Appl Environ Microbiol 75:3818–22

McMillan JD (1993) Xylose Fermenta to Ethanol: A Review

Metzger MH, Hollenberg CP (1995) Amino acid substitutions in the yeast *Pichia stipitis* xylitol dehydrogenase coenzyme binding domain affect the coenzyme specifiity. Eur J Biochem 228(1):50–54

Mishra P, Singh A (1993) Microbial pentose utilization. Adv Appl Microbiol 39:91–152

Ni H, Laplaza M, Jeffries TW (2007) Transposon mutagenesis to improve the growth of recombinant *Saccharomyces cerevisiae* on d-xylose. Appl Environ Microbiol 73:2061–2066

Nilsson A, Gorwa-Grauslund MF, Hahn-Hägerdal B, Lidén G (2005) Cofactor dependence in furan reduction by *Saccharomyces cerevisiae* in fermentation of acid-hydrolyzed lignocellulose. Appl Environ Microbiol 71(12):7866–7871

Ostergaard S, Olsson L, Nielsen J (2000) Metabolic engineering of *Saccharomyces cerevisiae*. Microbiol Mol Biol Rev 64:34–50

Pasha C, Kuhad RC, Rao LV (1997) Strain improvement of thermotolerant *Saccharomyces cerevisiae*VS3 strain for better utilization of lignocellulosic substrates. J Appl Microbiol 103:1480–1489

Petschacher B, Leitgeb S, Kavanagh KL, Wilson DK, Nidetzky B (2005) The coenzyme specifiity of *Candida tenuis* xylose reductase (AKR2B5) explored by site-directed mutagenesis and X-ray crystallography. Biochem J 385:75–83

Rodriguez-Peña JM, Cid VJ, Arroyo J, Nombela C (1998) The YGR194c (XKS1) gene encodes the xylulokinase from the budding yeast Saccharomyces cerevisiae. FEMS Microbiology Letters 162(1):155–160

Rouhollah H, Iraj N, Giti E, Sorah A (2007) Mixed sugar fermentation by *Pichia stipitis*, *Saccharomyces cerevisiae* and an isolated xylose-fermenting *Kluyveromyces marxianus* and their cocultures. Afr J Biotechnol 6:1110–1114

Shi NQ, Davis B, Sherman F, Cruz J, Jeffries TW (1999) Disruption of the cytochrome c gene in xylose-utilizing yeast *Pichia stipitis* leads to higher ethanol production. Yeast 15:1021–1030

Shi NQ, Cruz J, Sherman F, Jeffries TW (2002) SHAM-sensitive alternative respiration in the xylose-metabolizing yeast *Pichia stipitis*. Yeast 19:1203–1220

Skoog K, Hahn-Hägerdal B (1990) Effect of oxygenation on xylose fermentation by *Pichia stipitis*. Appl Env Microbiol 56(11):3389–3394

Sonderegger M, Sauer U (2003) Evolutionary engineering of *Saccharomyces cerevisiae* for anaerobic growth on xylose. Appl Environ Microbiol 69:1990–8

Sonderegger M, Jeppsson M, Hahn-Hägerdal B, Sauer U (2004a) Molecular basis for anaerobic growth of *Saccharomyces cerevisiae* on xylose, investigated by global gene expression and metabolic flux analysis. Appl Environ Microbiol 70:2307–2317

Sonderegger M, Jeppsson M, Larsson C, Gorwa-Grauslund MF, Boles E, Olsson L, Spencer-Martins I, Hahn-Hägerdal B, Sauer U (2004b) Fermentation performance of engineered and evolved xylose-fermenting *Saccharomyces cerevisiae* strains. Biotechnol Bioeng 87(1):90–98

Sreenath HK, Jeffreis TW (2000) Production of ethanol from wood hydrolysate by yeasts. Bioresour Technol 72:253–260

Stephanopoulos G, Aristidou AA, Nielsen J (1998) Metabolic engineering: principles and methodologies Academic press, p 17

Stephanopoulos (2007) Challenges in engineering microbes for biofuels production. Science 315 (5813):801–4

Tantirungkij M, Izuishi T, Seki T, Yoshida T (1994) Fed-batch fermentation of xylose by a fast-growing mutant of xylose-assimilating recombinant *Saccharomyces cerevisiae.* Appl Microbiol Biotechnol 41:8–12

Toivari MH, Aristidou A, Ruohonen L, Penttila M (2001) Conversion of xylose to ethanol by recombinant *Saccharomyces cerevisiae*: importance of xylulokinase (XKS1) and oxygen availability. Metab Eng 3:236–249

Toivari MH, Salusjärvi L, Ruohonen L, Penttilä M (2004) Endogenous xylose pathway in *Saccharomyces cerevisiae* Applied and environmental microbiology 70(6):3681–3686

van Vleet JH, Jeffries TW (2009) Yeast metabolic engineering for hemicellulosic ethanol production. Curr Opin Biotechnol 20:300–306

Wahlbom F, Cordero Otero RR, van Zyl WH, Hahn-Hägerdal B, Jonsson LJ (2003) Molecular analysis of a *Saccharomyces cerevisiae* mutant with improved ability to utilize xylose shows enhanced expression of proteins involved in transport, initial xylose metabolism, and the pentose phosphate pathway. Appl Environ Microbiol 69:740–746

Watanabe S, Kodaki T, Makino K (2005) Complete reversal of coenzyme specificity of xylitol dehydrogenase and increase of thermostability by the introduction of structural zinc. J Biol Chem 280:10340–10349

Watanabe S, Saleh AA, Pack SP, Annaluru N, Kodaki T, Makino K (2007) Ethanol production from xylose by recombinant *Saccharomyces cerevisiae* expressing protein engineered NADP+-dependent xylitol dehydrogenase. J Biotechnol 130:316–319

Webb SR, Lee H (1990) Regulation of D-xylose Utiization by Hexoses I Pentose-fermenting Yeasts. Biotechnolog Adv 8:685–697

Wisselink HW, Toirkens MJ, Wu Q, Pronk JT, van Maris AJ (2009) A novel evolutionary engineering approach for accelerated utilization of glucose, xylose and arabinose mixtures by engineered *Saccharomyces cerevisiae.* Appl Environ Microbiol 75:907–914

Yan F, Bai F, Tian S, Zhang J, Zhang Z, Yang X (2009) Strain construction for ethanol production from dilute-acid lignocellulosic hydrolysate. Appl Biochem Biotechnol 157(3):473–482

Yu S, Jeppsson H, Hahn-Hägerdal B (1995) Xylulose fermentation by *Saccharomyces cerevisiae* and xylose-fermenting yeast strains. Appl Microbiol Biotechnol 44:314–320

Zhang W, Geng A (2012) Improved ethanol production by a xylose—fermenting yeast strain constructed through a modified genome shuffling method. Biotechnol Biofuels 5:46

Zhu J-J, Yong Q, Xu Y, Chen S-X, Yu SY (2009) Adaptation fermentation of *Pichia stipitis* and combination detoxification on steam exploded lignocellulosic prehydrolyzate. Nat Sci 1:47–54

Pretreatment of Paddy Straw to Improve Biogas Yield

Rupali Mahajan, Harmanjot Kaur, Raman Rao and Sachin Kumar

Abstract Biofuels have been considered as an alternative energy source to cope up the future energy demand. There are various biomass-based energy sources including biogas, bioethanol, biobutanol, etc. Biogas is one of the biofuels which can be produced from the lignocellulosic biomass. Cellulose, hemicellulose, lignin, and several inorganic materials constitute lignocelluloses. Lignocelluloses are often the sole component of different agricultural waste such as paddy straw. Paddy straw is one of the lignocellulosic biomass that can be used as feedstock for biogas production because of its great profusion, sustainable supply, and low price. This type of biomass is not fully biodegraded in anaerobic digestion process at industrial scale due to their intricate physical and chemical structure that consequently leads to lower energy resurgence in terms of biogas yield. The biodegradability of these lignocelluloses waste can be increased by a pretreatment. Therefore, breakdown of complex materials to simpler ones or to make cellulose accessible to hydrolytic enzymes is the key step for better digestion to biogas. The aim of this paper is to review promising pretreatment technologies which have greatly improved the production of biogas. Effective parameters in paddy straw pretreatment involve increased fortification by lignin and hemicelluloses. Various pretreatment methods such as mechanical, thermal, chemical, biological, and their advantages and disadvantages as well as effects on the improvement of biogas production are discussed.

Keywords Biogas · Pretreatment · Lignocelluloses · Paddy straw

Rupali Mahajan · Harmanjot Kaur · Raman Rao · S. Kumar (✉)
Biochemical Conversion Division, Sardar Swaran Singh National Institute of Bio-Energy (Formerly Sardar Swaran Singh National Institute of Renewable Energy), Kapurthala 144601, Punjab, India
e-mail: sachin.biotech@gmail.com

S. Kumar et al. (eds.), *Proceedings of the First International Conference on Recent Advances in Bioenergy Research*, Springer Proceedings in Energy,
DOI 10.1007/978-81-322-2773-1_4

1 Introduction

With advancement in the technology, the energy requirement for the urbanized industries has been tremendously increased all around the world. Moreover, the easily accessible fossil fuels have also been decreased during almost two centuries and have resulted in the increased prices. Therefore, supply of energy becomes one of the major causes of concern for the globe (Cheremisinoff 2002). Conversely the incineration of fossil fuel carriers such as petrol, natural gas, etc., liberates CO_2, SO_x, and NO_X which causes gigantic environmental problems and adversely affects the ecosystem along with their negative impact on health. Currently, there is a real apprehension about the environment due to its significance to a sustainable future. The stipulation for a better way of living, as well as alleviation of environment obliges us to search about new kinds of technologies, habits, and also to develop way of thinking with least negative impacts (Mussgnug et al. 2010). As Richard Wright says "In the brilliance of universal climate change, poignant in the path of renewable energy sources seems indispensible".

The agricultural sector has long been acknowledged as one of the most fascinating activities for effectual execution of large scale use of renewable energy especially conversion of feedstock from the plant or animal origin. Therefore agricultural waste such as wheat straw and paddy straw can play an integral role in meeting the increasing societal demands of energy in sustainable manner. Specifically lignocellulosic waste has enormous potential for energy generation (Chandra et al. 2012a). Rice is ranked as third largest agricultural crop grown in the world with total cultivated land of 161, 42 million hectare with a gross yield production of 678.69 million tons. It is estimated that about 905 million tons per annum of lignocellulosic biomass is produced from cultivated rice. In order to obtain better yield from these lignocellulosic materials pretreatment plays an imperative role. Further for the better production of biogas the biodigestibility of the waste can be enhanced by pretreatment. Moreover, it has been observed that in anaerobic digestion most of the contents of biomass such as carbohydrates, proteins, and fats are converted into simple derivatives and at last transformed into biogas with the assistance of different types of anaerobic microorganisms. Anaerobic digestion (AD) of biodegradable waste and other agricultural residues is extensively used as it yields biogas rich in methane and carbon dioxide and is suitable for production of energy. A range of pretreatment technologies have been developed in the topical years in order to amplify the accessibility of AD of sugars and other small molecules in biogas substrates predominantly in lignocellulosic waste materials such as paddy straw (Fredriksson et al. 2006). These pretreatment technologies basically aim to

- Increase the yield of biogas
- To make anaerobic digestion faster
- To avert processing tribulations such as high electricity requirement for amalgamation and development of floating layers
- To make use of new substrates.

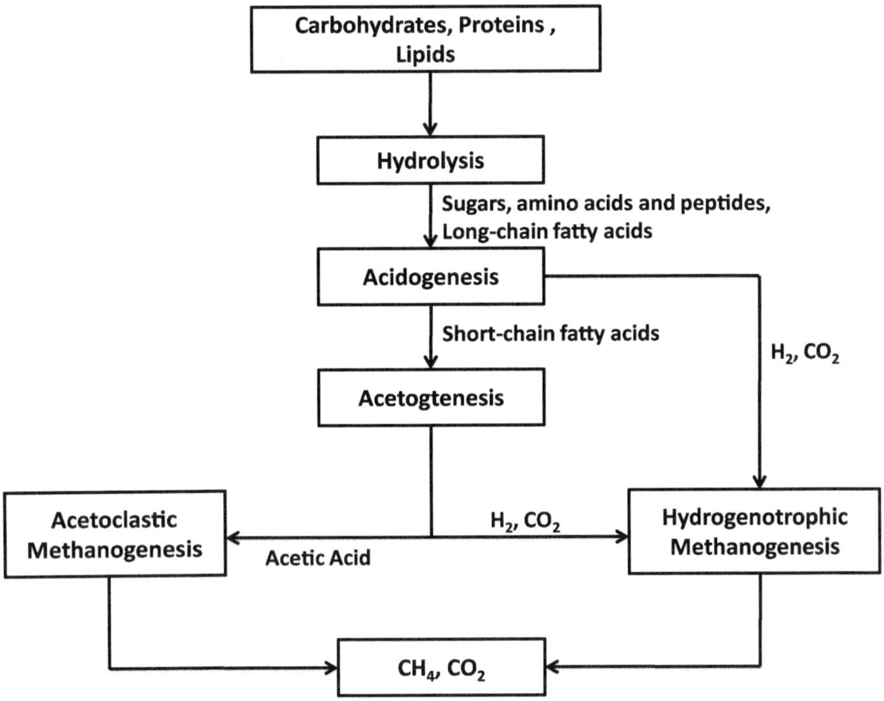

Fig. 1 Anaerobic digestion process

Anaerobic digestion generally involves four steps (as shown in Fig. 1), i.e.,

I. Hydrolysis: This step occurs at the beginning of the process as lignocelluloses have a recalcitrant structure which is naturally designed to prevent the enzymatic degradation; hydrolysis occurs as extracellular enzymes, produced by hydrolytic microbes that decompose complex organic polymers into simple monomers.
II. Acidogenesis: These small molecules are converted to the mixture of volatile fatty acids (VFAs) and alcohol by microorganisms of anaerobic and facultative group called acidogenic bacteria.
III. Acetogenesis: The VFAs are further converted to acetate, carbon dioxide, and hydrogen by Acetogenic bacteria and further act as substrate for methanogens.
IV. Methanogenesis: In this step, the anaerobic bacteria called as methane formers converts organic acid into biogas. The main constituents of biogas are methane and carbon dioxide with traces of H_2S, H_2, N_2, etc. These methanogens are quite sensitive to the pH changes (Zheng et al. 2014).

The degradation of lignocelluloses is a complicated process as lignocellulosic waste material such as paddy straw has a recalcitrant composition and often contain high amount of lignin which is difficult to degrade. Therefore, consecutively a pretreatment process is quite indispensable to disrupt the naturally intractable lignin shield that impairs the accessibility of enzymes and microbes to cellulose and hemicelluloses to improve the yield of biogas from paddy straw (Teghammar et al. 2014). The present paper paves the way for integrating the waste obtained from rice, by reviewing different methods for their pretreatment and production of biogas with its positive and negative aspects.

2 Material Composition of Paddy Straw

2.1 *Cellulose*

It is the main component of lignocellulosic biomass. It is a polysaccharide that consists of a linear chain of D-glucose linked by $\beta-1$, 4 linkages. The cellulose strains are allied together to make cellulose fibrils (Li et al. 2010). Cellulose fibers are connected by a number of intra and intermolecular hydrogen bonds. Therefore cellulose has high solubility in water and the majority of organic solvents (Swatloski et al. 2002).

2.2 *Hemicellulose*

Hemicelluloses are assorted branched polymers containing pentoses and hexoses. They are located in the secondary cell walls of the lignocellulosic biomass (Girio et al. 2010). They are easy to hydrolyze because of its amorphous nature and branched structure as well as due to its low molecular weight. As hemicelluloses covers the cellulose fibrils and affects its digestibility and availability for enzymatic hydrolysis (Agbor et al. 2011). The accessibility to celluloses can be increased by the removal of hemicelluloses.

2.3 *Lignin*

Lignin is an aromatic polymer synthesized from propanoid precursors. The cellulose and hemicelluloses are creased by lignin. It is accountable for structural stringency and integrity. The major chemical phenylpropane units of lignin make a complicated matrix consisting primarily of syringyl, guaiacyl, and p-hydroxy phenol. Therefore, the pretreatment of lignocellulosic biomass results in improved rate of hydrolysis. By the removal of lignin, the pretreatment provides better surface accessibility for enzymes by increasing the pore volume (Demirbas 2008). Figure 2 shows the percentage material composition of lignocellulosic biomass.

Fig. 2 Composition of paddy straw

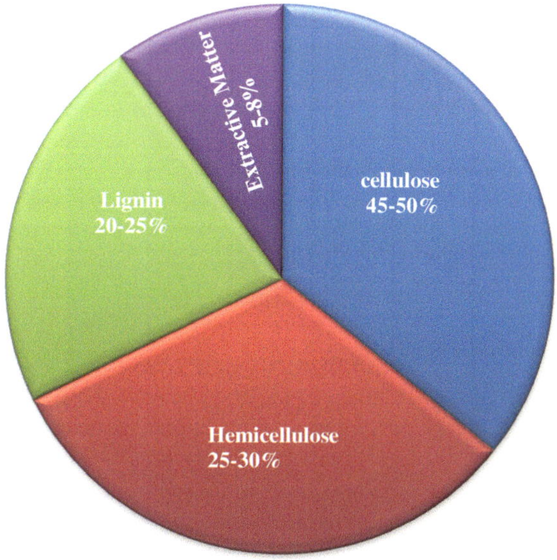

3 Pretreatment Methods

The basic idea behind pretreating a particular substrate for biogas process is to increase the yield of methane and also to minimize the required time in order to get the maximum yield. If the degradation process is fast, it will result in the higher total degradation of the substrate in the biogas process and therefore would result in the better yield of methane.

3.1 Physical Pretreatment

Physical pretreatment can result in increased pore size and accessible surface area. It further helps in reducing the crystallinity and degree of polymerization of cellulose. Among physical pretreatments diverse types of techniques such as mechanical, thermal, as well as other techniques such as freeze pretreatment and hydrothermal pretreatment can be used to improve the biodegradability or enzymatic hydrolysis of lignocellulosic wastes.

3.1.1 Mechanical Pretreatment

The main intent of mechanical pretreatment is to bring reduction in particle size as well as crystallinity of lignocellulosic materials. The specific surface area increases with the reduction in particle size and also reduction in the degree of polymerization. This effect can be obtained by grinding or milling, finally depending upon the particle size of the material. Further depending up on the nature of biomass, type of milling, and also the extent of milling; hydrolysis yield can be increased by 5–25 % (Delgenes et al. 2002). One of the disadvantages of using mechanical pretreatment is that the power constraint of this method is comparatively high depending upon the characteristics as well as the final particle size of the biomass.

3.1.2 Thermal Pretreatment

Thermal pretreatment is basically responsible for structural alteration of the insoluble fraction to make it more amenable to biodegradability. In this the substrate is held at high temperature of 125–190 °C under pressure for up to 1 h. The high temperature is accountable for disrupting the hydrogen bonds that hold together crystalline cellulose and lignocelluloses complexes. Thermal pretreatment is effective up to certain temperatures. The maximum range of temperature depends upon the nature of substrate and also on pretreatment retention time (Garrote et al. 1999). Although various studies have revealed that thermal pretreatment is used to increase the yield of biogas up to certain range of temperature, above which biogas yield starts decreasing. Moreover at high temperatures, certain dark colored lignin and xylose breakdown products are produced. These compounds comprise of phenolic and heterocyclic compounds such as furfural. These compounds act as inhibitory compounds and inhibit the activity of AD microorganisms and hence affect the yield of biogas.

Liquid Hot Water (LHW) Pretreatment

Liquid hot water pretreatment is used to make cellulose accessible to AD microorganisms by solubilizing the hemicelluloses. Further the formation of inhibitors can also be avoided by maintaining the pH range between 4 and 7. The pH range between 4 and 7 minimizes the formation of monosaccharides and other degradation products that can further catalyze hydrolysis of cellulosic material during pretreatment. Therefore, by maintaining the pH range of 4 and 7 the autocatalytic formation of fermentation inhibitors is avoided during pretreatment (Kohlmann et al. 1995; Mosier et al. 2005; Weil et al. 1997). The major advantage of using liquid hot water is that the solubilized hemicelluloses and lignin products are present in lower concentrations due to higher water input. Due to the lower concentrations of hemicelluloses and lignin the risk on degradation products like furfural is reduced. Further it also helps reducing condensation and precipitation of lignin compounds.

3.1.3 Hydrothermal Pretreatment

In hydrothermal pretreatment water is used under high temperature and pressure so that it can penetrate into the biomass, hydrate cellulose, and remove hemicelluloses and part of lignin. The major advantages of this pretreatment are

- No addition of chemicals
- No requirement of corrosion resistant materials for hydrolysis reactors
- No need for size reduction of biomass (Taherzadeh and Karimi 2008).
- Enhances accessible and susceptible surface area of the cellulose and makes it more accessible to hydrolytic enzymes (Zeng et al. 2007).

Pretreatments of paddy straw biomass were investigated by Chandra et al. (2012b) hydrothermal pretreatment was done which was further followed by addition of 5 % NaOH at 200 °C for 10 min, in order to maintain the pH suitable for the fermentation of methane and to circumvent the formation of inhibitor. Various experimental studies have also revealed that addition of NaOH to paddy straw results in enhanced biodegradability of paddy straw biomass which is apparent from the augmented production of biogas. The result showed that biogas and methane production yield as 315.9 and 132.7 L/kg VS, respectively, increase of 225.6 % in biogas production and 222.0 % in methane production relative to untreated paddy straw. Hydrothermal pretreatment resulted into enhanced biogas and methane production by accelerated prehydrolysis of biomass contents during the treatment process. Moreover, it has taken longer hydraulic retention time as compared to untreated substrates. The reason for taking longer HRT to convert the available volatile degradable matter into biogas is that inadequate bacterial population were present in the substrate during the process of degradation. Further an increase in HRT may provide ample time for methanogens to convert organic matter to methane and carbon dioxide (Kaparaju et al. 2010). Therefore, it is quite apparent that the hydrothermal pretreatment is potentially effective for pretreatment to enhance biogas yield from paddy straw biomass followed by proper amount of NaOH addition.

3.1.4 Freeze Pretreatment

This pretreatment is requisite to amend the structural and chemical composition of feedstock sequentially to provide efficient hydrolysis. For efficient hydrolysis the cell structures need to be broken down and the pore size must be increased to convert carbohydrates to fermentable sugars. Freeze pretreatment is one of the unique approaches that boost the enzyme digestibility of paddy straw (Keshwani and Cheng 2009). According to the study of Chang et al. (2011), with the help of freeze pretreatment the enzymatic hydrolysis yield was increased by 84 %. In this pretreatment the paddy straw was pretreated by freezing at −20 °C for 2 h and the frozen straw was thawed at room temperature for 1 h. The result showed that freeze pretreatment can reduce sugar yields from 226–77 to 417–27 g/kg and 93–84 to

138–77 g/kg substrate for 150 U cellulose and 100 U xylanase. Further it was reported that acetate buffer was used to preincubate paddy straw for 1 h and the solute of acetate buffer penetrated in the pores of paddy straw. The pore size and the accessible surface area expanded as the buffer solute freeze. This further results in breakdown of some structures and enhanced the rate of enzymatic hydrolysis (Tuankriangkrai and Benjakul 2010).

3.2 Chemical Pretreatment

Chemical pretreatment has been investigated under diverse conditions using the series of different chemicals, mainly acids, and bases of diverse strengths

3.2.1 Alkali Pretreatment

As discussed earlier, lignocellulosic materials due to their structure and composition are defiant to hydrolysis. Lignocellulosic materials swell by the addition of alkali and further results in partial solubilization of lignin. Additionally, alkali pretreatment due to its low price, high recoverability, safe handling, and minor environmental effects have been extensively used in improving the biodegradability. There have been numerous reports which show that alkali pretreatment is quite effective for anaerobic digestion. According to He et al. (2008) noteworthy increment was seen in the biogas yield in batch tests using paddy straw. The paddy straw used was pretreated with 6 % NaOH for 3 weeks at ambient temperature. The biogas yield was increased by 27.3–64.5 %. The enhancement in yield was accredited to NaOH pretreatment that increased biodegradability of paddy straw.

$Ca(OH)_2$ is one of the alkalis that has been comprehensively used in improving the biodegradability due to its high recoverability, low cost, safe handling, and inconsequential environmental effects. Song et al. (2012) reported that the lignin, cellulose, and hemicelluloses content of paddy straw were significantly degraded when pretreated with 9–81 % $Ca(OH)_2$ (w/w TS) for 5–89 day pretreatment time and 45–12 % inoculums content. Further the methane yield was also increased up to 225–300 ml/g VS. These studies verified that alkali pretreatment can act as a potential pretreatment option to increase the gas yield from lignocellulosic substrates. Further it is important to note that pretreatment time considerably affects the degradation of core components of paddy straw and the extensive treatments increased the diminution in cellulose and hemicelluloses contents under certain conditions. In general, this pretreatment technology is economically not so attractive due to high costs of alkalis but it may be useful for acidic substrates or substrates rich in lignin content that could otherwise not be anaerobically digested (Chang et al. 1997).

3.2.2 Acidic Pretreatment

Acid pretreatment disparately from alkali pretreatment does not result in lignin disruption but is considered to break down hemicelluloses and interrupt ether bonds between lignin and hemicellulose. Further for lignocellulosic biomass dilute acid is particularly favored over concentrated acid for acid pretreatment (Kalnappert et al. 1981) Dilute acid primarily hydrolyzes up to 100 % of hemicelluloses into its constituent sugars for example galactose, arabinose, and xylose depending on the pretreatment conditions (Zheng et al. 2014). Zhao et al. (2010) reported that dilute acid pretreatment with mixture of acetic-propionic acid significantly enhanced 35–84 % production of methane. Further it was also revealed from the study that acid pretreatment at concentration of 0–75 mol/L for 2 h at solid to liquid ratio of 1:20 (w/v), was able to remove about 34–19 % lignin and 21–15 % hydrolysis rate was obtained. 0–75 mol/L of acid concentration can enhance the biodegradability of paddy straw and can remove the hemicelluloses and lignin to some extent. This was represented by the increased rate of hydrolysis because it could contribute more hydrogen ions which help solubilizing hemicelluloses and lignin. As a result, the cellulose can be more easily converted into fermentable sugars due to increased accessibility caused by disruption of barriers by lignin and hemicelluloses. The removal of lignin is not associated with the longer pretreatment time. It is more likely that alteration in the structure of lignin results in removal of hemicelluloses, in so doing enhancing the cellulose digestibility in the residue solids. The lignin removal depends up on the certain fixed value of acid concentration, i.e., the quantity of lignin removed depends on certain amount of hydrogen ion in the reaction liquid. It can be concluded that hemicelluloses can be effectively removed with the help of dilute acid pretreatment but it proves to be ineffective for lignin removal. Another study by Hsu et al. (2010) reported that when paddy straw was pretreated with 1 % (w/w) sulfuric acid for 1–5 min at 160 °C, followed by enzymatic hydrolysis, the maximal sugar yield of 83 % was achieved. The pore size of pretreated solid residues increased as a result of complete release of sugar such as glucose and xylose. This further resulted in efficient enzymatic hydrolysis of 70 %.

3.2.3 Oxidative Pretreatment

Oxidative pretreatment is one of the effective pretreatments as it helps in breakdown of lignin with hydrogen peroxide. Hydrogen peroxide (H_2O_2) is a strong oxidant and has been used for biomass pretreatment for biogas production (Hendriks and Zeeman 2009). It degrades biomass into oxygen and water without leaving any residues. Further it also prevents the formation of any secondary products (Teixeira et al. 1999). H_2O_2 treatment is a nonselective oxidation process and partially breaks down lignin and hemicelluloses and releases a cellulose fraction with high degradability to the anaerobic microorganisms. Further the inhibitors might be generated as lignin is oxidized to form soluble aromatic compounds. The H_2O_2 concentration should also be considered because high concentration of hydrogen peroxide was found to inhibit the AD process due to toxicity of an excess of hydroxyl ions to methanogens (Song et al. 2013a).

Song et al. (2013b) reported that methane yield of 288 mL/g VS was obtained when paddy straw was pretreated with 2–68 % H_2O_2 concentration and 1–08 substrate/inoculum for 6–18 days at ambient temperature of 25 ± 2 °C. The cellulose and hemicelluloses content in the pretreated paddy straw decreased remarkably as compared to untreated paddy straw. There was considerable decrease in the cellulose, hemicellulose, and lignin concentrations pretreated with H_2O_2 ranging from 0.9 to 22.0 %, 3.7 to 60.3 %, and 0.4 to 12.5 %, respectively. The results showed that H_2O_2 was particularly effective in the hemicelluloses decomposition of paddy straw.

There is great effect of pretreatment time (PT) on the degradation of the main constituents of agricultural wastes. The extensive treatments could certainly contribute to increased reductions in the contents of cellulose and hemicellulose under firm conditions. At the higher H_2O_2 concentration (HC) it has been observed that there is higher reduction in cellulose, hemicellulose, and lignin contents. The higher H_2O_2 augmented the accessibility of contents to the anaerobic microorganisms. This further facilitates the usage of low molecular weight soluble compounds by the microorganisms and thus resulting in increased biodegradability. One possible disadvantage of this pretreatment is that the proportion of CO_2 increases in the produced biogas by the increased volume of oxygen introduced in the system. Moreover this pretreatment is not usually carried out, apparently due to its high cost.

3.2.4 Sodium Carbonate–Sodium Sulfite Pretreatment

Sodium carbonate (Na_2Co_3) and sodium sulfide are the main constituents of green liquor which shows promising enzymatic hydrolysis (Yang et al. 2012a). Even though sodium sulfide is precisely good for delignification but the elemental sulfur leads to the difficulty of handling process of spent liquor as well as environmental contagion (Yang et al. 2012b). The considerable amounts of hemicelluloses can be degraded by sulfite. Moreover the sulfite can reduce the degree of polymerization of cellulose. It can further also increase the lignin hydrophilicity by sulfonation (Zhu et al. 2009). Na_2SO_3 and Na_2CO_3 have low environmental risks and barriers and are considered potentially advantageous.

Yang et al. (2012b) used different ratios of Na_2SO_3 and Na_2CO_3 at 140°C, chemical charge of 12 % for 60 min to pretreat paddy straw. It was observed that when pretreated with only Na_2CO_3 at ratio of 1:0, the increasing Na_2CO_3 charge may strengthen the alkaline degradation of polysaccharides. Further the lignin macromolecules are degraded by breaking down ester bonds between lignin, hemicelluloses, and cellulose (Gaspar et al. 2007). The paddy straw at the ratio of 0:1 is pretreated with only Na_2SO_3, the sulfite (SO_3^{2-}), or bisulfite (HSO_3^{1-}) leads to the lignin sulfonation to increase its hydrophilicity and may subsequently promote the process of enzymatic hydrolysis. Better delignification was observed with the higher ratio of Na_2SO_3 as compared to Na_2CO_3 at the same chemical charge. The delignification capacity of Na_2SO_3 was remarkable even at low chemical charge. While in case of Na_2CO_3 chemical charge resulted in momentous impact on its improved delignification. The pretreated paddy straw showed the highest sugar recovery of 74.5, 82.7, and 59.5 % for total sugar, glucan, and xylan, respectively. The Na_2CO_3 pretreatment at the ratio of 1:0 showed sugar recovery of 67.1, 74.4, and 53.7 % for total sugar, glucan, and xylan, respectively. Its analogous delignification ratio was 40.3 % and polysaccharides conservation ratio was 88.9 %.

3.2.5 NMMO Pretreatment

N-methylmorpholine-N-oxide is an environmental friendly organic solvent used for pretreatment. More than 98 % of this solvent can be recovered without the production of any noxious waste pollutants and chemical derivatization (Adorjan et al. 2004). Compared with other volatile organic cellulose solvents, NMMO possess the advantages of low toxicity, low hydrophobicity, low viscosity, low volatility with potentially negligible environmental impact, nonflammable properties, thermal stability, enhanced electrochemical stability and wide assortment of anion and cation combinations, and high rate of reaction (Zheng et al. 2009).

Teghammar et al. (2012) reported that 328 Nml CH_4/gCH when paddy straw was pretreated with NMMO. During this pretreatment the paddy straw was digested with 50 % w/w NMMO solution at 130 °C. As a result the produced methane was seven times more than the methane production of untreated paddy straw samples. This further corresponds to theoretical yield of 79 %. Additionally, it was also reported that 1 to 3 h of pretreatment is considered quite efficient for paddy straw. Extensive pretreatments resulted in less production of biogas. Furthermore, even though the experiments went on for six weeks, the paddy straw can be efficiently digested in about one week of anaerobic digestion. This means that just a couple of days would have been enough to convert the pretreated paddy straw to biogas provided if the digesting bacteria need no lag phase. This improvement is presumably due to the breakdown of the crystalline structure of the lignocelluloses. In addition, the NMMO solvent converts the crystallinity of the dissolved cellulose, which further increases the biodegradability (Jeihanipour et al. 2009). The best results were obtained when paddy straw was digested for 1 h. Further the extensive pretreatments resulted in lower preliminary digestion rate and also less production of methane. The chemical pretreatment unlike physical and biological pretreatment, may leave chemical residues which may perhaps influence the downstream anaerobic digestion process. For NMMO pretreatment, it is possible to recover nearly 100 % of the used NMMO liquid, so chemical residues might not be a problem (Dadi et al. 2006).

3.3 *Biological Pretreatment*

It is one of the alternative methods that is cost effective and does not cause any environmental pollution. One of the advantages of biological pretreatment is that it can take place at lower temperature. Further as compared to other pretreatments it does not require any chemicals. One of the disadvantages of biological pretreatment is that it could be slower than other nonbiological methods.

3.3.1 Fungal Pretreatment

In context of anaerobic digestion, quite a lot of fungi are known for their ability to remove lignin from lignocellulosic biomass. This predominantly includes white rot fungi. The lignins are essential to remove for efficient production of biogas as lignins could either cause problems during digestate use or can impede anaerobic digestion (Barr and Aust 1994). Fungal pretreatment is one of the methods which has been investigated to confiscate phenolic toxins before the process of anaerobic digestion (Dhouib et al. 2006). Sahni and Phutela (2013) reported pretreatment of paddy straw with standard culture namely *C. versicolor* MTCC 138 and isolated lignocellulolytic culture, that is, Fusarium sp. In the case of control, there was very little or insignificant change in almost all components, that is, cellulose, hemicellulose, lignin, and silica. In case, when paddy straw is pretreated with *C. versicolor* MTCC 138 and *Fusarium* species, the lignin content decreased with the increase of incubation period with maximum reduction of 17.5 and 27.1 % in 20 days, respectively. These observations clearly indicate that *C. vesicolor* MTCC 138 and *Fusarium* sp. are lignocellulolytic fungus.

3.3.2 Enzyme Addition

Enzymes required to breakdown lignocellulosic biomass are produced by microorganisms of anaerobic digestion and therefore, they are already present in the digester. Further to enhance the process of breakdown, an addition of enzyme mixture can be done. This enzyme mixture may include cellulose, hemicellulose, pectin, and starch degrading enzymes. Practically the enzyme additives can be applied in three different ways:

- By direct addition to a single stage anaerobic digester,
- By addition to hydrolysis and acidification vessel of a two-stage system or
- By addition to a dedicated enzymatic pretreatment vessel.

Various studies have been conducted to analyze the addition of enzymes to anaerobic digester (Rintala and Ahring 1994). It was suggested in some of the studies that there was no significant effect when the enzymes were directly added to the biogas reactor. Moreover, in such case enzymes are quickly degraded after addition (Binner et al. 2011). A number of studies have indicated that higher substrate solubilisation is obtained by the addition of enzymes to first stage of a two-stage anaerobic digestion process and consequently leads to higher yield of biogas for example addition of cellulosic enzyme cocktails on paddy straw leads to better yield of biogas (Quemeneur et al. 2012). Some studies showed that higher substrate solubilisation or biogas yields in batch AD tests can be achieved by enzymatic pretreatment in a dedicated vessel for example with various agriculture residues with cellulolytic enzyme cocktail. Continuous anaerobic digestion of different agricultural residues pretreated in a dedicated vessel leads to small increases in biogas yield (Quinones et al. 2011).

3.3.3 Anaerobic Microbial Pretreatment

This pretreatment is also acknowledged as two-stage digestion or dark fermentation, pre-acidification. The pretreatment technology is quite simple kind of in which first step of anaerobic digestion is alienated from methane production. Although the pH during methane production must be between 6.5 and 8, the pH value of the first digester must lie between 4 and 6, which inhibits production of methane and causes accumulation of volatile fatty acids (Deublein and Steinhauser 2010). This pretreatment method is having a positive effect on the methane concentration in biogas. During the pre-acidification step, CO_2 is formed in addition to H_2 and volatile fatty acids. CO_2 formed can be present in three forms: in the form of carbonate ion CO_3^{2-} at higher pH values, in the form of bicarbonate ion HCO^{3-} at neutral pH and in acidic environment; it is present in the form of carbon dioxide. Due to low pH most of the carbonate is present in the form of CO_2. The CO_2 thus produced is released into hydrolysis gas produced from the pre-acidification step due to its volatility. Therefore, the concentration of CH_4 obtained is much higher which shows that there is less CO_2 in the gas phase of methanogenesis. The advantage of using two-stage digestion is that the microorganisms are less responsive to many chemicals in the first stage as compared to second stage. Thus in the first step itself large number of inhibiting chemicals can be broken down. On the whole, the two-stage digestion is quite useful for the array of substrates and added permanence of feed with a steady pH.

3.3.4 Aerobic Microbial Pretreatment

This type of pretreatment can be conceded in naturally occurring mixed cultures. Aerobic pretreatment is basically carried out to solubilise the substrate as some aerobic microorganisms produce large amounts of cellulose, hemicelluloses, and lignin degrading enzymes. While in case of anaerobic microbial pretreatment, the chemicals that might otherwise hinder the step of methanogenesis in the anaerobic digester are wrecked by microorganisms. Moreover, the pH and temperature are also quite suitable for hydrolysis enzymes. There have been no studies showing the comparison of aerobic microbial pretreatment to anaerobic microbial pretreatment. Broadly, the advantage of an aerobic process is that it is significantly faster. While one of the disadvantages is that a lot of the organic matter is despoiled to CO_2 which could otherwise be degraded to CH_4 if the duration of pretreatment is too long (Jagadabhi et al. 2010).

3.4 Combined Pretreatment

These are complex but are typically most effective than the other processes using one mechanism.

3.4.1 Chemical + Microwave Irradiation

Microwave irradiation leads to considerable energy savings due to high heating efficiency and also it reduces the reaction time and eventually results in increase in reaction rate. However the processing by microwave irradiation may lead to certain side effects such as production of heat-induced inhibitors such as phenolic compounds and furfural (Hsu et al. 2010). Therefore, it becomes essential to control these pretreatment conditions to evade the formation of these inhibitors. Thus the microwave treatment has not been used individually but usually it has been used to provide heat for assisting acid or alkaline pretreatment at relatively low temperatures without compromising pretreatment effects (Liu and Cheng 2009; Cheng and Liu 2010). Zhu et al. (2005) used microwave and 1 % alkali (NaOH) at 700 W for 30 min. The results indicated that higher rate of hydrolysis was obtained when paddy straw was pretreated with microwave and NaOH. Additionally, the higher glucose content was observed in the hydrolysate as compared to paddy straw pretreated with alkali alone. Gong et al. (2010) also reported that sugar yield of paddy straw can be effectively improved by the combination of microwave and chemical pretreatment with organic acid. It was obtained from the results that sugar yields of 71.41 and 80.08 % with 46.1 and 51.54 % removal of lignin was achieved at optimal conditions of 230 W microwave intensity, 1:15 solid liquid ratio, 25 % acid concentration and 5 min irradiating time.

3.4.2 Steam Explosion

The combination of heating and sudden pressure referred to as steam explosion, increases the accessibility of the substrate. During steam explosion the substrate is heated up to a temperature ranging between 160 °C and 220 °C, in a closed vessel to further cause a rise in pressure. The pressure is abruptly released after a certain time period of around 5 to 60 min. The sudden pressure drop results in rapid evaporation of intracellular water and thus causing a phenomenon known as steam explosion. In one of the studies carried out by Bauer et al. (2009) the methane yield was found to be increased by 20 % in the paddy straw pretreated with steam explosion than the untreated paddy straw.

4 Advantages and Disadvantages

There is no distinct pretreatment technology that is suitable for all substrates and systems of anaerobic digestion. The diverse pretreatment technologies discussed above may be suitable for one or the other economic situation of particular region.

Table 1 Advantages and disadvantages of different pretreatment technologies (Taherzadeh and Karimi 2008; Hendriks and Zeeman 2009)

Pretreatment process	Advantages	Disadvantages
Mechanical	• Increased surface area • Makes substrate easier to handle • Improves fluidity in digester	• Augmented demand of energy • Elevated maintenance cost.
Thermal	• Possibility to ensure anaerobic process stabilization • High efficiency in improving organic matter solubilisation	• Possible formation of compounds which are difficult to degrade with overall reduction of methane yield • High energy consumption
Liquid hot water	• Increases the accessibility of enzyme	• Elevated heat demand • Only effectual up to certain temperature
Oxidative	• Strong oxidizing power ensuring short reaction time • High solubilisation improvement • No addition of chemical to the substrate	• High maintenance cost • Possible formation of less biodegradable byproducts
Acid	• Solubilises hemicellulose	• High cost of acids • oxidization problems • Formation of inhibitors
Alkali	• Breaks down lignin	• High concentration of alkali in digester • High cost of chemical
Microbial	• Low energy expenditure • No chemical addition	• No lignin breakdown • Slow in process
Enzymatic	• Low energy utilization • No addition of chemicals	• High cost of enzymes • Continuous addition required • High reaction time
Steam explosion	• Breaks down lignin and solubilises hemicelluloses	• Effective up to certain temperature • High heat and electricity demand

Table 2 The influence of different pretreatment methods on the breakdown of lignocelluloses

Pretreatment method	Cellulose decrystallisation	Hemicellulose degradation	Lignin degradation	Increase in specific surface area
Mechanical	+			+
Liquid hot water (LHW)	ND	+	−	+
Concentrated acid		+	+	+
Diluted acid		+		+
Alkaline		−	+	+
Oxidative	ND		+/-	+
Biological				+
Steam explosion		+	+	+

Here, + = major effect
− = minor effect
ND = Not determined
No symbol means that it is unclear if there is effect or not

Further the selection of pretreatment method depends on the composition of the substrate. The greatest challenge for the pretreatment technologies is to increase the bioavailability of substrate. This could only be achieved if the right pretreatment technology is combined with right substrate. Energy balance and cost effectiveness are the most important parameters for selecting the right choice. The erroneous choice can make the process extravagant. Further, to obtain better gas yield, it is quite essential to make the process economically feasible. The Table 1 given below summarizes all the advantages and disadvantages of different pretreatment methods.

In the recent years, the influence of pretreatment technologies on anaerobic digestion process has been investigated but still these technologies need to be optimized for the better yield of biogas. The Table 2 given below summarizes the influence of different pretreatment methods on the breakdown of lignocelluloses.

5 Conclusion

Pretreatment in paddy straw is basically employed to remove the recalcitrance of lignocellulosic biomass for increased biogas yield. Pretreatment can further potentially reduce the lignin content and thus results in increased accessible surface area. A successful pretreatment method should be able to

(1) Improve the feedstock's digestibility for microbes in anaerobic digestion
(2) Avoid formation of inhibitors
(3) Require low energy input
(4) Avoid loss of carbohydrates

(5) Avoid the need of waste disposal
(6) Be cost effective and environmentally friendly.

Pretreatment is predominantly apt for substrates where there is nonavailability of sterically degradable biomass for enzyme attack. Finally, it should be noted that it is not realistic to treat all the substrates with one single pretreatment technology.

References

Adorjan I, Sjoberg J, Rosenau T, Hofinger A, Kosma P (2004) Kinetic and chemical studies on the isomerization of monosaccharides in N-methylmorpholine-N-oxide (NMMO) under lyocell conditions. Carbohydr Res 339:1899

Agbor VB, Cicek N, Sparling R, Berlin A, Levin DB (2011) Biomass pretreatment: fundamentals toward application. Biotechnol Adv 29:675–685

Barr DP, Aust SD (1994) Pollutant degradation by white rots fungi. In: Ware, GW (ed) Reviews of Environmental Contamination and Toxicology, Reviews of Environmental Contamination and Toxicology. Springer, New York, pp 49–72

Bauer A, Bosch P, Friedl A, Amon T (2009) Analysis of methane potentials of steam-exploded wheat straw and estimation of energy yields of combined ethanol and methane production. J Biotechnol 142:50–55

Binner R, Menath V, Huber H, Thomm M, Bischof F, Schmack D, Reuter M (2011) Comparative study of stability and half-life of enzymes and enzyme aggregates implemented in anaerobic biogas processes. Biomass Convers Biorefin 1:1–8

Chandra R, Takeuchi H, Hasegawa T, Kumar R (2012a) Improving biodegradability and biogas production of wheat straw substrates using sodium hydroxide and hydrothermal pretreatments. Energy 43(1):273–282

Chandra R, Takeuchi H, Hasegawa T (2012b) Hydrothermal pretreatment of rice straw biomass: A potential and promising method for enhanced methane production. Appl Energy 94:129–140

Chang V, Burr B, Holtzapple M (1997) Lime pretreatment of switchgrass. Appl Biochem Biotechnol 63–65:3–19

Chang KL, Thitikorn-amorn J, Hsieh J-F, Ou BM, Chen SH, Ratanakhanokchai K, Chen ST (2011) Enhanced enzymatic conversion with freeze pretreatment of rice straw. Biomass Bioenergy 35(1):90–95

Cheng XY, Liu CZ (2010) Enhanced biogas production from herbal-extraction process residues by microwave-assisted alkaline pretreatment. J Chem Technol Biotechnol 85:127–131

Cheremisinoff N (2002) Handbook of water and wastewater treatment technologies. Butterworth-Heinemann, USA

Dadi AP, Schall CA, Varanasi S (2006) Enhancement of cellulose saccharification kinetics using an ionic liquid pretreatment step. Biotechnol Bioenergy 95:904–910

Delgenes JP, Penaud V, Moletta R (2002) Pretreatments for the enhancement of anaerobic digestion of solid wastes Chapter 8: Biomethanization of the Organic Fraction of Municipal Solid Wastes. IWA Publishing, pp. 201–228

Demirbas A (2008) Heavy metal adsorption onto agro–based waste materials: a review. J Hazard Mater 157:220–229

Deublein D, Steinhauser A (2010) Biogas from waste and renewable resources: an introduction. Wiley

Dhouib A, Ellouz M, Aloui F, Sayadi S (2006) Effect of bioaugmentation of activated sludge with white-rot fungi on olive mill wastewater detoxification. Lett Appl Microbiol 42:405–411

Fredriksson H, Baky A, Bernesson S, Nordberg A, Noren O, Hansson PA (2006) Use of on-farm produced biofuels on organic farms-evaluation of energy balances and environmental loads for three possible fuels. Agric Syst 89(1):184–203

Garrote G, Dominguez H, Parajo JC (1999) Hydrothermal processing of lignocellulosic materials. Holz Roh Werkst 57:191–202

Gaspar M, Kalman G, Reczey K (2007) Corn fiber as a raw material for hemicellulose and ethanol production. Process Biochem 42:1135–1139

Girio FM, Fonseca C, Carvalheiro F, Duarte LC, Marques S, Bogel-Lukasik R (2010) Hemicelluloses for fuel ethanol: a review. Bioresour Technol 101:4775–4800

Gong G, Liu D, Huang Y (2010) Microwave-assisted organic acid pretreatment for enzymatic hydrolysis of rice straw. Biosyst Eng 107(2):67–73

He Y, Pang Y, Liu Y, Li X, Wang K (2008) Physicochemical characterization of rice straw pretreated with sodium hydroxide in the solid state for enhancing biogas production. Energy Fuels 22:2775–2781

Hendriks ATWM, Zeeman G (2009) Pretreatments to enhance the digestibility of lignocellulosic biomass. Bioresour Technol 100:10–18

Hsu TC, Guo GL, Chen WH, Hwang WS (2010) Effect of dilute acid pretreatment of rice straw on structural properties and enzymatic hydrolysis. Bioresour Technol 101(13):4907–4913

Jagadabhi PS, Kaparaju P, Rintala J (2010) Effect of microaeration and leachate replacement on COD solubilization and VFA production during mono-digestion of grass-silage in one-stage leach-bed reactors. Bioresour Technol 101:2818–2824

Jeihanipour A, Karimi K, Taherzadeh MJ (2009) Enhancement of ethanol and biogas production from high-crystalline cellulose by different modes of NMO pretreatment. Biotechnol Bioenergy 105:469

Kaparaju P, Serrano M, Angelidaki I (2010) Optimization of biogas production from wheat straw stillage in UASB reactor. J Appl Energy 87:3779–3783

Keshwani DR, Cheng JJ (2009) Switchgrass for bioethanol and other value-added applications: a review. Bioresour Technol 100:1515–1523

Kalnappert D, Grethlein H, Converse A (1981) Partial acid hydrolysis of poplar wood as a pretreatment for enzymatic hydrolysis. Biotechnol Bioeng Symp 11:67–77

Kohlmann KL, Westgate PJ, Sarikaya A, Velayudhan A, Weil J, Hendrickson R, Ladisch MR (1995) Enhanced enzyme activities on hydrated lignocellulosic substrates. BTEC paper 127. In: 207th American Chemical Society National Meeting, ACS Symposium series No. 618. Enzymatic Degradation of Insoluble Carbohydrates, pp. 237–255

Li MF, Fan YM, Xu F, Sun RC, Zhang XL (2010) Cold sodium hydroxide/urea based pretreatment of bamboo for bioethanol production: characterization of the cellulose rich fraction. Ind Crops Prod 32:551–559

Liu CZ, Cheng XY (2009) Microwave-assisted acid pretreatment for enhancing biogas production from herbal-extraction process residue. Energy Fuel 23:6152–6155

Mosier N, Hendrickson R, Ho N, Sedlak M, Ladisch MR (2005) Optimization of pH controlled liquid hot water pretreatment of corn stover. Bioresour Technol 96:1986–1993

Mussgnug JH, Klassen V, Schluter A, Kruse O (2010) Microalgae as substrate for fermentative biogas production in a combined biorefinery concept. J Biotechnol 150(1):51–56

Quemeneur M, Bittel M, Trably E, Dumas C, Fourage L, Ravot G, Steyer JP, Carrere H (2012) Effect of enzyme addition on fermentative hydrogen production from wheat straw. Int J Hydrogen Energy 37:10639–10647

Rintala JA, Ahring BK (1994) Thermophilic anaerobic digestion of source-sorted household solid waste: the effects of enzyme additions. Appl Microbiol Biotechnol 40:916–919

Quinones TS, Plochl M, Budde J, Heiermann M (2011) Enhanced methane formation through application of enzymes: results from continuous digestion tests. Energy Fuels 25:5378–5386

Sahni N, Phutela UG (2013) Comparative profile of paddy straw pretreated with standard and isolated lignocellulolytic fungal cultures, pp 92–97

Song ZL, Yang GH, Guo Y, Zhang T (2012) Comparison of two chemical pretreatments of rice straw for biogas production by anaerobic digestion. Bioresources 7:3223–3236

Song Z, Yag G, Feng Y, Ren G, Han X (2013a) Pretreatment of rice straw by hydrogen peroxide for enhanced methane yield. J Integr Agric 12(7):1258–1266

Song Z, Yang G, Han X, Feng Y, Ren G (2013b) Optimization of the alkaline pretreatment of rice straw for enhanced methane yield. BioMed Res Int 9:686–692

Swatloski RP, Spear SK, Holbrey JD, Rogers RD (2002) Dissolution of cellulose with-ionic liquids. J Am Chem Soc 124(18):4974–4975

Taherzadeh MJ, Karimi K (2008) Pretreatment of lignocellulosic wastes to improve ethanol and biogas production: a review. Int J Mol Sci 9:1621–1651

Teghammar A, Karimi K, Sarvari Horvath I, Taherzadeh MJ (2012) Enhanced biogas production from rice straw, triticale straw and softwood spruce by NMMO pretreatment. Biomass Bioenergy 36:116–120

Teghammar A, Forgacs G, Sarvari Horvath I, Taherzadeh MJ (2014) Techno-economic study of NMMO pretreatment and biogas production from forest residues. Appl Energy 116:125–133

Teixeira LC, Linden JC, Schroeder HA (1999) Alkaline and peracetic acid pretreatments of biomass for ethanol production. Appl Biochem Biotechnol 77:19–34

Tuankriangkrai S, Benjakul S (2010) Effect of modified tapioca starch on the stability of fish mince gels subjected to multiple freeze-thawing. J Muscle Foods 21:399–416

Weil JR, Brewer M, Hendrickson R, Sarikaya A, Ladisch MR (1997) Continuous pH monitoring during pretreatment of yellow poplar wood sawdust by pressure cooking in water. Appl Biochem Biotechnol 6:21–40

Yang L, Jin Y, Han Q, Chang HM, Jameel H, Phillips R (2012a) Green liquor pretreatment for improving enzymatic hydrolysis of corn stover. Bioresource Technology

Yang L, Cao J, Jin Y, Chang HM, Jameel H, Phillips R, Li Z (2012b) Effects of sodium carbonate pretreatment on the chemical compositions and enzymatic saccharification of rice straw. Bioresource Technology

Zhao R, Zhang Z, Zhang R, Li M, Lei Z, Utsumi M, Sugiura N (2010) Methane production from rice straw pretreated by a mixture of acetic-propionic acid. Bioresour Technol 101(3):990–994

Zeng M, Mosier NS, Huang CP, Sherman DM, Ladisch MR (2007) Microscopic examination of changes of plant cell structure in corn stover due to hot water pretreatment and enzymatic hydrolysis. J Biotechnol Bioenergy 97:265–278

Zheng Y, Pan Z, Zhang RH (2009) Overview of fuel ethanol production from lignocellulosic biomass. Int J Agric Biol Eng 2:51–68

Zheng Y, Zhao J, Xu F, Li Y (2014) Pretreatment of lignocellulosic biomass for enhanced biogas production. Prog Energy Combust Sci 42:35–53

Zhu S, Wu Y, Yu Z, Liao J, Zhang Y (2005) Pretreatment by microwave/alkali of rice straw and its enzymic hydrolysis. Process Biochem 40(9):3082–3086

Zhu JY, Pan XJ, Wang GS, Gleisner R (2009) Sulfite pretreatment (SPORL) for robust enzymatic saccharification of spruce and red pine. Bioresour Technol 100:2411–2418

Bioethanol Production from Fermented Cashew Apple Juice by Solar Concentrator

Y.P. Khandetod, A.G. Mohod and H.Y. Shrirame

Abstract In the present study, it is used for bioethanol production from locally available cashew apple fruit juice (variety Vengurla-4) grown in the Konkan region of Maharashtra, India. The objective of this study is to show the ways of ethanol production. Fermentation followed by distillation are the processes for ethanol production. For fermentation, *Saccharomyces cerevisiae* was the cheapest strain available for the conversion of substrate. Raw fermented cashew apple juice after fermentation is used for ethanol production. Distillation process using concentrating solar cooker having the per day rate of distillation of the system was 1170 ml with distillation efficiency 33.41 %. The concentration of ethanol increased from initial 12 to 18.6 % in first distillation, which was further increased to 35.5 % after second distillation. As the boiling point of ethanol is 78.5 °C, the temperature obtained was higher, i.e., 109.2 °C. Same amount of water also got evaporated, which diluted distillated product. Therefore, the higher concentration of ethanol was not obtained. The specific gravity and acid value of double distilled ethanol at 35.5 % concentration were 0.947 and 1.044 mg KOH/g, respectively.

Keywords Cashew apple juice · *Saccharomyces cerevisiae* · Fermentation · Distillation · Ethanol

List of Abbreviations and Symbols

%	Percent
°C	Degree centigrade
A_a	Aperture area of concentrator
A_c	Collector area required
A_{cont}	Total surface area of the container
A_r	Area of receiver
A_S	Surface area of paraboloid
Cc	Cubic centimeter

Y.P. Khandetod · A.G. Mohod · H.Y. Shrirame (✉)
Department of Electrical and Other Energy Sources, College of Agricultural Engineering and Technology, DBSKKV, Dapoli, Maharashtra, India
e-mail: hemantshrirame@gmail.com

S. Kumar et al. (eds.), *Proceedings of the First International Conference on Recent Advances in Bioenergy Research*, Springer Proceedings in Energy,
DOI 10.1007/978-81-322-2773-1_5

C_c	Specific heat of fermented cashew apple juice
C_{cont}	Specific heat of the material of container
C_e	Specific heat ethanol
C_w	Specific heat of water
C_{pr}	Specific heat of the material of receiver
CR	Concentration ratio
D	Aperture diameter of paraboloid
E	East
d_c	Diameter of cylinder
et al.	And others
etc.	Et cetra
FAO	Food and Agriculture Organization
Fig.	Figure
F'_{UL}	Overall heat loss coefficient
$F'_{\eta o}$	Optical efficiency factor
F	Focal length of paraboloid
G.I.	Galvanized iron
H_s	Total incident radiation on reflecting surface
H_c	Convective heat coefficient
h_c	Height of the container
h	Hour
hrs/day	Hours per day
i.e.	That is
I_b	Solar insolation
$^{\circ}$K	Degree Kelvin
K	Thermal conductivity
H	Convective heat coefficient
J/kg K	Jules per kilogram per Kelvin
kcal/m^2 h	Kilo calorie per square meter per hour
kg/m s	Kilogram per meter per second
kg/m^3	Kilogram per cubic meter
W/m K	Watt per meter per Kelvin
W/m^2 K	Watt per square meter per Kelvin
kW/m^2	Kilo watt per square meter
L	Length of the circumference of paraboloid
m	Meter
Ml	Milli-lit
m/s	Meter per second
m^3	Cubic meter
(MC)$'_w$	Total heat capacity
M_c	Mass of cashew apple juice
M_{cont}	Mass of the empty container with lid
M_r	Mass of receiver
M_w	Mass of water

N	North
N_{Gr}	Grashof number
N_{Nu}	Nusselt number
N_{Pr}	Pradalt number
n_{ref}	Reference density of water
n_t	Thermal efficiency
P_{evp}	Evaporation rate of product
Q_a	Heat available kcal
Q_f	Average energy falling at Dapoli
Q_d	Energy delivered by paraboloid
Q_{loss}	Total heat loss Kcal
Q_r	Energy required for vaporization
q_{conv}	Heat transfer rate by convention
q_{rad}	Heat transfer rate by radiation
R	Aperture radius of parabolic
S	Arc length of paraboloid
Sr. No.	Serial number
TSS	Total soluble solid
T_a	Ambient temperature
T_r	Surface temperature of receiver
T_d	Distillation point of ethanol
T_f	Final temperature of fermented cashew apple juice
T_i	Initial temperature of fermented cashew apple juice
T_V	Vaporization temperature of cashew apple juice
T_w	Water temperature in the container
V_c	Volume of container
W	Watt
W_u	Weight of ethanol sample
W/m^2	Watts per square meter
λ_V	Latent heat of vaporization of cashew apple juice
P	Density of air
ρ_c	Density of cashew apple juice
Z	Depth of paraboloid
T	Specular reflectance of reflector
A	Absorptance of absorber
B	Volumetric coefficient of expansion of the fluid
Φ	Intercept factor
M	Viscosity
E	Emissivity
ΔT	Positive temperature difference between the surface of container and fluid
Π	Pie

1 Introduction

Energy plays a crucial role in modern life. The main problem of the world is the increasing demand for fossil fuels but the storage strata are limited. Nowadays, alternative fuel like ethanol has more demand in developed and developing countries for the fulfillment of requirement for fuel. However due to increasing petroleum shortage, renewable resources can be used for production (Ames 2001). Glucose rich substrates can be fermented to produce ethanol which is a clean energy source (Yu and Zhang 2004). Demand for ethanol as an alternative renewable fuel sources has steadily increased (Sheoran et al. 1998) which helps the world to secure its future supply of energy by reducing its dependency on fossil fuels. Biomass with higher sugar content can produce ethanol by fermentation and distillation. Ethanol is a clear, colorless liquid with sweet flavor, but at higher concentration it has burning taste. In India, research and development efforts for using ethanol blended transport fuels started in 1980 (Linoj Kumar and Prakash 2006). Nowadays, fermentation technology produces nearly 80 % ethanol as clean fuel and for which Saccharomyces cerevisiae is considered to be the most important yeast strain because of its biotechnological application in the field of fermentation technology (Sreenath and Jeffries 1996). Ethanol produced by fermentation has been found to serve considerably as transportation fuel for cars, trucks, and trains. In India and Brazil, ethanol is blended with petrol at 5–10 and 24 % by volume, respectively. India is the second largest fruit producer in the world after China (Costa et al. 2009). In the Konkan region, cashew is a traditional crop mainly grown on hill slope as rained permissible horticulture crop. Cashew (*Anacardium occidentale* L.) belongs to family Anacardiaceae. North Brazil is the origin of cashew. In the world, a total of 32 countries produced the cashew fruit, and the major cashew apple producing countries, based on the FAO report, their production are approximately, Vietnam 8.4, Nigeria 5, India 4, Brazil 1.6, and Indonesia 10 lakh tons, respectively. In India, Maharashtra, Goa, Karnataka, and Kerala are leading states along the west coast and Tamil Nadu, Andhra Pradesh, Orissa, and West Bengal along the east coast for cashew growing (Hubballi 2006).

The total cashew plantation area in Konkan region of Maharashtra is about 1,73,601 hectares with annual production of about 1.5 lakh metric tones (Haldankar et al. 2007). The region is endeavored with the average solar energy availability of 450–600 W/m^2 for 7–8 h in a day and 250 days in a year. Solar energy is the cheapest, inexhaustible, and ample source of energy, which is a direct form of energy. Utilization of solar energy for thermal applications like distillation, cooking, heating, and drying is well recognized in tropical and semitropical regions. Ethanol obtained using distillation process using concentrating solar cooker provides a solution for energy conservation in ethanol production.

2 Materials and Methods

2.1 Cashew Apple Fruit

Cashew apple fruits of variety Vengurla-4 were used which is grown in the Konkan region of Maharashtra, India. The cashews are collected/harvested from cashew nut orchard of the Central Experiment Station, Wakawali (in the year 2009) of Konkan region, Maharashtra, India. Cashew apple seems as an alternative raw material for ethanol production, due to its vast availability and high concentration of reducing sugars (Pinheiro et al. 2008).

2.2 Processing

Cashew apple (*Anacardium occidentale* L.) fruits were picked at mature and ripe stages Plate 1. Basket press and mechanical juice extractor were used for the extraction of cashew apple juice. Basket press juice extractor is a manually operated press as given in Plate 2. Cashew apple juice obtained from basket press had good quality as compared to mechanical juice extractor, therefore juice obtained from basket press was used and the overlook of the extracted juice is shown in Plate 2.

2.3 Physical Properties of Cashew Apple

The Vengurla-4 variety of cashew apple was used for ethanol production. An average weight, length, diameter, and volume of Vengurla-4 variety is reported in Table 1. The cashew apple had an average length of 4.9 cm and diameter of 2.7 cm. The cashew apple was red in color and it had an average volume 56.66 ml. The average juice recovery was 67.30 % w/w.

Plate 1 Cashew apple fruits of variety Vengurla-4

Plate 2 Overlook of the extracted cashew apple juice

Table 1 Physical properties of cashew apple

Sr. No.	Physical property	Particulars
1	Weight (g)	54.24
2	Length (cm)	4.9
3	Diameter (cm)	2.7
4	Volume (ml)	56.66
5	Color	Red
6	Juice recovery (%)	67.30 w/w

2.4 *Chemical Properties of Cashew Apple Juice*

The data on chemical properties of cashew apple are reported in Table 2.

2.4.1 Determination of TSS

In cashew apple, TSS was 11.8 °Brix, it indicates the percentage of sucrose by weight in grams per 100 ml of water, which is determined with the help of hand refractometer of Erma (Japan) manufacturing and values were corrected at 20 °C

Table 2 Chemical properties of cashew apple

Sr. No.	Chemical property	Particulars
1	Total soluble solid (°Brix)	11.8
2	pH	4.4
3	Sugar (%)	11.3
4	Acidity (%)	0.3

using temperature correction chart with the range 0–90 %, scale (1) 0–42 %, (2) 42–90 % (A.O.A.C 1975).

2.4.2 Determination of PH

The pH of cashew apple juice was measured using digital pH meter and the pH of apple was 4.4.

2.4.3 Determination of Sugar

The monosaccharides sugars include glucose, fructose, and galactose, which are derived from different sources. First 50 ml sugar solution will be taken in 250 ml conical flask, then 5 g of citric acid and 50 ml of water will be added in it. For the complete inversion of sucrose, solution will be boiled for 10 min and after that it will be kept for cooling, then this solution will be transferred to 250 ml volumetric flask to neutralize it with 1 N NaOH using phenolphthalein indicator. For inversion at room temperature, 50 ml of aliquot of clarified and deleaded solution will be transferred to 250 ml flask, then 10 ml of HCl will be added in it and solution will be allowed to stand at room temperature for 24 h. Finally solution will be neutralized with concentrated NaOH solution. For the determination of total sugar aliquot will be taken for testing. Results will be expressed on percent basis (Ranganna 2009). In cashew apple juice sugar was found to be 11.3 %.

2.4.4 Determination of Acidity

A known quantity of sample (cashew apple juice) was titrated against 0.1 N sodium hydroxide (NaOH) solution using phenolphthalein as an indicator. A known quantity of sample was blended in pestle and mortar with 20–25 ml distilled water and then it was transferred to 100 ml volumetric flask and filtered. A known volume of aliquot was titrated. The results were expressed as a percentage of anhydrous citric acid (Ranganna 2009). The acidity was found to be 0.3 %.

The chemical analysis of cashew apple revealed that the cashew apple is rich in sugar content. The acid sugar blend in the cashew apple was good and indicated that the apples could be successfully utilized for the production of alcohol.

2.5 Microorganism

The yeast *Saccharomyces cerevisiae* was used for the fermentation process (Pinheiro et al. 2008). For the fermentation, the juice was supplemented with diammonium hydrogen phosphate (DAHP) as 1 g/l and potassium metabisulfite (KMS) as 0.03 g/l. The juice was inoculated with pure culture of *Saccharomyces cerevisiae* var. ellipsoideus (NCIM-3315) as 0.30 g/l.

2.6 Solar Concentrating Cooker

Different types of concentrating cookers like Flat absorber with flat reflectors, parabolic concentrator, compound parabolic concentrator, Fresnel lens, and cylindrical parabolic concentrator are used for various applications with lower temperatures to higher temperatures (Garge and Prakash 2005). For this study, the parabolic solar cooker (SK-14), which concentrated the incoming solar radiation at the focal point, was used as heat source for the distillation of fermented cashew apple juice. The concentrating system followed the sun, so that maximum sun rays were always focused on reflecting surface. A solar concentrating collector SK-14 having aperture diameter of 1.4 m and depth 0.35 m with focal length of 0.30 m was used for distillation. The distillation of cashew apple juice was carried out on solar concentrating collector by placing the container at focal point to get maximum concentration of ethanol.

3 Results and Discussion

3.1 Fermentation Process for Cashew Apple Juice

Known volume of cashew apple juice was taken and TSS and pH content of cashew apple juice was adjusted to 24 °Brix and 3.5, respectively, by addition of sugar. The acidity was adjusted by addition of citric acid. The juice was supplemented with diammonium hydrogen phosphate (DAHP) as 1 gm/l and potassium metabisulfite (KMS) as 0.03 g/l. After adjusting TSS, pH, and acidity, the measured volume of juice was taken in fermentation flask. The juice was inoculated with pure culture of *Saccharomyces cerevisiae* as 0.30 gm/l. The fermentation flask was incubated at 28 ± 2 °C. The start of fermentation was indicated by evolution bubbles. The end of fermentation was indicated by cessation of foaming and bubbling after fermentation assembly was dismantled.

First observation was taken after 24 h and then observations were continuously taken at alternate days till total soluble solid of juice was found to be constant. The fermentation was allowed to continue for 10 days at the end of which the total

Table 3 Changes in total soluble solids TSS, pH and acidity in cashew apple juice during fermentation process

Sr. No.	Time (h)	TSS (°Brix)	pH	Acidity (%)
1	0	24	3.50	0.6144
2	24	18.4	3.47	0.7936
3	72	12.2	3.53	0.9472
4	120	8.8	3.63	0.6656
5	168	8.6	3.65	0.704
6	216	8.0	3.70	0.768
7	264	8.0	3.68	0.793
7	312	8.0	3.79	0.768
9	360	8.0	3.66	0.704

soluble solid was found to be 8.0 °Brix. There was no further decrease in total soluble solid after 8 days of fermentation. Table 3 shows changes in TSS, pH, and acidity in cashew apple juice during fermentation process.

3.2 Performance of Solar Concentrating Collector for Distillation of Fermented Cashew Apple Juice

The parabolic solar cooker SK-14, which concentrated the incoming solar radiation at the focal point, was used as heat source for distillation of fermented cashew apple juice. The angle of focus to attend the required temperature for distillation was fixed based on preliminary testing. The present study on distillation of cashew apple juice was carried out to get maximum concentration of ethanol which is shown in Plate 3.

The performance of solar parabolic concentrator was firstly evaluated without load (i.e., without juice in the container). The solar pyranometer is used for the measurement of solar radiation. Measurement of solar radiation was made when

Plate 3 Solar parabolic concentrator for distillation of ethanol

pyranometer faces the sky. It was observed that the maximum temperature of the container was 225.7 °C reached at 13.00 h when solar radiation was 594 W/m^2 with 36.4 °C ambient temperature. Then there was a gradual decrease in the temperature along with time and solar intensity which is presented in Fig. 1.

Solar parabolic concentrator with load was used for the distillation of ethanol. When container was fully loaded with cashew apple juice with its designed capacity, the maximum temperature of inside container was 109.2 °C. The maximum solar insolation was 611 W/m^2 at 13.00 h. during the test Fig. 2 shows the temperature profile at full load test.

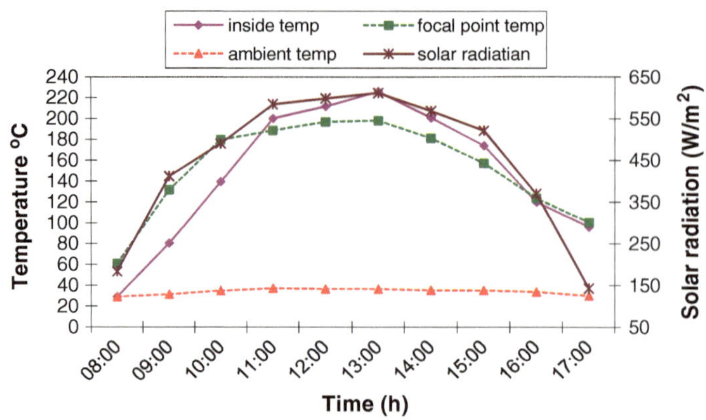

Fig. 1 Performance of parabolic solar concentrator for without juice in the container

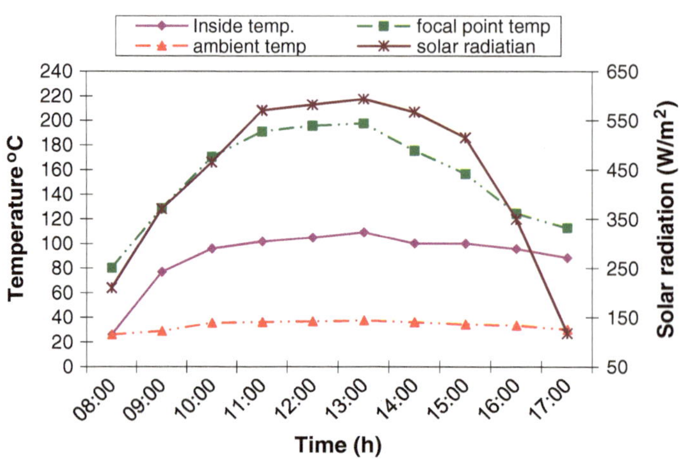

Fig. 2 Performance of parabolic solar concentrator for with juice in the container

3.2.1 Distillation Efficiency of Solar Parabolic Concentrator

The solar parabolic concentrator was manually adjusted towards the sun as the sun inclination changed. The condensation was provided to collect ethanol. The receiver was mounted on parabolic solar concentrator with its full load capacity, i.e., 3.00 kg of fermented cashew apple juice for production of ethanol. During the test, actual distillation was obtained between 9.00 and 16.00 h. The maximum distillation rate was obtained between 11.00 and 13.00 h. when a maximum solar intensity ranged from 450 to 600 W/m^2 was available and the distillation rate decreased after 13.00 h, as solar intensity decreased. The distillation efficiency of parabolic solar concentrator with time for different solar insolations is given in Table 4. Higher efficiency of the system was found to be 33.4 % at 13.00 h when solar intensity was 500 W/m^2 and ambient temperature 31.5 °C.

The relationship between cumulative distillation rate and solar radiations versus time is shown in Fig. 3. The maximum distillation observed was 580 ml when solar radiation was 620 W/m^2 at 13.00 h and minimum distillation rate was measured at 16.00 h. The maximum range of total distillation rate calculated during the whole day, i.e., from 8.00 to 17.00 h was 2230 ml from 3.00 kg of fermented cashew apple juice. Same trend was observed which is shown in Figs. 4 and 5 with minimum difference in solar radiation and ethanol distillation rate.

Water heating and cooling tests are recommended and both test performed with a load of water in aluminum made container. The first test covered the water heating test using parabolic concentrator till the steady state temperature of water was reached under clear sky condition. Then the test of water cooling was done by shading the concentrator. The water heating and cooling curves were plotted between difference in temperature of water and ambient temperature versus time, i.e., (T_w–T_a) is given in Fig. 6 which show the time required for water boiling. For this test, water was required 55 min to reach the steady state temperature at 9.30 h when average solar insolation was 563 W/m^2. The analysis was done for both the

Table 4 Performance of parabolic concentrator from water heating test water heating and cooling test

Time (h)	Water temperature °C initial	final	τ_o (min)	F'_{UL} (W/m2 K)	F'_{UL}/C	$F'no$	Avg. $F'no$	Thermal efficiency (ηt)
09.30	27.2	36.5	25	12.83	0.32	0.0946	0.1202	33.41
09.40								
09.40	36.5	47.7	25	12.83	0.32	0.1242		
09.50								
09.50	47.7	61.2	25	12.83	0.32	0.1267		
10.00								
10.00	61.2	80.6	25	12.83	0.32	0.1274		
10.10								

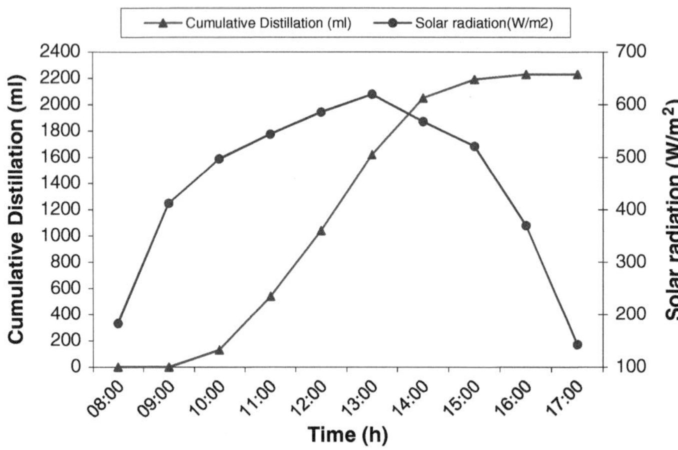

Fig. 3 Cumulative distillation rate and solar radiations versus time

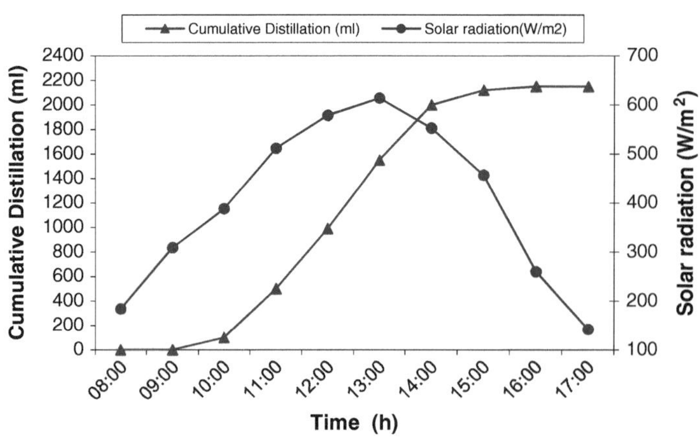

Fig. 4 Cumulative distillation rate and solar radiations versus time

curves of water heating and cooling. The water cooling curves give overall heat loss coefficient (F'_{UL}) and optical efficiency factor (F'_{no}) was computed by water heating test.

3.2.2 Overall Heat Loss Coefficient (F'_{UL})

Cooling Curve Analysis

The calculated value of [ln ($T_w - T_a$)] for each set of the data points that plotted with this value on Y-axis and time on X-axis was drawn (Fig. 7). Different points of

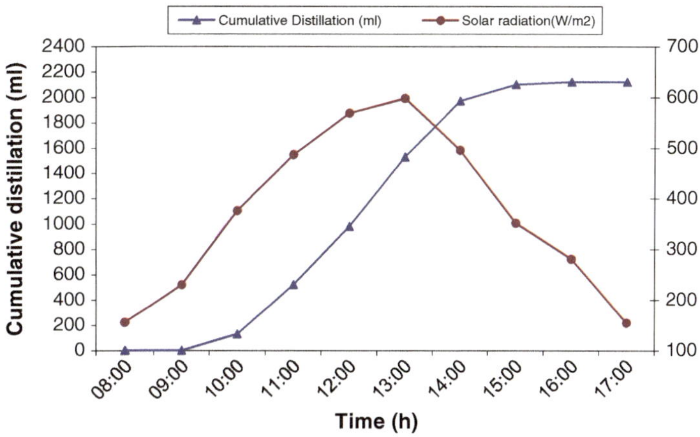

Fig. 5 Cumulative distillation rate and solar radiations versus time

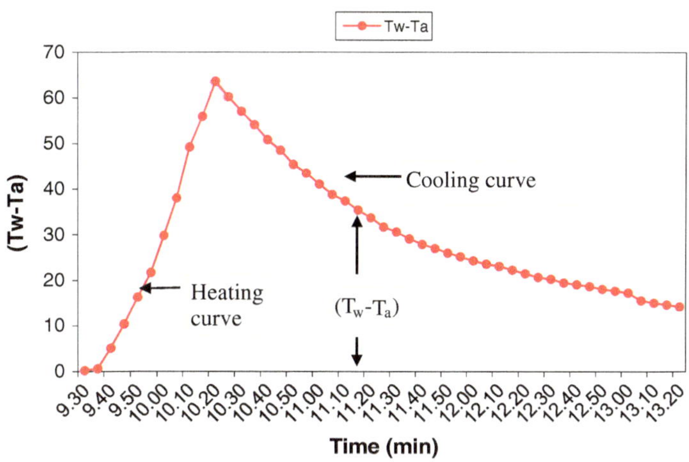

Fig. 6 Water heating and cooling curve test

the plot were fitted to a least square linear regression equation. The slope of the line equals to $(-1/\tau_0)$, where τ_0 is defined as the time constant for cooling.

Substituting known values of time constant for cooling (τ_0) in Eq. 1, surface area of the container (A_{cont}), and total thermal capacity of the container in the Eq. 1 and the value of the heat loss factor (F'_{UL}) of the cooker was evaluated.

$$F'_{UL} = \frac{(MC)'_W}{A_{cont}\tau_0} \tag{1}$$

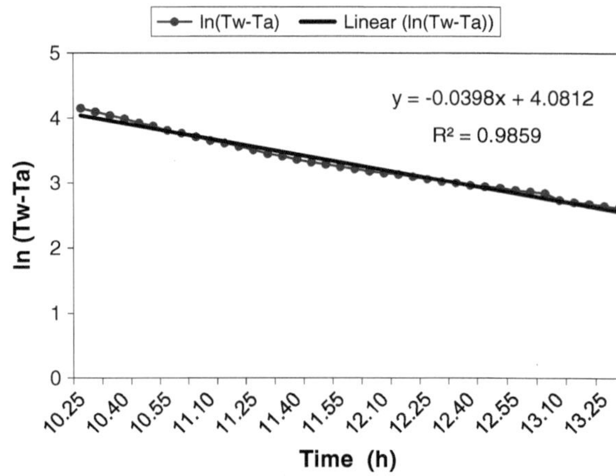

Fig. 7 Graph for cooling curve

where,

A_{cont} = Surface area of the container, m^2
$(\mathrm{MC})'_{\mathrm{w}}$ = Heat capacity, J/K
C_{cont} = Specific heat of container, J/kg K
C_{w} = Specific heat of water, J/kg K
M_{cont} = Mass of the container with cover, kg
M_{w} = Mass of water, kg
τ_{o} = Sensible cooling constant

By regression analysis of the test, the value of the time constant (τ_{o}) is 25 min at the average ambient temperature of 31.5 °C. With the cylinder surface area 0.0791 m^2 and values of cooling time constant (τ_{o}), overall heat loss factor (F'_{UL}) was found to be 12.83 W/m^2. The overall heat loss factor (F'_{UL}) was found higher, because of increase in thermal resistance between the receiver cylinder and the surrounding.

3.2.3 Optical Efficiency

The optical efficiency factor indicates the theoretical upper limit of the overall efficiency of solar paraboloid concentrator which is related to the performance of reflecting surface and its reflectance. The determination of optical efficiency factor (F'_{no}) is by using water heating test. This is found after analyzing the sensible heating curve as shown in Fig. 6.

$$F'_{\eta\mathrm{o}} = \frac{\frac{(F'_{\mathrm{UL}})A_{\mathrm{cont}}}{A_{\mathrm{aperture}}}\left[\left(\frac{T_{\mathrm{wf}}-T_{\mathrm{a}}}{I_{\mathrm{b}}}\right) - \left(\frac{T_{\mathrm{wi}}-T_{\mathrm{a}}}{I_{\mathrm{b}}}\right)\right]e^{\tau/\tau_o}}{1 - e^{-\tau/\tau_o}} \quad (2)$$

where

T_{wi} = Temperature of water at the beginning of the interval, C
T_{wf} = Temperature of water at the end of the interval, C.
τ = Duration of the interval (e.g. 10 min or 600 s), s.
I_b = Average intensity of beam radiation incident on the aperture of concentrator during the interval, W/m^2.
T_a = Average ambient air temperature during the interval, C.
A_{cont} = Total surface area, m^2.
$A_{aperture}$ = Aperture area of the cooker, m^2.

Equation 2 was used for the determination of optical efficiency factor (F'_{no}). The result of the above test is given in Table 4. The optical efficiency factor (F'_{no}) was found 0.1202 for average solar insolation of 500 W/m^2. The major factor, which affects this value, i.e., reflectance of the reflector and absorptance of receiver cylinder.

3.3 Characteristic Components of Ethanol

After distillation, the value of ethanol concentration was calculated by using standard ethanol curve shown in Fig. 8. Percentage of ethanol available in raw cashew apple juice was 12 %. After first distillation process, it was increased up to 18.6 %. The value of ethanol concentration solution obtained from second distillation was 35.5 %. Ethanol production of 35.5 % by volume is possible by using solar concentrating collector. Table 5 shows ethanol concentration during first distillation and second distillation and Plate 4 shows ethanol samples of first distillation and second distillation.

After second distillation, the analysis of ethanol–water mixture was done. The characteristic components of ethanol, i.e., the specific gravity and acid values were obtained 0.947 and 1.044 mg KOH/g, respectively. The efficiency of distillation unit for the production of 35.5 % ethanol is about 33.4 %. The temperature of the

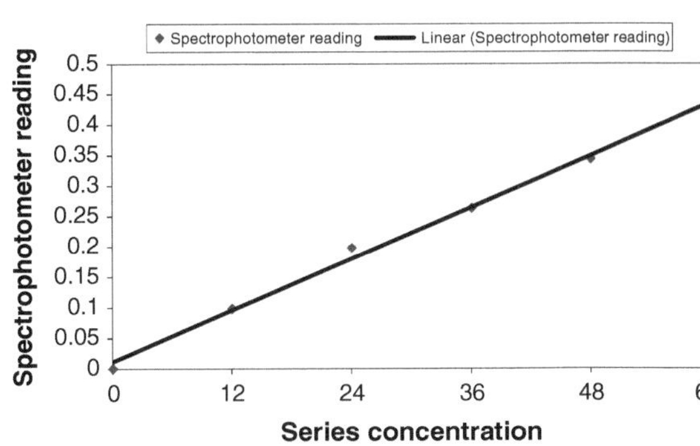

Fig. 8 Standard curve of ethanol

Table 5 First distillation, second distillation and ethanol concentration

Sr. No.	Time (h)	1st distillation (ml)	2nd distillation (ml)
1	8.00	0	0
2	9.00	0	0
3	10.00	130	50
4	11.00	410	240
5	12.00	500	310
6	13.00	580	380
7	14.00	430	170
8	15.00	140	20
9	16.00	40	0
10	17.00	0	0
	Total	2230	1170
	Ethanol concentration	18.6 %	35.5 %

Plate 4 Ethanol samples of first distillation and second distillation

fermented juice was much higher than boiling point of ethanol, i.e., 78.5 °C. Due to higher temperature, same amount of water also got evaporated, which diluted distillated product. Therefore, the higher concentration of ethanol was not obtained by using solar parabolic cooker. It was possible to get 95 % ethanol by further distillation in laboratory. For obtaining 100 % ethanol concentration, the benzene would be used as a dehydrating agent.

4 Conclusions

This study could establish that cashew apple juice commercially for any industrial application. Ethanol production from cashew apple juice was carried out by using solar concentrating collector/cooker. The two processes, i.e., fermentation and distillation are involved in ethanol production. *Saccharomyces cerevisiae* was used as inoculation for better fermentation (Pinheiro et al. 2008). The total distillation rate per day of the system was obtained 2230 ml. The average distillation efficiency of the system was found to be 33.4 with 35.5 % ethanol concentration.

References

Ames BN (2001) DNA damage from micronutrient deficiencies is likely to be a major cause of cancer. Mut Res 475:7–20

A.O.A.C. (1975) Official Methods of Analysis. Association of official analytical chemists, 12th edn. Washington, DC

Costa JMC, Felipe FMF, Maia GJ, Hernandez FFF, Brasil IM (2009) Production and characterization of the cashew apple (Anacardium occidentale L.) and guava (Psidiumguajava L.) fruit powders. J Food Process Preserv 33:299–312

Garge HP, Prakash J (2005) Solar concentrating collectors. Solar energy fundamentals and applications, Tata McGraw-Hill Publishing Co. Ltd., New Delhi, pp 118–119

Haldankar PM, Haldvnkar PC, Govekar MS, Mali PC (2007) Cashew research and development in Konkan region of Maharashtra. National seminar on research, development and marketing of cashew, 20th–21st November, held at Association for Coastal Agricultural Research, ICAR Research Complex for Goa, Ela, Old Goa, pp 33–35

Hubballi VN (2006) Indian cashew-cashew development strategies in next decade. The cashew, vol XX, no 2, pp 20–26

Linoj Kumar NV, Prakash J (2006) Alternative feedstock for bio-ethanol production in India. In: PP Bhojvaid (ed) Biofuels towards a greener and secure energy future, TERI Press Publication, p 90

Pinheiro ADT, Rocha MVP, Macedo GR, Goncalves LRB (2008) Evaluation of cashew apple juice for the production of fuel ethanol. Appl Biochem Biotechnol 148:227–234

Ranganna S (2009) Handbook of analysis and quality control for fruit and vegetable products. Tata McGraw Hill Publishing, New Delhi. ISBN 0-07-451851-8

Sheoran A, Yadav BS, Nigam P, Singh D (1998) Continuous ethanol production from sugarcane molasses using a column reactor of immobilized *Saccharmyces cerevisiae* HAU-1. J Basic Microbiol 38:123–128

Sreenath KH, Jeffries TW (1996) A variable-tilt fermentation rack for screening organisms in micro tubes. Biotechnol Tech 24:239–242

Yu Z, Zhang H (2004) Ethanol fermentation of acid-hydrolyzed cellulosic pryolysate with *Saccharmyces cerevisiae*. Biores Technol 93:199–204

Potential Role of Xylose Transporters in Industrial Yeast for Bioethanol Production: A Perspective Review

Nilesh Kumar Sharma, Shuvashish Behera, Richa Arora and Sachin Kumar

Abstract Sustainable development in lignocellulosic bioethanol production has major challenge due to high cost of production. There are several issues such as efficient utilization of pentose sugars present in lignocelluloses, economical production of lignocellulolytic enzymes with high specificity and economical product recovery, etc. In line, genetically modified yeast strains have been approached to utilize pentose and hexose sugars for bioethanol production. However, these strains showed limited xylose consumption. For efficient utilization of xylose, it is necessary to provide efficient molecular transportation of xylose to the yeast cells. The yeast strains which have been found prominent are *Saccharomyces cerevisiae, Candida intermedia, C. tropicalis, Kluyveromyces marxianus* and *Scheffersomyces stipitis* which have been engineered and examined for xylose transporter genes for xylose utilization. Several transporter genes of interest have been targeted; however, there are major bottleneck in this approach such as xylose transporters which are significantly inhibited by glucose and other hexose sugars. Hence, there are several molecular approaches that have been applied to engineer the yeasts which could improve the xylose transportation. This review has been focused to discuss the molecular advancements in xylose transporter genes and its complexity.

Keywords Genetic engineering · Xylose transporter · Industrial yeasts · Xylose

1 Introduction

For the last three decades, biofuels produced from renewable feedstocks have received much publicity because of their potential to replace conventional fossil fuels. However, the use of readily fermentable agricultural feedstocks like sugar-

N.K. Sharma · Shuvashish Behera · Richa Arora · S. Kumar (✉)
Biochemical Conversion Division, Sardar Swaran Singh National Institute of Bio-Energy (Formerly Sardar Swaran Singh National Institute of Renewable Energy), Kapurthala 144601, Punjab, India
e-mail: sachin.biotech@gmail.com

S. Kumar et al. (eds.), *Proceedings of the First International Conference on Recent Advances in Bioenergy Research*, Springer Proceedings in Energy,
DOI 10.1007/978-81-322-2773-1_6

cane juice and corn for the production of bioethanol is less desirable because of its negative connotation in the food versus fuel debate. Instead, lignocellulosic biomass from hardwood, softwood, and agricultural residues is generally considered as a more sustainable source of feedstocks for biofuels (Kumar et al. 2008; Solomon 2012; Behera et al. 2014a; Sharma et al. 2016). Pentose sugars derived from the hydrolysis of lignocelluloses make up nearly one-third of the reducing sugars with xylose as the major component (Rydholm 1965; Lee et al. 1979; Sharma et al. 2014). Hemicellulose, upon hydrolysis, yields D-xylose as the major component and a number of other sugars, somewhat depending on the source of the hemicellulose, including glucose, D-mannose, D-galactose, and L-arabinose (Wenzl 1970; Jeffries 1983) (Fig. 1). Yeasts capable of fermentation of xylose, thus producing ethanol, have been identified by some researchers (Maleszka and Schneider 1982; Behera et al. 2014b). *Pachysolen tannophilus* (Maleszka 1982), *Candida shehatae,* and *Scheffersomyces stipitis* (Preez and Bosch 1986) appear to generate the highest yields of ethanol from xylose, and these yields compare favorably with that of ethanol from glucose in *Saccharomyces cerevisiae* (Skoog and Hahn-Hägerdal 1988). Efficient conversion of xylose to ethanol occurs only under aerobic conditions in some yeast and under anaerobic conditions in others (Does and Bisson 1989). The explanation for this finding appears to lie in the specificity of the cofactor required for xylose reductase activity. In *C. utilis*, the cofactor for xylose reductase is NADPH, whereas NAD^+ is required for xylitol dehydrogenase activity, resulting in an imbalance of cofactor production and consumption in the absence of an endogenous electron acceptor such as molecular oxygen (Bruinenberg et al. 1983). Xylose reductase utilizes either NADPH or NADH as cofactor in the more efficient xylose-fermenting yeasts such as *C. shehatae, P. tannophilus,* and *S. stipitis*, thus permitting fermentation of xylose under anaerobic conditions (Does and Bisson 1989). Uptake of xylose into the cell may also be the cause of rate limiting for catabolism. The kinetics of xylose uptake has been investigated in *C. shehatae*, *S. stipitis,* and *Rhodotorula gracilis*. In all cases, both high- and low-affinity xylose transport systems were observed (Does and

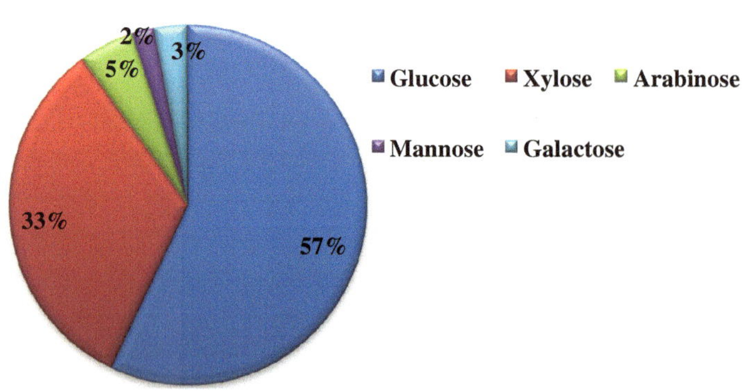

Fig. 1 Percentage sugar content in lignocellulosic biomass

Bisson 1989). Xylose fermentation and subsequent ethanol production have been studied in several other yeast species, including *S. stipitis*, *C. shehatae*, and *P. tannophilus*; however, their ethanol productivities are very poor as compared to *S. cerevisiae* when glucose-based substrates are used (Prior et al. 1989; Hahn-Hägerdal et al. 1994; Nobre et al. 1999). Efforts to establish a xylose-utilizing pathway in *S. cerevisiae* by insertion of the genes encoding xylose reductase and xylitol dehydrogenase from *S. stipitis* or other organisms have resulted in only poor ethanol production from xylose (Hamacher et al. 2002; Sharma et al. 2014). Various steps, including the uptake of xylose, have been suggested to limit the metabolism of xylose in metabolically engineered *S. cerevisiae* (Eliasson et al. 2000). Although a number of research articles have been published that deals with monosaccharide transport in baker's yeast, little study has been done in the field of specificity of the sugar-transporting system (Cirillo 1961, 1963).

Therefore, this review has been focused to discuss the molecular advancements in xylose transporter genes and its complexity in yeast cells.

2 Transport of Sugars

There are number of machineries that are available in industrial yeast for assisting sugar transport. It covers a complete range from energy-independent to energy-dependent transport system. However, adaptation of machinery depends on a large number of factors including substrate availability, type of strain, growth environment, physical conditions, and availability of cofactors (Weusthuis et al. 1994) (Fig. 2).

2.1 Glucose

Glucose transport research studies on industrial yeasts including *S. cerevisiae* and *Kluyveromyces*, *Scheffersomyces*, *Candida,* and *Schizosaccharomyces* species show the existence of different machineries present in all the yeast to facilitate glucose transportation inside the cell. *S. cerevisiae* has high-affinity glucose transport system including energy-independent carrier-mediated facilitated diffusion; however, energy-dependent proton symport system is not available in *S. cerevisiae* (Jansen et al. 2002). Glucose transport is facilitated by carrier-mediated facilitated diffusion together with phosphorylation by glycolytic kinases such as Hxk1p and Hxk2p and Glk1p in *S. cerevisiae*. *S. cerevisiae* has the highest glucose influx rate as compared to other yeasts (Teusink et al. 1998). In higher concentration of glucose, the glucose permeates freely with the help of low-affinity system (Gamo et al. 1995; Naftalin 2008). In other yeasts, glucose accumulation, arisen from a proton electrochemical

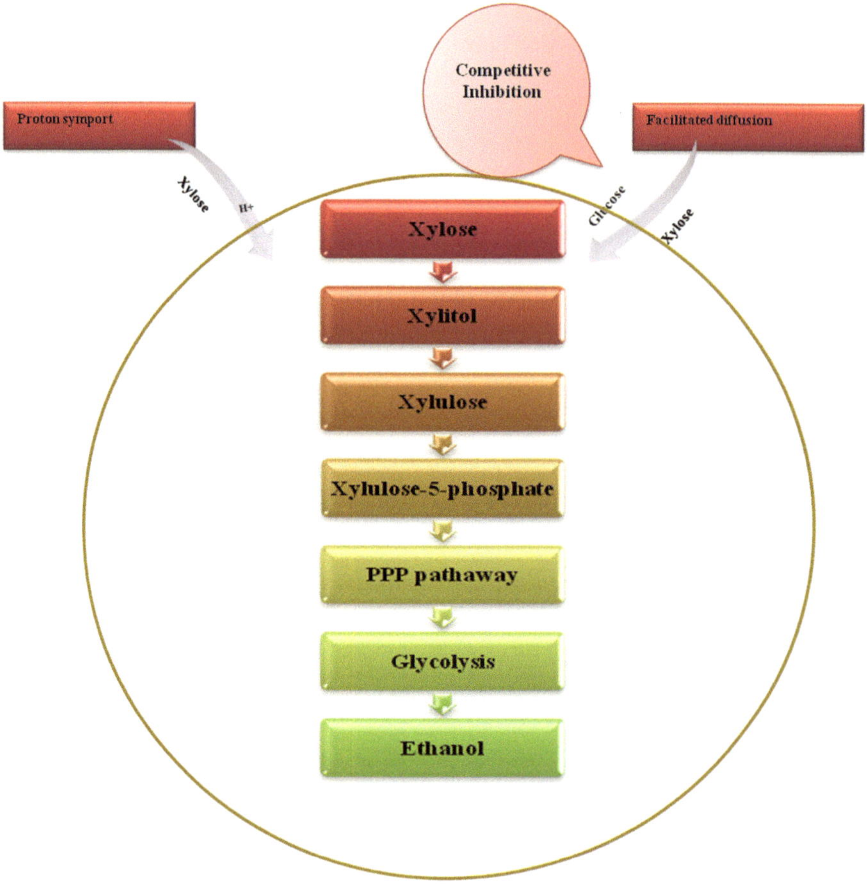

Fig. 2 Sugar transportation mechanism in yeast

gradient across the cell membrane against a concentration gradient, is driven by energy. In fact, the energy cost is quite high for active proton symport of glucose (Stambuk et al. 2003).

With the help of molecular experiments, transportation of glucose and other sugars in the yeasts has identified several genes encoding high- and low-affinity transporters. The huge number of putative glucose transporters has been identified including Hxt1p, Hxtl4p, and Snf3, belonging to a sugar permease family (Boles et al. 1997). The HXT genes affect glucose (HXT2), galactose (GAL2), glucose and mannose (HXTI), or glucose, fructose and galactose (HXT4) uptake. However, none of the genes has been identified in *S. cerevisiae* specifically fructose uptake (Reifenberger et al. 1997; Boles and Hollenberg 1997) (Table 1).

Table 1 Glucose transport systems in some biotechnologically industrial yeast (Walker 1998)

S/N	Yeast	Mechanism	Control
1	*C. utilis*	Proton symport	Repressible
2	*C. shehatae*	Proton symport	Repressible
		Facilitated diffusion	Non-repressible
3	*K. marxianus*	Proton symport	Repressible
		Proton symport	Repressible
4	*S. stipits*	Proton symport	Non-repressible
5	*S. cerevisiae*	Facilitated diffusion	Repressible
		Facilitated diffusion	Constitutive

Notes

- Control refers to whether or not the transport system is repressible by glucose or is constitutively expressed
- Facilitated diffusion is an energy-independent system mediated by protein carriers, whereas proton symport refers to active energy-dependent ATPase-mediated proton motive force

2.2 *Xylose*

Studies concerning to pentose sugar transport by yeasts again reveal that different transport systems exist in different yeasts. Pentose transportation in yeasts generally occurs by membrane potential due to proton symport, while sometimes facilitates diffusion through low-affinity transport systems (Singh and Mishra 1995). Different machineries for sugar uptake exist in different yeast species such as *K. marxianus*, *S. stipitis*, *C. shehatae,* and *Rhodotorula glutinis*. These yeasts have been used as a model organism in the study of pentose transport (Young et al. 2014). Under aerobic environment, a bioelectrogenic proton symporter is perceptible, during high-affinity D-xylose uptake in *R. glutinis*. A low-affinity system involving a proton symport mechanism may also exist in this yeast (Singh and Mishra 1995). In *S. stipitis*, xylose transport is a rate-limiting step under aerobic growth conditions and uptake of this sugar is mediated by both low- and high-affinity carriers (Du et al. 2010). In *C. shehatae*, both machineries exist including facilitated diffusion and proton symport mechanisms for D-xylose transport. The former mechanism is operative under conditions of repression but it is absent in starving cultures of this yeast (Walker 1998). *S. cerevisiae* is not able to utilize xylose but sometimes in starvation xylose uptake occurs under aerobic conditions by a low-affinity system (Toivari et al. 2004). In general, xylose uptake is inhibited in the presence of glucose in most of the xylose-fermenting yeasts (Wang et al. 2014), which reveals the preferential utilization of glucose in cells grown on complex sugar mixtures of hemicellulosic hydrolysates. The sequential uptake pattern represents a major

Table 2 Xylose transport systems in some biotechnologically industrial yeast (Walker 1998)

S/N	Yeast	Mechanism	Comments
1	*C. shehatae*	Proton symport	Xylose both respired and fermented
		Facilitated diffusion	–
2	*C. utilis*	Proton symport	Only high-affinity system operates in glucose de-repressed cells
3	*M. reukaufii*	Proton symport	Xylose only respired
4	*P. stipitis*	Proton symport	Xylose both respired and fermented
5	*R. glutinis*	Proton symport	Xylose only respired
		Facilitated diffusion	–
6	*S. cerevisiae*	Facilitated diffusion	Xylose non-metabolizable
7	*K. marxianus*	Facilitated diffusion	Xylose metabolizable
8	*S. pombe*	Facilitated diffusion	Xylose non-metabolizable

factor, which limits the biotechnological exploitation of pentose-fermenting yeasts in the bioconversion of wood hydrolysates (Spencer-Martins 1994; Nijland et al. 2014) (Table 2).

3 Factor Affecting Xylose Transporters

Xylose transporters have its own importance in the xylose uptake in the yeast cells. However, a lot of efforts including the cloning of the transporter genes of other species have been made to sort out the transportation issues. However, several other factors as discussed below play the major role in inhibition of xylose transporter (Nevoigt 2008).

3.1 Co-expression of GXF1 and GXS1

When glucose/xylose transporter genes from *C. intermedia* were cloned in *S. cerevisiae,* i.e., GXF1 (glucose/xylose facilitator) and GXS1 (glucose/xylose proton symporter), Gxf1p was observed fully functional *in S. cerevisiae*. However, the symporter Gxs1p exhibits very low-glucose/xylose transport activity, which could not be able to produce the sufficient protein. Therefore, it is evident that co-expressions of GXS1 and GXF1 inhibit the transportation of pentose sugars in *S. cerevisiae* (Leandro et al. 2006, 2008; Young et al. 2011).

3.2 Extracellular Glucose

Similar transport system is accompanied for xylose and glucose in *S. stipitis* as *S. cerevisiae* (Farwick et al. 2014). However, xylose at 100 mM concentration did not inhibit glucose uptake, while glucose inhibited the xylose uptake. Thus, xylose did not compete with glucose for uptake of glucose in *P. heedii* which also shows the involvement of different carrier systems (Does and Bisson 1989; Barnett et al. 2008).

3.3 Acetic Acid

Acetic acid is fungistatic, act by dissipating plasma membrane proton gradients and depressing cell pH when they dissociate into ions in the yeast cytoplasm. Acetic acid disrupts proton gradients by diffusing from the acidified external medium into the cells and dissociating in the more neutral environment of the cytoplasm (Walker and White 2011). Therefore, proton gradient looses to drive nutrient uptake. Higher concentration of acetic acid also changes growth and thermal death profile. *C. shehatae* could tolerate up to 0.4 ml/l acetic acid at pH 4.5, and shifted temperature range from 5–34 °C to 21–27 °C and resulted in decreasing cell mass yield by 64 % on D-xylose (Rodrigues et al. 1992; Singla et al. 2012).

4 Role of Transporter Genes

Several transporter genes have been identified for both glucose and xylose transportations in different yeast strains. Each transporter gene has a different role in the system. Hence, many genetic engineering approaches have been applied to increase the xylose and glucose uptake in industrial yeasts (Young et al. 2011). Some of the functions of major transporter genes have been summarized in Table 3.

4.1 AT5G17010 and AT5G59250

The directed movement of carbohydrate into, out of or within a cell, or between cells occurs by means of some agents such as a transporter or pore. The genes AT5G17010 and AT5G59250 are D-xylose-proton symporter-like protein 2 and protein 3 of *A. thaliana*, respectively (Büttner 2007), which facilitates the translocation of a specific substance or associated substances from out to in and vice versa. These genes have been cloned in *S. cerevisiae* to improve xylose transporter

Table 3 List of transporter genes and their role

S/N	Transporter genes	Function	Source organism	References
1	AT5G17010	D-xylose-proton symporter-like 2	*Arabidopsis thaliana*	Hector et al. (2008)
2	AT5G59250	D-xylose-proton symporter-like 3	*A. thaliana*	Hector et al. (2008)
3	HXT1	Hxt1p (MFS transporter, SP family)	*S. cerevisiae*	Sedlak et al. (2004)
4	HXT2	Hxt2p (MFS transporter, SP family)	*S. cerevisiae*	Sedlak et al. (2004)
5	HXT5	Hxt5p (MFS transporter, SP family)	*S. cerevisiae*	Sedlak et al. (2004)
6	HXT7	Hxt7p (MFS transporter, SP family)	*S. cerevisiae*	Sedlak et al. (2004)
7	GXF1	Glucose/xylose facilitator 1	*C. intermedia*	Leandro et al. (2006)
8	GXS1	Glucose/xylose symporter 1	*C. intermedia*	Leandro et al. (2008)
9	XUT4	High-affinity xylose transporter (Putative) (HGT3)	*S. stipitis*	Young et al. (2011)
10	XUT5	Putative xylose transporter	*S. stipitis*	Young et al. (2011)
11	XUT6	Sugar transporter, putative (STL12)	*S. stipitis*	Young et al. (2011)
12	XUT7	Xylose transporter, high affinity, putative similarity to STL13	*S. stipitis*	Young et al. (2011)
13	RGT2	Glucose transporter/sensor	*S. stipitis*	Hamacher et al. (2002)
14	SUT1	Hexose transporter	*S. stipitis*	Weierstall et al. (1999)
15	STP2	sugar transport protein 2	*A. thaliana*	Hamacher et al. (2002)
16	STP3	Sugar transport protein 3	*A. thaliana*	Hamacher et al. (2002)
17	TRXLT1	–	*Trichoderma reesei*	Saloheimo et al. (2007)
18	HUP1	H(+)/hexose cotransporter 1	*Chlorella kessleri*	Wolf et al. (1991), Hamacher et al. (2002)
19	CNBC3990	Uncharacterized protein	*Cryptococcus neoformans*	Young et al. (2011)

(continued)

Table 3 (continued)

S/N	Transporter genes	Function	Source organism	References
20	DEHA0D02167	–	*Debaryomyces hansenii*	Young et al. (2011)
21	ALI0C06424	–	*Yarrowia lipolytica*	Young et al. (2011)
22	YALI0B06391	–	*Y. lipolytica*	Young et al. (2011)

efficiency and significant improvement, i.e., xylose uptake rate increased by 45 % in comparison of isogenic control strain lacking the *A. thaliana* transporters (Hector et al. 2008).

4.2 HXTs Family

Several studies have been carried out to resolve the problems related to co-expression of glucose and xylose (Boles and Andre 2014). An approach has been applied in which xylose uptake properties recombinant xylose-utilizing *S. cerevisiae* has been modified by expression of heterologous and homologous permease-encoding genes. In this mutant strain main hexose transporter genes have been deleted and engineered for xylose utilization (Young et al. 2011). After screening in xylose plates, the enhanced growth was shown, whereas, when the clone was retransformed into the host, it did not support significant growth on xylose. After long-term adaptive mutation it was revealed that Hxt1p, Hxt2p, Hxt4p, and Hxt7p had the capability to uptake of the xylose (Saloheimo et al. 2007).

In *S. cerevisiae*, HXT2 gene was identified by its ability to complement the defect in the glucose transport of a snf3 mutant, when it was present on the multicopy plasmid pSC2 (Kruckeberg and Bisson 1990). It has 12 highly hydrophobic regions that can form transmembrane domain. It has been observed that an hxt2 null mutant strain lacks a high-affinity glucose transport when under derepressing (low-glucose) conditions. However, the hxt2 null mutation did not incur a major growth defect on glucose-containing media. Biochemical studies of this protein suggest that wild-type levels of high-affinity glucose transport require the products of both the HXT2 and SNF3 genes; these genes are not linked (Kruckeberg et al. 1990, Jansen et al. 2002).

To reveal glucose transport properties of HXT family transporters, HXT1, HXT2, HXT3, HXT4, HXT6, and HXT7 have separately expressed in a HXT1–7 null strain. The Km values from high to low affinity determined from counter-transport and initial-uptake experiments are Hxt6p 0.9 and 1.4 mM, Hxt7p 1.3 and 1.9 mM, Hxt2p 2.9 and 4.6 mM, Hxt4p 6.2 and 6.2 mM, Hxt3p 28.6 and 34.2 mM, and Hxt1p 107 and 129 mM. Same Km values were obtained for each transporter from both the methods counter-transport and initial uptake, which show the essential facilitated diffusion transporters for transporting glucose in *S. cerevisiae* (Maier et al. 2002).

4.3 STP2

The AtSTP2 gene (sugar transport protein 2) of *Arabidopsis thaliana* encodes a high-affinity, low-specificity monosaccharide carrier that can transport a number of hexoses and pentoses at similar rates. AtSTP2 has 12 putative transmembrane helices and a molecular mass of 55.0 kDa (Truernit et al. 1999).

5 Conclusion

There are different biotechnological applications like synthetic biology, systems biology, and metabolic engineering which have been followed to tackle the problems ensuring the stable supply of transportation fuels for the future. All genetic engineering methodologies must go beyond the different levels of genetic information. The transport physiology of the recipient organism is a very important part which depends on the expression of that genetic information, although the recombinant yeasts can be utilized to convert xylose into ethanol. However, the xylose consumption rate is very slow, which is not sufficient for efficient production of ethanol from lignocellulosic materials. There are some other problems such as inefficient co-fermentation of glucose and xylose using recombinant strains. Further, it is also necessary to consider the kinetics and energetics of sugar transport systems in relation to the overall metabolism of engineered organisms. The efforts are also required to study the expression pattern of different transporter genes affecting xylose uptake.

Acknowledgments Authors are very much thankful to Ministry of New and Renewable Energy, New Delhi, Govt. of India for the financial support and providing all research facilities to carry out research work. One of the authors (Nilesh Kumar Sharma) also acknowledges Sardar Swaran Singh National Institute of Bio-Energy, Kapurthala for providing Junior Research Fellowship and I. K. Gujral Punjab Technical University, Jalandhar for his Ph.D. registration (Pro. reg. 1422002).

References

Barnett JA (2008) A history of research on yeasts 13 active transport and the uptake of various metabolites. Yeast 25(10):689–731

Behera S, Arora R, Sharma NK, Kumar S (2014b) Fermentation of glucose and xylose sugar for the production of ethanol and xylitol by the newly isolated NIRE-GX1 yeast. Recent Advances in Bio-energy Research, SSS-NIRE, Chapter 16, pp. 175–182

Behera S, Arora R, Nandhagopal N, Kumar S (2014a) Importance of chemical pretreatment for bioconversion of lignocellulosic biomass. Renew Sust Energy Rev 36:91–106

Boles E, André B (2014) Role of transporter-like sensors in glucose and amino acid signalling in yeast. Molecular mechanisms controlling transmembrane transport. Top Curr Genet 9:155–178

Boles E, Hollenberg CP (1997) The molecular genetics of hexose transport in yeasts. FEMS Microbiol Rev 21(1):85–111

Bruinenberg PM, de Bot PHM, van Dijken JP, Scheffers WA (1983) The role of redox balances in the anaerobic fermentation of xylose by yeasts. Eur J Appl Microbiol Biotechnol 18:287–292

Büttner M (2007) The monosaccharide transporter(-like) gene family in Arabidopsis. FEBS Lett. 581:2318–2324

Cirillo VP, Wilkins PO, Anton J (1963) Sugar transport in a psychrophilie yeast. J Baeteriol 86 (6):1259–1264

Cirillo VP (1961) Sugar transport in microorganisms. Annual Rev Mierobiol 15:197–218

Does AL, Bisson LF (1989) Characterization of Xylose Uptake in the Yeasts Pichia heedii and Pichia stipitis. Appl Environ Microbiol 55(1):159–164

Du J, Li S, Zhao H (2010) Discovery and characterization of novel d-xylose-specific transporters from Neurospora crassa and Pichia stipitis. Mol BioSyst 6(11):2150–2156

Du Preez JC, Bosch M, Prior BA (1986) The fermentation of hexose and pentose sugars by Candida shehatae and Pichia stipitis. AppI Microbiol Biotechnol 23:228–233

Eliasson A, Christensson C, Wahlbom CF, Hahn-Hägerdal B (2000) Anaerobic xylose fermentation by recombinant Saccharomyces cerevisiae carrying XYL1, XYL2, and XKS1 in mineral medium chemostat cultures. Appl Environ Microbiol 66:3381–3386

Farwick A, Bruder S, Schadeweg V, Oreb M, Boles E (2014) Engineering of yeast hexose transporters to transport d-xylose without inhibition by d-glucose. PNAS 111(14):5159–5164

Gamo FJ, Moreno E, Lagunas R (1995) The low-affinity component of the glucose transport system in Saccharomyces cerevisiae is not due to passive diffusion. Yeast 11:1393–1398

Hahn-Hägerdal B, Jeppsson H, Skoog K, Prior B (1994) Biochemistry and physiology of xylose fermentation by yeasts. Enzyme Microb Technol 16:933–943

Hamacher T, Becker J, Gardony M, Hahn-Hägerdal B, Boles E (2002) Characterization of the xylose-transporting properties of yeast hexose transporters and their influence on xylose utilization. Microbiology 148:2783–2788

Hector RE, Qureshi N, Hughes SR, Cotta MA (2008) Expression of a heterologous xylose transporter in a Saccharomyces cerevisiae strain engineered to utilize xylose improves aerobic xylose consumption. Appl Microbiol Biotechnol 80(4):675–684

Jansen ML, De Winde JH, Pronk JT (2002) Hxt-carrier-mediated glucose efflux upon exposure of Saccharomyces cerevisiae to excess maltose. Appl Environ Microbiol 68(9):4259–4265

Jeffries TW (1983) Utilization of xylose by bacteria, yeasts and fungi. Adv Biochem Eng Biotech 27:1–31

Kruckeberg AL, Bisson LF (1990) The HXT2 gene of Saccharomyces cerevisiae is required for high-affinity glucose transport. Mol Cell Biol 10(11):5903–5913

Kumar R, Singh S, Singh OV (2008) Bioconversion of lignocellulosic biomass: biochemical and molecular perspectives. J Ind Microbiol Biotechnol 35:377–391

Leandro MJ, Spencer-Martins I, Goncalves P (2008) The expression in Saccharomyces cerevisiae of a glucose/xylose symporter from Candida intermedia is affected by the presence of a glucose/xylose facilitator. Microbiology 154:1646–1655

Leandro MJ, Goncalves P, Spencer-Martins I (2006) Two glucose/xylose transporter genes from the yeast Candida intermedia: first molecular characterization of a yeast xylose-H + symporter. Biochem J 395(3):543–549

Lee YY, Lin CM, Johnson T, Chambers RP (1979) Selective hydrolysis of hardwood hemicellulose by acids. Biotechnol Bioeng Symp 8:75–88

Maier A, Volker B, Boles E, Fuhrmann GF (2002) Characterisation of glucose transport in Saccharomyces cerevisiae with plasma membrane vesicles (countertransport) and intact cells (initial uptake) with single Hxt1, Hxt2, Hxt3, Hxt4, Hxt6, Hxt7 or Gal2 transporters. FEMS Yeast Res 2:539–550(2002)

Maleszka R, Schneider H (1982) Fermentation of D-xylose, xylitol and D-xylulose by yeasts. Can J Microbiol 28:360–363

Naftalin RJ (2008) Alternating carrier models of asymmetric glucose transport violate the energy conservation laws. Biophys J 95(9):4300–4314

Nevoigt E (2008) Progress in metabolic engineering of Saccharomyces cerevisiae. Microbiol Mol Biol Rev 72(3):379–412

Nijland JG, Shin HY, de Jong RM, de Waal PP, Klaassen P, Driessen AJ (2014) Engineering of an endogenous hexose transporter into a specific D-xylose transporter facilitates glucose-xylose co-consumption in Saccharomyces cerevisiae. Biotechnol Biofuels 7(1):168

Nobre A, Lucas C, Leão C (1999) Transport and utilization of hexoses and pentose in the halotolerant yeast Debaryomyces hansenii. Appl Environ Microbiol 65(8):3594

Prior BA, Kilian SG, du Preez JC (1989) Fermentation of D-xylose by the yeasts Candida shehatae and Pichia stipitis. Process Biochem 89:21–32

Reifenberger E, Boles E, Ciriacy M (1997) Kinetic characterization of individual hexose transporters of Saccharomyces cerevisiae and their relation to the triggering mechanisms of glucose repression. Eur J Biochem 245:324–333

Rodrigues AA, Morais JM, Lopes MA (1992) Effects of acetic acid on the temperature range of ethanol tolerance in Candida she hate growing on D-xylose. Biotechnol Lett 14:1181–1186

Rydholm SA (1965) Pulping processing. Interscience Publishers, Inc., New York, pp 95–98

Saloheimo A, Rauta J, Stasyk OV, Sibirny AA, Penttila M, Ruohonen L (2007) Xylose transport studies with xylose-utilizing Saccharomyces cerevisiae strains expressing heterologous and homologous permeases. Appl Microbiol Biotechnol 74:1041–1052

Sedlak M, Ho NW (2004) Characterization of the effectiveness of hexose transporters for transporting xylose during gluco se and xylose co-fermentation by a recombinant Saccharomyces yeast. Yeast 21:671–684

Sharma NK, Behera S, Arora R, Kumar S (2016) Enhancement in xylose utilization using *Kluyveromyces marxianus* NIRE-K1 through evolutionary adaptation approach. Bioprocess Biosyst Eng 1–9

Sharma NK, Behera S, Kumar S (2014) Genetic modification for simultaneous utilization of glucose and xylose by yeast. In: Kumar S, Sarma AK, Tyagi SK, Yadav YK (eds) Recent advances in bioenergy research, vol III. SSS-NIRE, Kapurthala, India. pp 194–207

Singh A, Mishra P (1995) Microbial pentose utilization: current applications in biotechnology. Elsevier Science Ltd, ISBN: 0444820396

Singla A, Paroda S, Dhamija SS, Goyal S, Shekhawat K, Amachi S, Inubushi K (2012) Bioethanol production from xylose: problems and possibilities. J Biofuels 3(1):39–49

Skoog K, Hahn-Hägerdal B (1988) Xylose fermentation. Enzyme Microb Technol 10:1–15

Solomon BD (2012) Biofuels and sustainability. Ann N Y Acad Sci 1185:119–134

Spencer-Martins I (1994) Transport of sugars in yeasts: implications in the fermentation of lignocellulosic materials. Bioresour Tech 50:51–57

Stambuk BU, Franden MA, Singh A, Zhang M (2003) D-Xylose transport by Candida succiphila and Kluyveromyces marxianus. Appl Biochem Biotechnol Spring 105–108:255–263

Teusink B, Diderich JA, Westerhoff HV, Dam KV, Walsh MC (1998) Intracellular glucose concentration in derepressed yeast cells consuming glucose is high enough to reduce the glucose transport rate by 50 %. J Bacteriol 180(3):556–562

Toivari MH, Salusjärvi L, Ruohonen L, Penttilä M (2004) Endogenous xylose pathway in Saccharomyces cerevisiae. Appl Environ Microbiol 70(6):3681–3686

Truernit E, Stadler R, Baier K, Sauer N (1999) A male gametophyte-specific monosaccharide transporter in Arabidopsis. Plant J. 17:191–201

Walker GM (1998) Yeast physiology and biotechnology. John Wiley & Sons Ltd

Walker GM, White NA (2011) Introduction to fungal physiology, in fungi: biology and applications. In: Kavanagh K (ed), Wiley, Ltd, Chichester, UK

Wang X, Jin M, Balan V, Jones AD, Li X, Li BZ, Dale BE, Yuan YJ (2014) Comparative metabolic profiling revealed limitations in xylose-fermenting yeast during co-fermentation of glucose and xylose in the presence of inhibitors. Biotechnol Bioeng 111(1):152–164

Wenzl HF (1970) The chemical technology of wood. Academic Press, New York and London, pp 119–123

Weusthuis RA, Pronk JT, van den Broek PJ, van Dijken JP (1994) Chemostat cultivation as a tool for studies on sugar transport in yeasts. Microbiol Rev 58:616–630

Wolf K, Tanner W, Sauer N (1991) The Chlorella H +/hexose cotransporter gene. Curr Genet 19 (3):215–219

Young E, Poucher A, Comer A, Bailey A, Alper H (2011) Functional survey for heterologous sugar transport proteins, using Saccharomyces cerevisiae as a host. Appl Environ Microbiol 77 (10):3311–3319

Young EM, Tong A, Bui H, Spofford C, Alper HS (2014) Rewiring yeast sugar transporter preference through modifying a conserved protein motif. PNAS 111(1):131–136

Part II
Chemical Conversion

Transesterification of Neem Oil Using Na Ion-Doped Waste Fishbone

Himadri Sahu and Kaustubha Mohanty

Abstract High cost of biodiesel produced from various known sources till date has focused the attention of researchers on low cost and renewable raw materials such as nonedible oils and waste cooking oils. Catalyst plays a significant role on the yield and quality of biodiesel. Heterogeneous catalysts are better over conventional homogeneous catalysts as they can be easily recovered from the reaction mixture by filtration and can be reused after activation, thereby making the process economically viable. Heterogeneous catalysts are mostly solid and can be prepared from various materials such as industrial wastes (e.g., red mud, aluminum dross, fly slag, blast furnace, and steel slag) and natural wastes (e.g., chicken egg shells, rice husks, fish skeletal, shells of crab, oyster, snail, shrimp, etc.). The use of such wastes to develop catalysts can be a win-win situation as this takes care of the waste disposal issues. Current focus is to develop catalysts from natural sources. In the present work, heterogeneous catalyst was developed from one natural hydroxyapatite, i.e., cooked fish bone and was doped with Na ion. Doping of metalloid ions and calcining not only results in higher surface area but also increases the ion exchange capability. The catalyst has been prepared employing wet impregnation method. This developed catalyst was used for transesterified neem oil to produce biodiesel with a acid value conversion closer to 96.7 %. The final transesterified product was analyzed and the parametric effects on the acid value conversion were studied.

Keywords Transesterification · Neem · Hydroxyapatite · Fishbone · Catalyst

Himadri Sahu · Kaustubha Mohanty (✉)
Department of Chemical Engineering, Indian Institute of Technology Guwahati, Guwahati 781039, India
e-mail: kmohanty@iitg.ernet.in

S. Kumar et al. (eds.), *Proceedings of the First International Conference on Recent Advances in Bioenergy Research*, Springer Proceedings in Energy,
DOI 10.1007/978-81-322-2773-1_7

1 Introduction

Globally in excess ninety one million tons of fish are caught and around 55 % of this are used for human consumption, while rest were discarded (Rustad 2003). The discard materials generate undesirable impact on environment. To reduce any adverse environmental effects, it is advisable to use these discard materials for high value-added products. Few fish species are exported after extracting the bones. These discarded bones are the cheap source of calcium phosphate. So this is also termed as hydroxyapatite, with stoichiometric formula $Ca_{10}(PO_4)_6(OH_2)$.

However, there are few differences between synthetic and raw hydroxyapatite. Natural hydroxyapatite has and better dynamic response to the environment, high catalytic activity, and good thermochemical stability than the synthetic hydroxyapatite (Obadiah et al. 2012; Best et al. 2008). As far as solubility and activity is concerned, crystallographic structure plays an important role. In line to synthetic hydroxyapatite, the natural hydroxyapatite has perturbed nanostructures and non-stoichiometric composition with low hydroxyl content. As physical and chemical properties depend on structure, natural hydroxyapatite will lead to good results (Leventouri 2006). Because of these properties, hydroxyapatite can be used as a heterogeneous catalyst. Although natural hydroxyapatite has favorable structure but the surface area is bit low, which can be increased by doping metal ions on to it.

In the present work, raw hydroxyapatite collected from waste fish bone was used as a base material for metal doping to conduct transesterification of neem (Azadirachta indica) oil. After the literature review, it was found that calcined Sodium silicate alone has given very good result in transesterification reaction (Fan et al. 2013; Long et al. 2011; Yin et al. 2010). So, sodium silicate metal ion was doped on to the raw hydroxyapatite by wet impregnation method to increase the surface area of hydroxyapatite. In this catalyst both the hydroxyapatite and Na/Si will act as an active source for the adsorption of reactants.

2 Materials and Detailed Methodology

2.1 Materials

From IIT Guwahati hostel mess-cooked waste fish bone was collected. Few chemicals namely, 25 % aqueous ammonia solution, methanol and sodium silicate ($Na_2O_3Si{\cdot}9H_2O$) were procured from Merk, India and were used as it is without any further treatment.

2.2 Na/Si Doping on Hydroxyapatite Methodology

Fish bone was washed properly to remove proteinaceous part and sun dried for 2 days. After that oily matters were removed completely by dipping in acetone for 24 h and dried at 60 °C. The fishbone was crushed and raw hydroxyapatite was obtained. Sodium Silicate was used as precursor for metal ions during wet impregnation (Wang et al. 2008). Following the method, 1:1 wt. proportions of raw hydroxyapatite and sodium silicate, at first 30 g of sodium silicate was added to 100 mL of water to get a solution of sodium silicate. To this solution, 30 g of fine powder of raw hydroxyapatite was added gradually and blended rigorously using a mechanical agitator under aggregate reflux time of 30 min. 25 % aqueous ammonia solution was used to maintain pH at 11. The entire solution was mechanically stirred for 2 h at 700 rpm at 60 °C. The final solution was kept for 24 h at room temperature so as to allow the deposition of sodium and silica ions on raw hydroxyapatite. After that, mother liquor was vacuum filtered and obtained sticky precipitate was dried in hot air oven. The dried mass was then calcined at 650 °C for 3 h and final Na/Si-doped hydroxyapatite was sent to lab for characterization.

2.3 Procedure for Transesterification of Neem Oil Using Raw, Calcined, and Na/Si-Doped Hydroxyapatite

Transesterification of purified neem oil was conducted in a batch reactor having conventional heating framework. For a standard analysis, 30 g of neem oil, methanol, and catalyst (2 wt%) were taken in a 3-necked glass reactor of 1 L volume and was heated up to 70 °C and then agitated for a particular time period. After that, the entire reaction mixture was transferred to a rotavapor for removing excess methanol. The reaction mixture was then centrifuged to separate the catalyst and the mixture was kept in a separating funnel overnight to separate the glycerol produced during the reaction. Then the acid value conversion was calculated by gas chromatography. Following this standard method, effect of oil to methanol ratio and catalyst loading on acid value conversion (%) with respect to reaction time were also studied.

2.4 Catalyst and Oil Characterization

2.4.1 Catalyst Characterization

Bucker X-ray D8 advance diffractometer was used to gather powder X-ray diffraction information for structural analysis. For X-ray radiation source, a Cu Kα ($\lambda = 0.154$ nm, 45 kV, 40 mA) anode was used, ranging 2θ between 2° and 90° with

0.5 s/step scan rate and 0.05 increment. Surface analyzer device (Beckman-Coulter; Model: SA3100) was used to measure BET surface area, average pore measurement, and pore volume (t-plot strategy) by physical adsorption–desorption of N_2 at the boiling point (77 K). Preceding the examination, raw hydroxyapatite and Na/Si-doped hydroxyapatite were subjected to a preheating for 90 min at 423 K under vacuum. A thermal analyzer (NETZSCH: STA449F3 Jupiter) was used for the thermo gravimetric analysis of the powder sample. Argon was passed through the instrument with a heating rate of 10 K min^{-1} using a ceramic crucible up to 1000 °C. A scanning electron microscope (JEOL-JSM-6390LV) furnished with an EDAX PV 9760 detector for energy dispersive X-ray spectroscopy (EDX) was used to investigate nearby chemical arrangement. The samples were dispersed in methanol and deposited on an Al foil before mounting on sample holder, then coated with gold–palladium to make the specimen fit for microstructures investigations.

2.4.2 Purified Neem Oil and Transesterified Neem Oil Characterization

The acid Number of purified neem oil (Azadirachta indica) and transesterified product was determined by colorimetric titration as per ASTM D974 (ASTM standard 2009) and saponification number of raw oil was calculated as per ASTM D5558 (ASTM standard 2006). The acid number and saponification value of purified neem oil was found to be 23.8 mg KOH/g oil and 190 mg KOH/g oil. The physical properties such as density, viscosity, flash point, and pour point were calculated by ASTM standards. Fourier transform infrared spectroscopy (FTIR) provides information with respect to various functional groups present in the sample. DRS connected with Excalibur Bio-Rad spectrophotometer (Model FTS 3500 GX) FTIR analyzer was utilized for the same. The IR range was between 400 and 4000 cm^{-1} at a scan rate of 40 and at a step size of 4 cm^{-1}.

3 Results and Discussion

3.1 Characterization of Na/Si-Doped Hydroxyapatite Catalyst

XRD patterns of raw hydroxyapatite and Na/Si-doped hydroxyapatite are delineated in Fig. 1. Well-resolved characteristic peak of most elevated intensity for hydroxyapatite was acquired at 2θ value of 31.77° corresponding to 211 planes. The phase formed was clear and matches well with standard pattern reported in

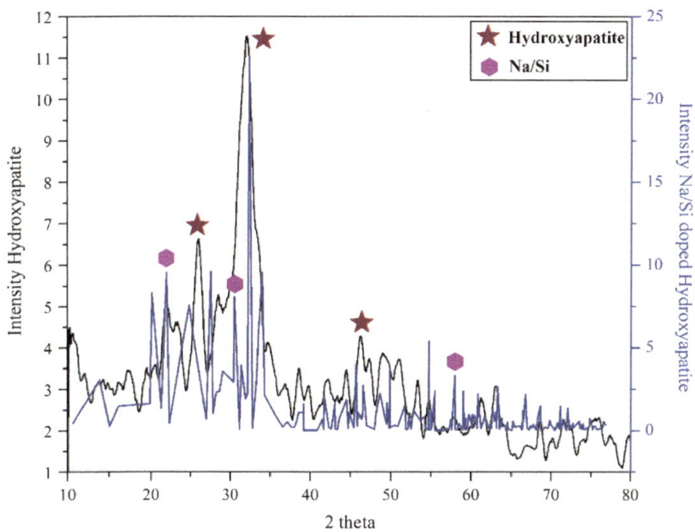

Fig. 1 X-ray diffraction patterns of raw hydroxyapatite and Na/Si-doped hydroxyapatite showing characteristic peaks

literature (Rocha et al. 2005). The standard corresponding plane for hydroxyapatite (namely, 100, 101, 200, 002, 211, 202, 301, 130, 131, 113, 203, 222, 132, 321, 004, 240, 241, 502, 323, and 511) was well observed in case of the synthesized catalyst. The diffraction peak at $2\theta = 21.8°$ with plane 2, 0, 0 corresponds to the Si support (Musić et al. 2011) as registered on the XRD patterns (Fig. 1).

The principal evidence for framing of raw hydroxyapatite was in the form of a strong complex broad FTIR band (Fig. 2) focused at about 1000–1120 cm^{-1} because of asymmetric stretching mode of vibration for PO_4 group (Varma and Babu 2005). The band at 570 cm^{-1} corresponds to P–O stretching vibration of the PO_4 group. As a major peak of phosphate group, the vibration peak could be distinguished in the region between 1120–960 cm^{-1} for both raw hydroxyapatite and Na/Si-doped hydroxyapatite, which were due to P–O asymmetric stretching of PO_4^{3-}. Many of the compounds can be acknowledged to be analogs of ethers, especially when an alkoxy group was available, featuring the P–O–C linkage at 1240 cm^{-1}. In case of Na/Si-doped hydroxyapatite the bands at 780–980 cm^{-1} and peak in the range of 900–1200 cm^{-1} are resulting to the peroxide formation (M–O–O–M) and the M–O–M bonding, respectively (Maensiri et al. 2006).

Figure 3 shows a typical nitrogen adsorption–desorption isotherm of raw, calcined and Na/Si-doped hydroxyapatite. The isotherms display a type IV curve with a hysteresis loop which corresponds to mesoporous materials. Raw, calcined, and Na/Si-doped hydroxyapatite have BET surface areas of 10.865, 167.86, and 57.124 $m^2\ g^{-1}$, respectively. The total pore volumes were found to be 0.0894, 0.4926, and 0.4943 cc g^{-1} for raw, calcined, and Na/Si-doped hydroxyapatite,

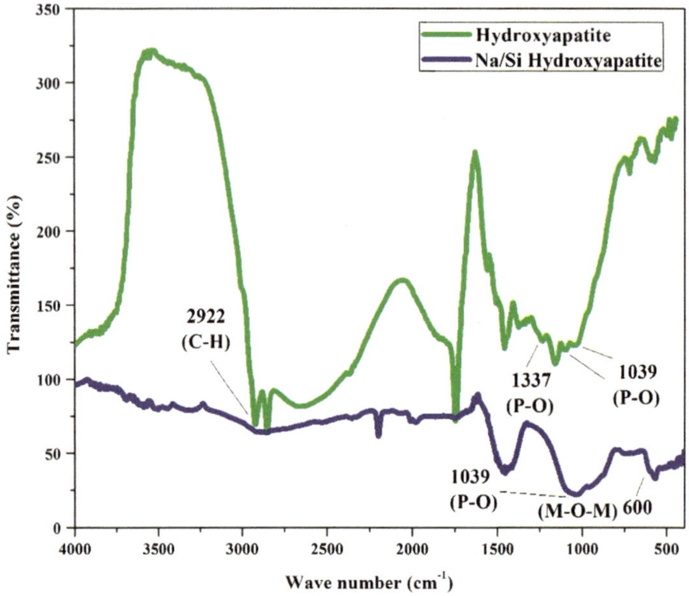

Fig. 2 FTIR spectrum of raw hydroxyapatite and Na/Si-doped hydroxyapatite

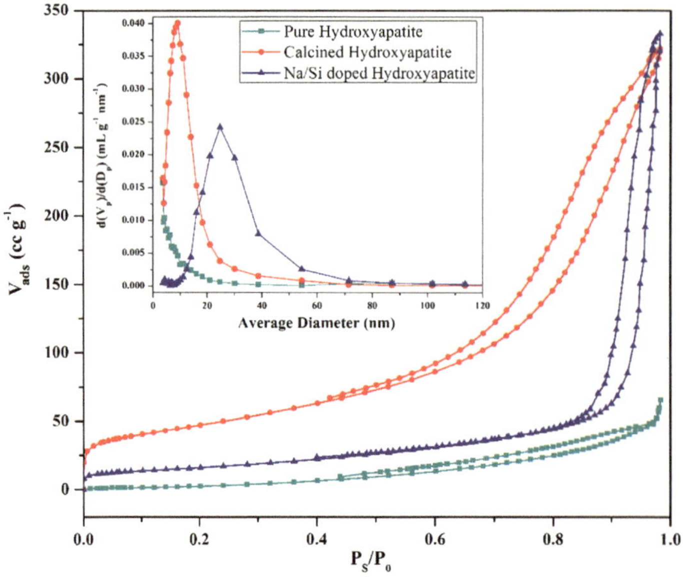

Fig. 3 Adsorption and desorption isotherms of N_2 for raw hydroxyapatite and Na/Si-doped hydroxyapatite

respectively. It can be seen from Table 1 that, the BJH pore volumes were close to the BET pore volumes for the samples.

Figure 4 demonstrates the TGA analysis of raw, calcined, and Na/Si-doped hydroxyapatite. In case of raw hydroxyapatite, initial weight loss of 6.84 % occurred in between the temperature regime of 90 and 150 °C because of the evaporation of water molecules. After that, the weight loss was found to be about 35 % which was due to the decomposition of carbonate as mentioned below.

$$CaCO_3 \rightarrow CaO + CO_2 \tag{1}$$

Table 1 Different surface area ($m^2\ g^{-1}$) measurements and pore volume ($cc\ g^{-1}$)

Parameter	Hydroxyapatite		
	Raw	Calcined	Na/Si
Langmuir surface area ($m^2\ g^{-1}$) at $P_S/P_0 = 0.9814$	5.7	164.7	55.9
One point BET surface area ($m^2\ g^{-1}$) at $P_S/P_0 = 0.3$	8.8	165.1	56.1
Adsorbed BET Surface area ($m^2\ g^{-1}$)	10.8	167.8	57.1
Adsorbed t-plot Surface area ($m^2\ g^{-1}$)	27.2	161.2	57.3
Total pore volume ($cc\ g^{-1}$) at $P_S/P_0 = 0.9814$	0.08	0.49	0.49
Total BJH pore volume ($cc\ g^{-1}$)	0.1	0.49	0.54

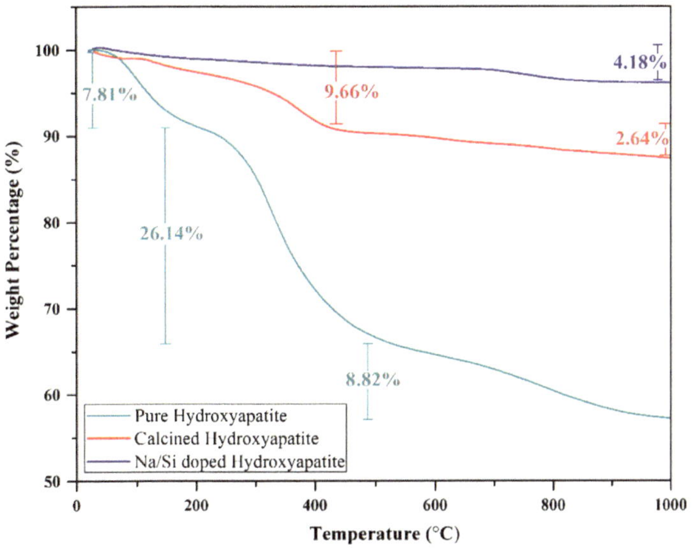

Fig. 4 Thermal gravimetric analysis of raw, calcined, and Na/Si-doped hydroxyapatite

Fig. 5 SEM micrograph of raw hydroxyapatite and Na/Si-doped hydroxyapatite

For calcined hydroxyapatite the total weight loss at 1000 °C was 12.3 % and after 450 °C the degradation was thought to be the aftereffect of continuous dehydroxylation. However, in case of Na/Si-doped hydroxyapatite, no crest was found within a temperature region of 30–1000 °C and the reported weight loss was only 4.18 %. It confirms the stability of Na/Si-doped hydroxyapatite.

The morphologies of raw and Na/Si-doped hydroxyapatite ares presented in Fig. 5. These SEM micrographs gave understanding into the structure with respect to pore sizes and distribution. As can be seen from the micrographs, surface morphologies of the Na/Si-doped hydroxyapatite appear as a polycrystalline material. For Na/Si-doped hydroxyapatite, the pores are uniformly distributed on the surface. The calcination at 650 °C resulted in decomposition of carbonate which helped in forming uniform pores. Hence, the surface area as well as the pore volume was higher in case of Na/Si-doped hydroxyapatite than raw hydroxyapatite that was evident from the BET results.

The EDX analysis (Fig. 6) of raw hydroxyapatite and Na/Si-doped hydroxyapatite showed the presence of C, O, Mg, P, Ca, Na, and Si and the corresponding weight percentages are tabulated in sight. The wt% analysis confirmed the oxide formation of Si.

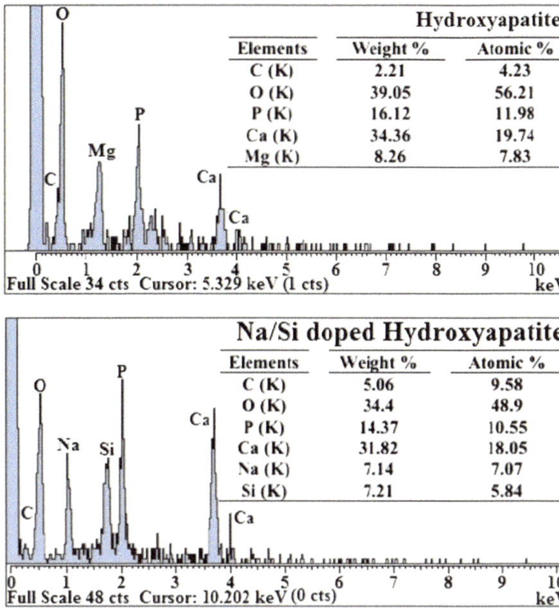

Fig. 6 EDX analysis of raw hydroxyapatite and Na/Si-doped hydroxyapatite)

3.2 Parametric Effects on Acid Value Conversion

Figure 7a shows the effect of neem oil to methanol molar ratio on acid value conversion (%) with respect to reaction time. It can be clearly observed that the time required to reach highest conversion (%) at lower concentration of Na/Si-doped hydroxyapatite 1:2 and lower oil to methanol ratio (1:10), was approximately 195 min. But with increase in oil to methanol ratio the time required to reach highest conversion gradually decreased to 135 min.

Further, the reduction in acid value (%) versus reaction time data was obtained at various Na/Si-doped hydroxyapatite, constant oil to methanol ratio (1:20), and reaction temperature (70 °C) are presented in Fig. 7b. It was evident from the analysis that the equilibrium conversion was achieved after three and half hour.

For Na/Si-doped hydroxyapatite 2:1 the stability over certain reaction time (160 min.) was studied at constant oil to methanol ratio (1:20) and reaction temperature (70 °C) and 2 wt% catalyst. The catalyst was regenerated after complete separation from the reacting mixture. It was first washed with methanol several times and then dried for 2 h at 110 °C. Then the dried sample was again calcined at 650 °C for 1 h and finally the regenerated catalyst was further used. Figure 8 depicting the acid value conversion of purified neem oil with respect to time (min.).

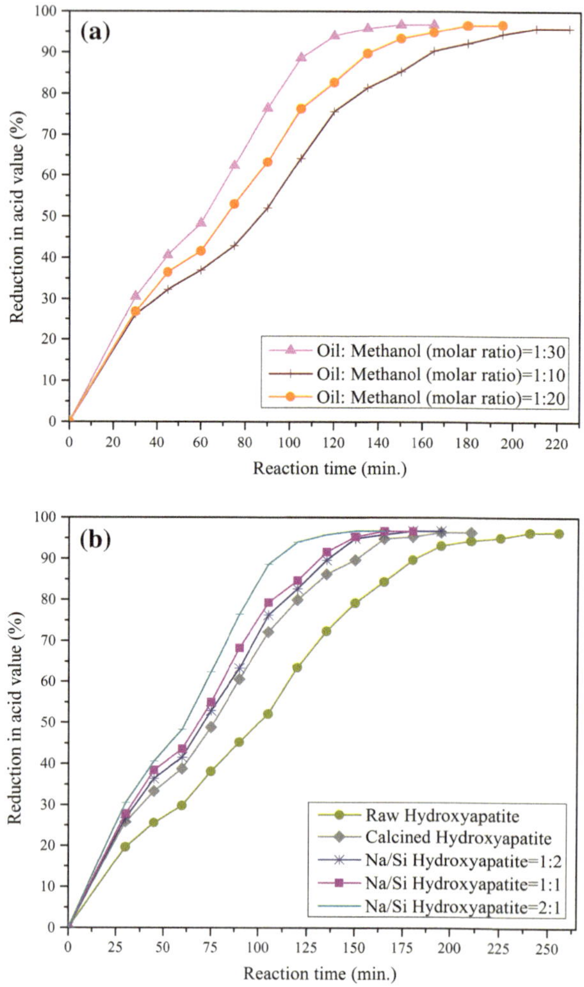

Fig. 7 Effect on reduction of acid value (%) with respect to reaction time by varying **a** oil: methanol and **b** different catalyst

It is clear that the conversion after regeneration does not vary significantly up to 4th regeneration. But after 5th regeneration, the conversion was decreased to 90.3 from 96.7 % after 160 min.

The physical properties of the transesterified product obtained at constant oil to methanol ratio (1:20) and reaction temperature (70 °C) and 2 wt% Na/Si-doped hydroxyapatite 2:1 catalyst was analyzed by following ASTM standards. The physiochemical properties of both neem oil and transesterified product are presented in Table 2.

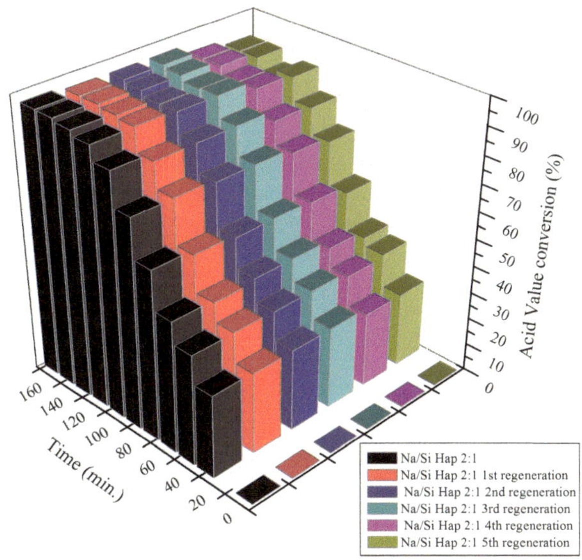

Fig. 8 Acid value conversion of neem oil with respect to time for Na/Si-doped hydroxyapatite 2:1 and the regenerated Na/Si-doped hydroxyapatite 2:1

Table 2 Physicochemical properties of neem oil and transesterified product

Properties	Unit	ASTM test method	Neem oil	Transesterified neem oil
Acid value	mg KOH/g oil	D 974	23.8	0.78
Density	g cm^{-3}	D 4052	910	872
Viscosity	mm^2 s^{-1}	D 445	52.1	4.82
Flash point	°C	D 93	–	165
Pour point	°C	D4419–90	3	−4

3.3 *FTIR Results of Purified and Transesterified Neem Oil*

Figure 9 presents the graph in between wave number (cm^{-1}) and transmittance (%) for transesterified neem oil produced at constant reaction temperature (70 °C), oil to methanol ratio (1:30) and Na/Si-doped hydroxyapatite 2:1. At first it is clear from figure that alkane is present without phosphorous and sulfur, which confirms that leaching of catalyst did not occur and it was separated clearly. Also, a C=O group is identified at 1742 cm^{-1} which refers to ester. Along with other wavenumbers, functional groups are also presented in Table 3.

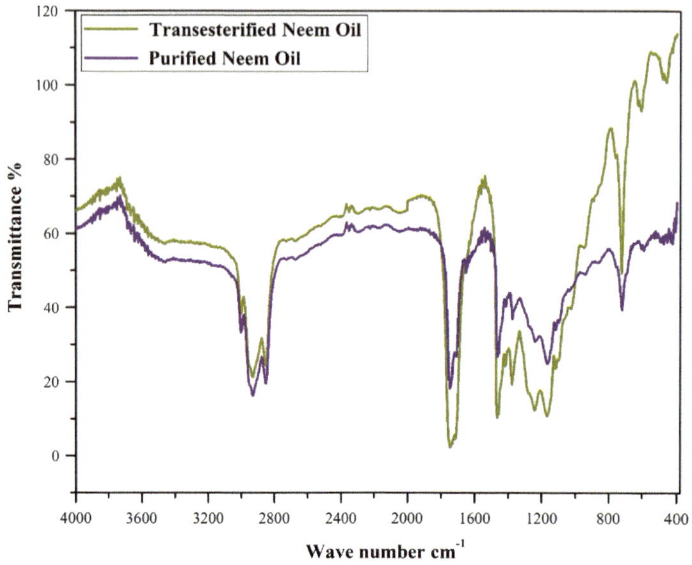

Fig. 9 FTIR of purified and transesterified neem oil

Table 3 FTIR frequencies of purified and transesterified neem oil

Frequency (cm^{-1})	Functional group	Frequency (cm^{-1})	Functional group
722	$–CH_2$	1742	–C=O
1163	C–O–C	2766	$–CH_2$
1375	$O–CH_2$	2912	$–CH_2$
1379	$O–CH_2$	3018	=C–H
1436	$(CO)–O–CH_3$	3422	–OH

4 Conclusion

An active, reusable solid catalyst was prepared from cooked waste fish bone. The method of reusing fishbone waste to prepare catalyst could help waste recycling, contaminants minimization, cost reduction, and making the catalyst environmentally viable. This high efficient and low-cost fishbone catalyst could make the process of transesterification economic and fully ecologically friendly. So, it can be anticipated that the low-cost catalyst could be utilized in a large-scale industrial process of biofuel, which will lead to cheap and ecologically viable process.

Acknowledgments Authors acknowledge the analytical facilities, i.e., FESEM, EDX, BET, and TGA analysis provided by Central Instruments Facility (CIF), IIT Guwahati. Use of XRD facility (procured through FIST Grant No. SR/FST/ETII–028/2010 from Department of Science and Technology, Government of India) at Department of Chemical Engineering, IIT Guwahati is also acknowledged.

References

ASTM Standard D5558−95 (2006) ASTM international, West Conshohocken, PA. doi:10.1520/D5558-95R11, www.astm.org

ASTM Standard D974-12 (2009) ASTM International, West Conshohocken, PA. doi:10.1520/D0974-12, www.astm.org

Best SM, Porter AE, Thian ES, Huang J (2008) Bioceramics: past, present and for the future. J Eur Ceram Soc 28:1319–1327

Fan F, Jia L, Guo XF, Lu X, Chen J (2013) Preparation of novel ethylene glycol monomethyl ether fatty acid monoester biodiesel using calcined sodium silicate. Energy Fuels 27:5215–5221

Leventouri T (2006) Synthetic and biological hydroxyapatites: crystal structure questions. Biomaterials 27:3339–3342

Long YD, Guo F, Fang Z, Tian XF, Jiang LQ, Zhang F (2011) Production of biodiesel and lactic acid from rapeseed oil using sodium silicate as catalyst. Bioresour Technol 102:6884–6886

Maensiri S, Laokul P, Promarak V (2006) Synthesis and optical properties of nanocrystalline ZnO powders by a simple method using zinc acetate dihydrate and poly (vinyl pyrrolidone). J Cryst Growth 289:102–106

Musić S, Filipović-Vinceković N, Sekovanić L (2011) Precipitation of amorphous SiO_2 particles and their properties. Braz J Chem Eng 28:89–94

Obadiah A, Swaroopa GA, Kumar SV, Jeganathan KR, Ramasubbu A (2012) Biodiesel production from palm oil using calcined waste animal bone as catalyst. Bioresour Technol 116:512–516

Rocha JHG, Lemos AF, Kannan S, Agathopoulosa S, Ferreira JMF (2005) Hydroxyapatite scaffolds hydrothermally grown from aragonitic cuttlefish bones. J Mater Chem 15:5007–5011

Rustad T (2003) Utilisation of marine byproducts. Electron J Environ Agric Food Chem 2:458–463

Varma HK, Babu SS (2005) Synthesis of calcium phosphate bioceramics by citrate gel pyrolysis method. Ceram Int 31:109–114

Yin JZ, Ma Z, Hu DP, Xiu ZL, Wang TH (2010) Biodiesel production from subcritical methanol transesterification of soybean oil with sodium silicate. Energy Fuels 24:3179–3182

Wang J, Nonami T, Yubata K (2008) Syntheses, structures and photophysical properties of iron containing hydroxyapatite prepared by a modified pseudo-body solution. J Mater Sci Mater Med 19:2663–2667

Assessment of the Seed Oils of *Persea americana* and *Melia dubua* for Their Potentialities in the Production of Biodiesel and Possible Industrial Use

Kariyappa S. Katagi, Ravindra S. Munnolli, Sangeeta D. Benni and Sneha S. Kulkarni

Abstract The seed oils of *Persea americana* and *Melia dubua* species are selected for this analytical study. These seed species yield 63.6 and 28 % of nonedible oil, respectively. The details of component fatty acids (CFAs) are collected from the literature. The necessary analytical data, viz., iodine value (IV) and saponification value (SV) of seed oils and major fuel properties of biodiesel, viz., cetane number (CN), higher heating value (HHV), and lower heating value (LHV) of the fatty acid methyl esters (FAMEs) are empirically computed. These parameters of biodiesel of seed oils under investigation are compared and evaluated in reference to existing biodiesels. This work reports the suitability of especially *Persea americana* for the biodiesel production.

Keywords Seed oil · *Persea americana* · *Melia dubua* · Biodiesel · Fatty acid methyl esters

1 Introduction

Sustainable development has been the issue of immense importance. Renewable raw materials add significantly to the sustainable progress. Plant oils cover the larger section of the existing utilization of renewable raw materials in the chemical industry (Liu 2013). Seed oils find various applications in production of surfactants, soaps, detergents, solvents, lubricants, paints, cosmetics, etc. The unusual fatty acids used in formulations of protective coatings, plastics, urethane derivatives,

K.S. Katagi (✉)
Department of Chemistry, Karnatak University's Karnatak Science College, Dharwad 580001, India
e-mail: kskatagi@gmail.com

R.S. Munnolli · S.D. Benni · S.S. Kulkarni
Department of Chemistry, Research Centre, KLS's VDR Institute of Technology, Haliyal 581329, India

S. Kumar et al. (eds.), *Proceedings of the First International Conference on Recent Advances in Bioenergy Research*, Springer Proceedings in Energy,
DOI 10.1007/978-81-322-2773-1_8

dispersants, cosmetics, lubricant additives, biolubricants, pharmaceuticals, textiles, variety of synthetic intermediates, and stabilizers in plastic formulations, etc. (FAO 2008; Kucuk 1994).

The fast depletion of natural resources of energy such as petroleum and coal has triggered the emergence of biodiesel as a potential substitute for energy. It is derived from animal fat or from plant seed oils. The advantages of biodiesel include its domestic origin, renewability, biodegradability, higher flash point, inbuilt lubricity, higher cetane number, low viscosity, improved heating value, etc. also with lowered emissions of particulates, CO_X and SO_X (Leduc et al. 2009; Mustafa and Havva 2010).

Presently, due to apprehension about energy security, fuel cost and safe environment, biodiesel is considered for part or full replacement of petro-diesel. The seed oil has 2.25 times higher energy density than starch or proteins (Mehta and Anand 2009; Banas et al. 2007). Currently, seed oils such as *Palm* seed oil, *Soybean* seed oil, *Sunflower* seed oil, *Coconut* oil, etc., which are edible are also used for biodiesel production this has resulted in a competing situation for plant seed oils being used for food and for possible industrial applications. Thus, due to increased obligation, biodiesel practice would continue to grow at a fast rate to meet the energy needs. As a result, the cost of edible seed oils led to escalating prices constraining to secure sufficient supply in various parts of the world. Consequently, there is a need to identify nonedible/alternative sources of plant oils to be used for energy-related applications. Thus, those plant crops which require only a minimum land for their crop production yielding significantly good amount of nonedible seed oils are to be endorsed (Dennis et al. 2010; Manjula et al. 2009).

Researchers in the recent times, are focusing on nonedible seed oils for the biodiesel production (Altun 2011; Balusamy and Marappan 2010; Banapurmath et al. 2008) to mitigate the consequences of usage of edible oils for the production of biodiesel which otherwise, would lead to adverse effect on food security.

1.1 *Persea americana*

The *Persea americana* (Avocado) belongs to *Lauraceae* plant family. It is a tree, native to Mexico and Central America, its fruit is shown in Fig. 1. It has a large berry with single seed. Fruits have a green skinned, fleshy body that is pear-shaped, or egg-shaped, or spherical. The tropical and subtropical regions are suitable climatic conditions for cultivation of this species.

It is one of the 150 varieties of avocado pear. Its parts such as, seeds, stems, leaves, roots, bark, etc., are used for the curing of various diseases affecting humans. The drugs used in medicine today are directly or indirectly derived from this plant due to its bioactive constituents such as; alkaloids, steroids, tannins, etc., (Glossary of Indian medicinal plants 1976).

The seed of *Persea americana* (avocado seed) has varied applications in ethno–medicine, like in the treatment of diarrhea, dysentery, toothache, intestinal parasites, dermatological treatments, and in beautification. The seed oil is used in the

Fig. 1 *Persea americana* fruit

treatment of obesity. The fruit possesses fatty, subtly flavored, smooth, and creamy texture. Fruits are consumed as milk shakes and intermittently added to ice cream in countries such as Mexico, Brazil, South Africa, and India (Ramos et al. 2004).

1.2 *Melia dubua*

Melia dubua belongs to *Meliaceae* family. It is an indigenous tree to India, Southeast Asia, and Australia. It has been cultivated as a source of firewood, also called as Maha Neem or Forest Neem (Glossary of Indian medicinal plants 1956a, b). The seeds are shown in Fig. 2.

1.3 Component Fatty Acids in Seed Oils

Listed below in Table 1 are the fatty acid components with their chemical formulae and molecular weights that make up the two seed oils under investigation. There are

Fig. 2 *Melia dubua* seeds

Table 1 Fatty acids that constitute the two seed oils under investigation

Fatty acid	Molecular formula	Molecular weight
Lauric acid	$CH_3(CH_2)_{10}COOH$	200.32
Myristic acid	$CH_3(CH_2)_{12}COOH$	228.38
Palmitic acid	$CH_3(CH_2)_{14}COOH$	256.43
Stearic acid	$CH_3(CH_2)_{16}COOH$	284.48
Oleic acid	$CH_3(CH_2)_7CH{=}CH(CH_2)_7COOH$	282.47
Linoleic acid	$CH_3(CH_2)_4CH{=}CHCH_2CH{=}(CH_2)_7COOH$	280.46
Arachidic acid	$CH_3(CH_2)_{18}COOH$	312.54

seven fatty acids out of which oleic acid and linoleic acid are the two unsaturated acids. Their content in higher proportion makes the oil as much more unfavorable for biodiesel synthesis.

2 Materials and Methods

2.1 *General Procedures*

2.1.1 Oil Extraction

The seeds are ground, powdered and the oil extracted using petroleum ether (boiling range: 40–60 °C) in Soxhlet extractor for 24 h. The organic extract is filtered and dried over anhydrous sodium sulphate. The petroleum ether is removed under vacuum.

2.1.2 Transesterification

Triglycerides in the seed oils on reaction with alcohol usually methanol or ethanol in presence of base or acid as a catalyst, yield FAMEs or fatty acid ethyl esters (FAEEs) along with glycerol as byproduct.

$$\underset{\text{(Seed oil)}}{\text{Triglyceride}} + \underset{\text{(Methanol/Ethanol)}}{\text{Alcohol}} \rightleftharpoons \underset{\text{(Biodiesel)}}{\text{FAMEs or FAEEs}} + \underset{\text{(Byproduct)}}{\text{Glycerol}}$$

The seed oil is treated with methanol or ethanol in 1:6 molar ratio in the presence of base or acid catalyst, then organic and aqueous phases are separated using a separating funnel. The lower layer containing glycerol, alcohol, and most of the catalyst is drained out. The upper layer containing alkyl esters, some alcohol, traces of the catalyst is washed thoroughly with deionized water. Then, the residual methanol is removed by rotary evaporation at around 70 °C. The product obtained is a mixture of FAMEs or FAEEs and is used as biodiesel (Katagi et al. 2011).

3 Computational Analysis of Biodiesel Properties of Seed Oils

The selected seed oils and the computed biodiesel properties are compared for fuel properties with those of ASTM, DIN D6751, and EN 14214 standards.

3.1 Saponification Value (SV) and Iodine Value (IV)

These parameters of the seed oils are calculated to establish their suitability for biodiesel synthesis. The SV and IV of seed oils are evaluated using Eqs. (1) and (2), respectively, based on composition of fatty acids (Kalayasiri et al. 1996). The results obtained are comparable with experimental values.

$$\mathrm{SV} = \sum \frac{560 \times A_i}{M_{wi}} \tag{1}$$

$$\mathrm{IV} = \sum \frac{254 \times N_{db} \times A_i}{M_{wi}} \tag{2}$$

where A_i is the percentage of component fatty acids, N_{db} is the number of carbon–carbon double bonds, and M_{wi} is the molecular mass of each component.

3.2 Cetane Number (CN)

Generally, biodiesel exhibits higher CN than conventional petro-diesel and results in higher combustion efficiency. CN is significant in knowing the suitability of biodiesel. The CN of FAMEs is computed from the following Eq. (3), which suits closely with the experimental values (Krisnangkura 1986).

$$\mathrm{CN} = 46.3 + \frac{5458}{\mathrm{SV}} - 0.225 \times \mathrm{IV} \tag{3}$$

3.3 Higher Heating Value (HHV)

It is known that the processed vegetable oils used in diesel engines are the complex chemical mixture of FAMEs. The HHV of biodiesel is calculated using Eq. (4) in accordance with regression model (Demirbas 1998).

$$\mathrm{HHV} = 49.43 - (0.015 \times \mathrm{IV}) - (0.041 \times \mathrm{SV}) \tag{4}$$

3.4 Lower Heating Value (LHV)

LHV of biodiesel is computed from Eqs. (5) and (6) based on bond energy values of different FAMEs. The methods deployed for the calculation of lower heating value are quite general and their predictability is reasonable (Mehta and Anand 2009).

For FAMEs,

$$\mathrm{LHV} = 0.0109\left(\frac{\mathrm{C}}{\mathrm{O}}\right)^3 - 0.3516\left(\frac{\mathrm{C}}{\mathrm{O}}\right)^2 + 4.2000\left(\frac{\mathrm{C}}{\mathrm{O}}\right) + 21.066 - 0.100\,N_{db} \quad (5)$$

$$\mathrm{LHV} = 0.0011\left(\frac{\mathrm{H}}{\mathrm{O}}\right)^3 - 0.0785\left(\frac{\mathrm{H}}{\mathrm{O}}\right)^2 + 2.0409\left(\frac{\mathrm{H}}{\mathrm{O}}\right) + 20.992 - 0.100\,N_{db} \quad (6)$$

where C, H, and O are the number of carbons, hydrogen, and oxygen, respectively. N_{db} has the same meaning as stated above.

4 Results and Discussion

Biodiesel property of selected seed oils of *Persea americana* and *Melia dubua* are computed based on the data from the literature (Shanbhag and Badami 1974; Thakkar 1983). Details of component fatty acids (CFAs) and other analytical values of seed oils and properties of biodiesel are depicted in Tables 2 and 3 respectively. The computed parameters of biodiesel from *Persea americana* are complimentary to quality biodiesel and make it a befitting candidate for the synthesis of biodiesel.

A comparison of the properties of different seed oils and corresponding biodiesels is made in Table 4. One nonedible seed oil and three edible seed oils are included for purpose of comparison.

Iodine value of seed oils should not exceed 120 mg I_2 g^{-1} which best fits especially for *Persea americana* as per the limitation laid by European standard organization EN 14214 for biodiesel. However, the IV is on the higher side for *Melia dubua* because of the higher proportion of unsaturated acids, oleic acid and linoleic acid. This is clear as is shown in Fig. 3a. There is consistency in the SV. This signifies that *Persea americana* is ideal for the biodiesel as per the conclusion derived by Knothe (Knothe 2008).

Literature reveals that, biodiesel standards of USA (ASTM D6751), Germany (DIN 51606), and European Organization (EN 14214) have set CN value as 47, 49, and 51 respectively (Krisnangkura 1986). The computed values of CN for the FAMEs of seed oils of *Persea americana* and *Melia dubua* are 59.8 and 41.5, respectively, whereas, the CN of petro-diesel is 42.0. The empirically calculated CN value of FAMEs of *Persea americana* meet the ASTM standards, however; *Melia dubua* exhibits relatively lesser CN than the expected value. Comparisons shown in Fig. 3b confirm the facts mentioned above.

Table 2 Analytical values of *Persea americana*[a] and *Melia dubua*[b] seed oils

Source/seed species	*Persea americana*	*Melia dubua*
Seed oil percentage	63.6	28.0
Molecular weight of oil (g mol^{-1})	762.2	880.2
Total saturated fatty acids (TSFAs) percentage	61.7	13.9
Total unsaturated fatty acids (TUSFAs) percentage	38.3	86.1
Saponification value (mg KOH g^{-1})	240.3	208.5
Iodine value (mg I_2 g^{-1})	40.2	140.0
Percentage of component fatty acids: 12:0 (Lauric)	36.3	NA
14:0 (Myristic)	14.3	NA
16:0 Palmitic)	7.9	9.4
18:0 (Stearic)	1.1	1.5
18:1 (Oleic)	29.6	20.5
18:2 (Linoleic)	8.7	65.6
20:0 (Arachidic)	1.1	NA

[a]Shanbhag and Badami (1974)
[b]Thakkar (1983)

Table 3 Computed biodiesel properties of *Persea americana* and *Melia dubua* plant species

Source/seed species	*Persea americana*	*Melia dubua*
Cetane number	59.8	41.5
Higher heating value (HHV; MJ kg^{-1})	39.1	39.0
Lower heating value C/O (LHV; MJ kg^{-1})	37.1	38.1
Lower heating value H/O (LHV; MJ kg^{-1})	37.3	38.0

Table 4 Comparison of properties of biodiesels from *Persea americana* and *Melia dubua* with existing biodiesels and petro-diesel

Fuel property	*Persea americana* (PA)	*Melia dubua* (MD)	*Pongamia pinnata*[a] (PP)	*Soybean*[a] (SB)	*Rapeseed*[a] (RS)	*Sunflower*[a] (SF)	*Petro diesel*[a] (PD)
SV (mg KOH g^{-1})	236.3	201.4	182.3	194.6	197.1	193.0	NA
IV (mg I_2 g^{-1})	42.9	141.6	58.4	120.5	108.1	132.3	NA
CN	59.8	41.5	63.1	45.0	59.0	49.0	42.0
HHV (MJ kg^{-1})	39.1	39.0	40.5	39.8	37.0	40.6	46.0
LHV (MJ kg^{-1})	37.1	38.1	34.3	33.5	32.8	33.5	43.1

NA Not applicable
[a]Indicates the data obtained from Ref. (Ramadas et al. 2005)

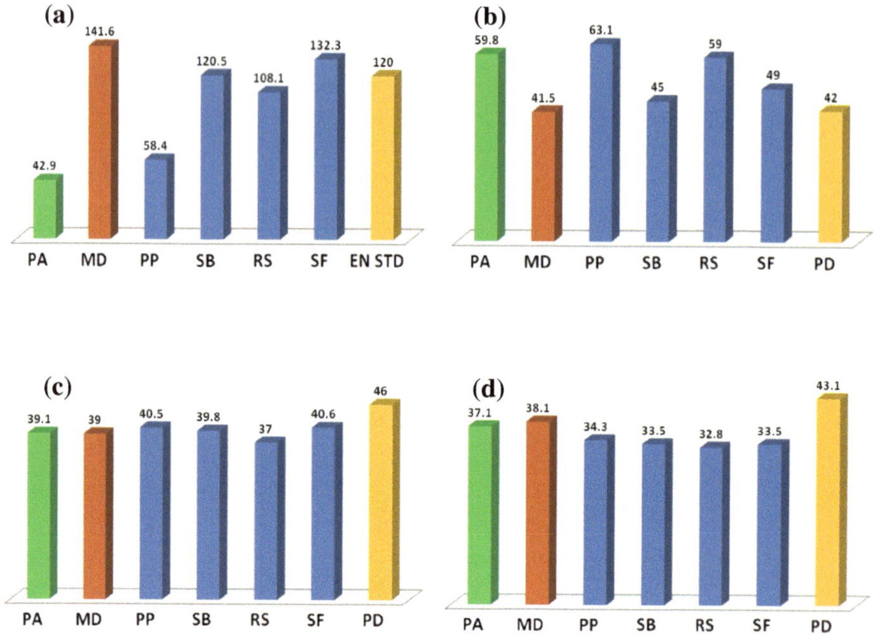

Fig. 3 Graphical comparison of biodiesel properties with standard fuel and petro-diesel. **a** IVs of biodiesels compared with that of EN standard. **b** CNs of biodiesels compared with that of petro-diesel. **c** HHVs of biodiesels compared with that of petro-diesel. **d** LHVs of biodiesels compared with that of petro-diesel

The carbon, hydrogen, and oxygen content of a fuel is directly proportional to its HHV and it is evaluated using the Eq. (4). The HHV is seen to be almost same at about 39.1 MJ kg^{-1} for the biodiesel derived from the seed oils under investigation are slightly lower than that of petro-diesel (43 MJ kg^{-1}) or petroleum (42 MJ kg^{-1}), but is higher than that of coal (32–37 MJ kg^{-1}). The LHVs are estimated using two Eqs. (5) and (6). The computed LHVs of respective biodiesels are in the range of 37–38 MJ kg^{-1}. The European Biofuels Technology Platform 2011 reported the LHV for biodiesel as 37.1 MJ kg^{-1} (Biofuels Technology Platform 2011). This is slightly lower than the LHV of petro-diesel (43 MJ kg^{-1}). It may be noted that heating values are a function of mainly carbon content in the fuel. Obviously, it is expected to be higher for petro-diesel because of the presence of mostly carbon and a little of hydrogen in petro-diesel. However, biodiesel is a mixture of fatty acid esters (oxygenates) reducing the carbon content per unit mass. It is also interesting to observe that the oxygen in biodiesel molecules promotes relatively better combustion efficiency and yields higher cetane number as is evident for most of the biodiesels including the two under investigation in this work. Figure 3b–d support the statements made above.

It may be noted that higher SV infers higher proportion of fatty acids per unit mass of oil and subsequent higher proportion of FAMEs in the biodiesel. Also, because, the FAMEs are oxygenates, improved combustion efficiency and CN is expected which is indicated by corresponding values for *Persea americana*. It is comparable with that of biodiesel from rape seed oil or *Pongamia pinnata*. Also, the computed value of LHV is more favorable in comparison to any other biodiesel listed above.

The analytical data for seed oils other than those investigated in this paper are obtained from the literature (Ramadas et al. 2005). It signifies, *Persea americana* to be an ideal candidate for biodiesel production with mostly favorable parameters especially with regard to IV, CN and heating values when compared to other seed oils/biodiesels.

5 Conclusion

The FAMEs of seed oil of *Persea americana* meet the major specifications of US biodiesel standard and European standards and can be utilized for the production of biodiesel and also has the possible industrial relevance. However, further research is required to evaluate these FAMEs for other properties like, tribological studies and long-term engine testing, etc.

Acknowledgments The authors acknowledge the support of authorities of Sri. S.S. Basavanal Library, Karnatak University, Dharwad for facilities provided during data collection. The authors also acknowledge Karnatak Science College, Dharwad and KLS's VDR Institute of Technology, Haliyal for extending encouragement in completing this work.

References

Altun S (2011) Fuel properties of biodiesels produced from different feed stocks. Energy Educ Sci Technol Part A 26:165–174

Balusamy T, Marappan R (2010) Effect of injection time and injection pressure on CI engine fuelled with methyl ester of Thevetia peruviana seed oil. Int J Green Energy 7(4):397–409

Banapurmath NR, Tewari PG, Hosmath RS (2008) Performance and emission characteristics of a DI compression ignition engine operated on Honge, Jatropha and Sesame oil methyl esters. Renewable Energy 33(9):1982–1988

Banas A, Debski H, Banas W, Heneen WK, Dahlqvist A, Bafor M, Gummeson PO, Martilla S, Ekman A, Carlsson AS, Stymne S (2007) Lipids in grain tissues of oat (Avena sativa): differences in content, time of deposition and fatty acid composition. J Exp Bot 58:2463–2470

Demirbas Ayhan (1998) Fuel properties and calculation of higher heating values of vegetable oils. Fuel 77(9–10):1117–1120

Dennis YC, Leung Xuan Wu, Leung MKH (2010) A review on biodiesel production using catalyzed transesterification. Appl Energy 87:1083–1095

FAO (2008) AOSTAT. http://faostat.fao.org

Glossary of Indian medicinal plants (1956a) CSIR, New Delhi p 34

Glossary of Indian medicinal plants (1956b) CSIR, New Delhi, p 164

Glossary of Indian medicinal plants (1976) vol 11. CSIR, New Delhi, p 108

Kalayasiri P, Jayashke N, Krisnangkura K (1996) Survey of seed oils for use as diesel fuels. J Am Oil Chem Soc 73:471–474

Katagi KS, Munnolli RS, Hosamani KM (2011) Unique occurrence of unusual fatty acid in the seed oil of Aegle marmelos Correa: screening the rich source of seed oil for bio-energy production. Appl Energy 88:1797–1802

Kucuk MM (1994) Recent Advances in Biomass Technology. Fuel Sci Technol Int 12(6):845–871

Knothe G (2008) "Designer" biodiesel: optimizing fatty ester composition to improve fuel properties. Energy Fuels 22:1358–1364

Krisnangkura KA (1986) Simple method for estimation of cetane index of vegetable oil methyl esters. J Am Oil Chem Soc 63:552–553

Leduc S, Natarajan K, Dotzauer E, Mc Callum I, Obersteiner M (2009) Optimizing biodiesel production in India. Appl Energy 86:125–131

Liu Z (2013) Preparation of biopolymers from plant oils in green media. Bioenergy Res 6:1320–1336

Manjula Siriwardhana GK, Opathella C, Jha MK (2009) Bio-diesel: initiatives, potential and prospects in Thailand. Energy Field Energy Policy 37:554–559

Mehta Pramod S, Anand K (2009) Estimation of a lower heating value of vegetable oil and biodiesel fuel. Energy Fuels 23:3893–3898

Mustafa Balat and Havva Balat (2010) Progress in biodiesel processing. Appl Energy 87:1815–1835

Ramadas AS, Muraleedharan C, Jayaraj S (2005) Performance and emission evaluation of a diesel engine fueled with methyl esters of rubber seed oil. Renewable Energy 30:1789–1800

Ramos MR, Jerz G, Villanueva S, Lopez-Dellamary F, Waibe R, Winterhalter P (2004) Two glucosylated abscisic acid derivatives from avocado seeds (Persea americana Mill. Lauraceae cv. Hass). Phytochemistry 65:955–962

Shanbhag MR, Badami RC (1974) Studies on chemistry of fatty acids. Ph.D. Thesis, Karnatak University, Dharwad, Karnataka, India, p 118

Thakkar JK (1983) Studies on chemistry of fatty acids. Ph.D Thesis, Karnatak University, Dharwad, Karnataka, India, p 103

www.biofuelstp.eu (2011) Fact sheet european biofuels technology platform

Investigation of CI Engine Fueled with Ethanol Nano Additives Blended Diesel

V. Karthickeyan, P. Balamurugan and R. Senthil

Abstract Experimental investigation on the performance and emission characteristics of ethanol–cerium oxide blended diesel in a single cylinder four-stroke diesel engine is presented. Solubility of cerium oxide in diesel and ethanol is investigated with diethyl ether. Four blended diesel test fuels have been prepared using ethanol, diethyl ether, and cerium oxide in diesel. A computerized CI engine with eddy current dynamometer test rig with gas analyzer has been used. Result shows that the prepared blended fuel is capable of replacing neat diesel as fuel for diesel engine with lower CO_2 emission. The blended fuel has higher thermal efficiency with nominal variations in fuel consumption, due to nano additives. NO_x emission is also observed lower when compared to neat diesel.

Keywords Diesel · Ethanol · Cerium oxide · Computerized diesel engine · Performance · Emission

1 Introduction

In the present scenario of energy and environment, the drive for research in the field of IC Engines is always targeted toward reducing the dependence on fossil fuels and the emission from the engines. These could be achieved with the exploration of different alternate fuels from renewable sources of energy. Investigation on preparation of different blended diesel fuels and their performance on engines are being carried out by researchers across the world.

V. Karthickeyan · P. Balamurugan (✉)
Department of Mechanical Sciences, Sri Krishna College of Engineering and Technology, Coimbatore 641008, Tamil Nadu, India
e-mail: balamuruganp@skcet.ac.in

R. Senthil
Department of Mechanical Engineering, University College of Engineering Villupuram, Villupuram 605103, Tamil Nadu, India

S. Kumar et al. (eds.), *Proceedings of the First International Conference on Recent Advances in Bioenergy Research*, Springer Proceedings in Energy,
DOI 10.1007/978-81-322-2773-1_9

Aksoy et al. (1990) carried out investigation on the properties of triglyceride oils obtained from seeds of fermented raisins of Turkish origin. The technical properties of the derivatives of pot still distillation have been determined, to be recommended as an alternative to conventional fuels. Gowen Marcia (1989) explored sectoral differences between biofuel and fossil fuels with regard to production costs, economies of scale, subsidies, and other economic incentives in developing countries. Malaya Naik et al. (2008) investigated the production of biodiesel from high FFA Karanja oil. The production of biodiesel was discussed using the mechanism of dual process. It was observed that high FFA Karanja oil yield 96.6–97 % biodiesel by dual step process.

Experimental investigation on the performance and emissions of a diesel engine fuelled with ethanol–diesel blends was conducted by Huang et al. (2009). The fuel consumptions of the engine fuelled by the blends were found to be higher that of pure diesel. However, it was observed that the blended fuel gave comparable thermal efficiency like pure diesel, with marginal deviations at different loads and speed. Also, the CO emissions of blends were found higher than those of diesel. However, it was found that the engine fuelled with blends emitted less NO_x at low engine speed. Also, smoke emissions from the engine with blended fuel were lower than that of diesel, the reductions ranging from 16.7 to 87.5 %.

The effect of thermal insulation of CI engine with ethanol was investigated by Karthickeyan and Srithar (2011). Yttria-stabilized zirconia was used as insulation to control the ignition delay of the low heat rejection engine. Benefits of insulation were investigated with ethanol and diesel operated in glow plug assisted engine. It was observed that highest efficiency was obtained with ethanol with insulated combustion chamber.

The effect of metal-based additives ($FeCl_3$ and FBC) in diesel engine with waste cooking palm oil is investigated by Kannan et al. (2011). BSFC decreased by 8.6 % and BTE increased by 6.3 %. CO, HC and smoke emission of FBC added biodiesel decreased by 52.6, 26.6 and 6.9 %, respectively, when compared without FBC.

Performance of high alcohols and diesel fuel in diesel engine was compared by Campos-Fernandez et al. (2012). The existence of oxygen in the molecular structure of 1-butanol and 1-pentanol offsets its reduced low heating value, showing better combustion and brake thermal efficiency, while power and torque were found similar to that of diesel fuel. With regard to BSFC, butanol blends exhibited slightly better behavior than pentanol blends and neat diesel fuel. Ramesh Kumar and Nagarajan (2012) experimentally investigated the effect of thermal barrier coating using alumina (Al_2O_3) in SI engine. It was found that the combustion chamber temperature increased resulting in limiting the preignition and knocking. It was also observed that the partially insulated SI engine and when fuelled with alternate fuel ethanol 20 % improved performance with a reduction of HC and CO emissions up to a maximum of 48 and 50 %, respectively.

Experimental investigation on tobacco seed oil methyl ester (TSOME) in diesel engine found that effective power and torque which were increased was analyzed by Parlak et al. (2013). However, on the contrary, as TSOME contains only 10 % of oxygen in its content, blended fuel was found to cause an increase of about 6 % in

NO_x emission. Golimowski et al. (2013) evaluated the performance of common rail diesel tractor engine using pure plant oil. The engine generated maximum power of 102 kW when fuelled with diesel, and 90 kW with raps oil. The engine operating on vegetable oil displayed better efficiency over diesel. Experiments with Pongamia oil methyl (PME) ester and diesel blends (B20 and B100) were investigated by Prabhahar and Rajan (2013) in a 4S direct injection diesel engine with and without coated piston. Ceramic coating material (TiO_2) was applied on the piston crown using plasma spray method. From the experiment it was observed that it is 100 % biodiesel. The efficiency of the engine was improved with a decrease in BSFC by about 10 % at full load. Also, the coated engine showed lower carbon monoxide and hydrocarbon emissions, however, with an increase of 15 % of nitrogen oxide emission. Kim and Choi (2013) carried out experiments using biodiesel fuel (BDF) and ethylene glycol mono-n-butyl ether (EGBE) to investigate the reduction of smoke and NO_x emission in IDI engines. Better smoke reduction was observed with the BDF and the mixture of BDF and EGBE when compared to neat diesel. Also, the simultaneous application of cooled EGR was found to have augmented reduction effects of emissions. The prepared alternate fuel was validated by comparing with the results available in the literature. From the literature survey, it has been found that even though different researches have been carried out on the alternate fuels, very few works have been carried out with respect to nano additives in fuel. In this context, this research work shall be highly significant and unique one.

In the present work, performance and emission characteristics are investigated experimentally in direct injection diesel engine using ethanol–cerium oxide blended diesel fuel and the results are presented.

2 Preparation of Biofuel Blends

The present experimentation is mainly intended to examine the characteristics of new blended diesel combinations at different loading conditions. The new fuel is prepared by blending ethanol and cerium oxide with diesel in different proportions. Ethanol addition to diesel fuel results in different physiochemical changes in diesel fuel properties, particularly reductions in cetane number, viscosity, heating value, etc. Since ethanol is having polar molecule and its solubility in diesel is prone to be affected by temperature and water content, high percentage addition of ethanol to diesel is difficult. In order to mix ethanol and diesel, an emulsifier or cosolvent like diethyl ether should be added. Cerium oxide nano particles as an additive (Sajith et al. 2010) could possibly exhibit high catalytic activity because of their large surface area per unit volume, leading to improvement in the fuel efficiency and reduction in the emissions. Hence by adding the cerium oxide nano particle the fuel reduces the NO_x emissions and prolongs combustion, which leads to a reduction in unburnt hydrocarbons (by up to 15 %) and a decrease in fuel consumption up to 5–9 %. In this, an ultrasonic bath has been used to mix diesel, ethanol, and diethyl

Table 1 Composition of four trial samples of test fuels

NEAT DIESEL	100 % diesel
D89E10DEE1	89 % diesel + 10 % ethanol + 1 % DEE
D89E10DEE1CEO10	89 % diesel + 10 % ethanol + 1 % DEE + 10 gm cerium oxide
D89E10DEE1CEO15	89 % diesel + 10 % ethanol + 1 % DEE + 15 gm cerium oxide
D89E10DEE1CEO20	89 % diesel + 10 % ethanol + 1 % DEE + 20 gm cerium oxide

Table 2 Properties of diesel and other trial sample prepared

Properties	Neat Diesel	D89E10DEE1	D89E10DEE1CEO10	D89E10DEE1CEO15	D89E10DEE1CEO20
Diesel content (%vol)	100	89	89	89	89
Ethanol content (%vol)	0	10	10	10	10
Density at 150 °C (kg/m^3)	843	831	833	836	838
Viscosity at 40 °C (Ns/m^2)	2.48	1.86	1.88	1.92	1.92
Flash point (°C)	52	89	86	86	85

ether with cerium oxide. The sonicator resulted in better mixing of components resulting ultimately with an essential solution of cerium oxide.

In order to carry out the investigation on performance and emission characteristics, four combinations of blended diesel fuels have been prepared using the cerium oxide solution, ethanol, and diesel. The compositions of individual components of the four trial samples are given in Table 1. The physiochemical properties of the prepared fuels namely, density, viscosity, calorific values, flash point, etc., have been determined and the values are presented in Table 2.

3 Experimental Setup

The schematic diagram of computerized single cylinder four-stroke diesel which is used for experiment is shown in Fig. 1.

The setup includes air chamber with U-tube manometer, digital temperature indicator, digital speed indicator, fuel measuring unit, and fuel tank. The system also includes engine jacket cooling water circuit with calorimeter temperature indicator and glass tube rotameter for measuring cooling water, valves, and instrumentation. Various measuring devices such as pressure sensors, flow meters, and temperature sensors are fitted at suitable locations. The specification of the engine used and uncertainties of various instruments are listed in Tables 3 and 4, respectively.

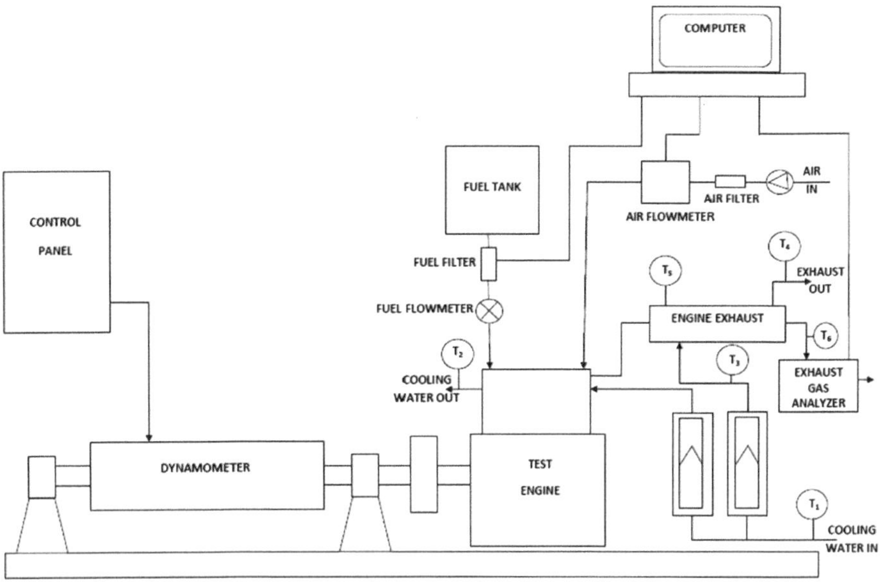

Fig. 1 Schematic diagram of computerized IC engine test rig

Table 3 Specifications of test engine

Make	Kirloskar
Cylinder number	1
Type	Four-stroke, stationary, constant speed, direct injection, air cooled, and diesel engine
Max. power/speed	4.4 kW/1500 rpm
Bore	87.5 mm
Stroke	110 mm
Compression ratio	17.5:1
Injection timing	23.4° bTDC

Table 4 List of measurement uncertainty

Measurement	Uncertainty (%)
Load	±0.2
Speed	±0.1
Time	±0.2
Manometer	±1
CO_2	±0.2
NO_x	±0.2

The experiment is carried out by varying the brake load using the eddy current dynamometer. During which the data of fuel consumption, engine temperature, engine speed, and exhaust gas temperature are recorded using NI USB-6210, 16 bit, 250 kS/d data acquisition system. U-tube manometer is used to measure the air intake. Exhaust gas has been investigated using AVL make 444 five gas analyser.

4 Experimental Procedure

Figure 2 shows the experimental setup of computerized engine test rig with eddy current dynamometer.

Initially, the diesel is taken in fuel tank. Before the engine is started at no load condition, the cooling water line is regulated to ensure continuous flow of water. Simultaneously the data acquisition system and the AVL gas analyzer, which are connected to engine, are switched ON to keep track of the performance and emission. After starting, the engine is given sufficient time to attain a constant steady state speed of 1500 rpm. Initially, the engine is investigated at no load condition to measure the necessary data. Consequently the load is increased to full load condition gradually with an increment of 25 %. At every loading, after reaching steady state condition, all required data are measured using data acquisition system and the gas analyzer. Then the entire steps are repeated with four blended biodiesel trial samples.

Fig. 2 Computerized engine coupled with eddy current dynamometer

5 Results and Discussion

Experimentation has been carried out on the computerized single cylinder four-stroke diesel engine connected to data acquisition system, with one fuel after another. There are totally five fuels (including, neat diesel) tested. This section compares the performance and emission characteristics of four combinations of blended diesel fuels with those of neat diesel. By the end of the present work, a suitable replacement to diesel engine shall be recommended.

Figure 3 shows the variation of brake thermal efficiency with brake power. As expected, it has been observed that the efficiency of the engine increases with the increase in load and brake power. As the blended diesel contains cerium oxide as one of the constituents, it facilitates the production of oxygen in the combustion chamber of engine.

Larger the quantity of cerium oxide, larger will be the oxygenation of fuel. This results in the higher efficiency of blended diesels with higher cerium oxide content. Due to this, it is found that D89E10DEE1CEO15 and D89E10DEE1CEO20 have higher brake thermal efficiency at every loading conditions. Also, these two fuels are found to have higher values of efficiency than that of the neat diesel at all operating conditions.

For the same reason, it has been observed that fuels with higher cerium oxide composition have lower brake specific fuel consumption than their counterpart. This has been substantiated in Fig. 4, showing the variation of brake specific fuel consumption with brake power. Figure 5 shows the variation of percentage of carbon dioxide emitted from the engine at different operating conditions investigated with the different test fuels. Being a green house gas, any recommended fuel shall have the lowest value of carbon dioxide emission. It has been found that the carbon dioxide percentage increases with increase in brake power. Also in all the operating conditions, the blended diesels are found to have lower emission when compared to that of neat diesel. Among the prepared test fuels, D89E10DEE1CEO20 is observed

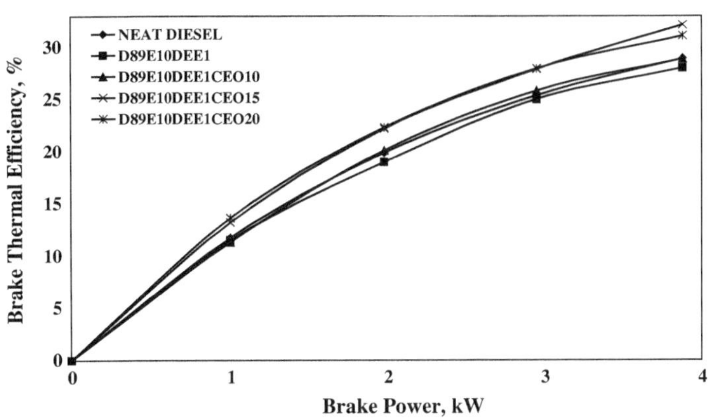

Fig. 3 Effect of brake power on brake thermal efficiency

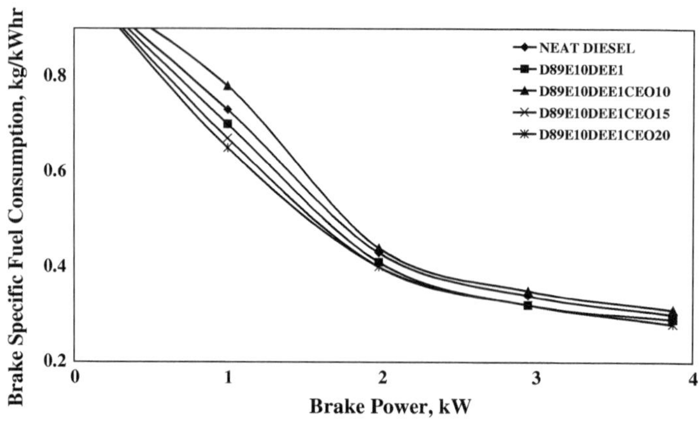

Fig. 4 Effect of brake power on brake specific fuel consumption

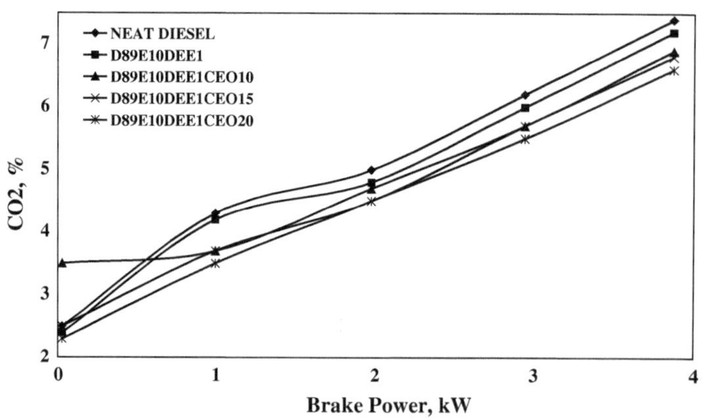

Fig. 5 Effect of brake power on carbon dioxide

to have the lowest carbon dioxide percentage at every load. However, the test fuel D89E10DEE1CEO15 is having only marginally higher emission than test fuel D89E10DEE1CEO20. The amount of cerium oxide present in the blended diesels prepared, contributes for the complete combustion due to the presence of more oxygen produced. The complete combustion ensures emission of less carbon dioxide from the engine. Hence the test fuel D89E10DEECEO20 with more cerium oxide has the lowest emission of carbon dioxide.

Figure 6 shows the effect of engine loading on NO_x emission from the engine. From the figure, it has been inferred that the NO_x emission increases with the increase in engine loading, due to the higher temperature of the cylinder attained (Campos-Fernandez et al. 2012).

In the entire range of operating conditions, the NO_x emission of blended diesels is found to be always less than that of neat diesel. Also, among the blended diesel

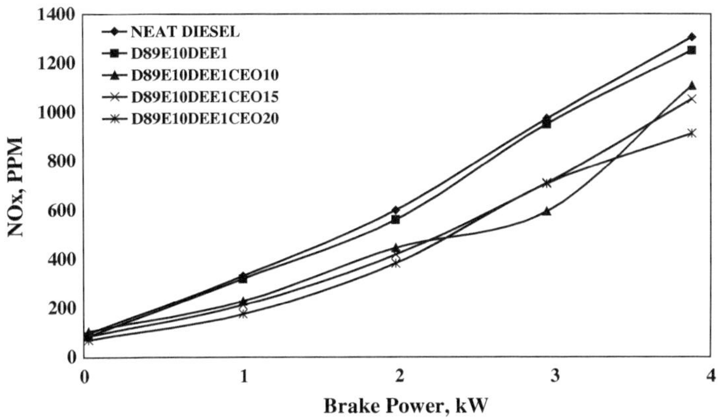

Fig. 6 Effect of brake power on oxides of nitrogen

fuels, D89E10DEE1CEO20 is found to have the lowest NO_x emission next to test fuel D89E10DEE1CEO15. Even though the high temperature of engine at higher engine loading condition causes the increase in NO_x emission, the presence of cerium oxide in the blended diesel acts favorable to facilitate complete combustion inside the cylinder. This results in less NO_x emission for test fuel D89E10DEE1CEO20 next to D89E10DEE1CEO15 at every engine loading condition.

With the foregoing discussions it has been observed that in the entire range of operating conditions, the test fuel D89E10DEE1CEO20 has a 3.2 % increase in efficiency and 0.6 % decrease in emission when compared to the neat diesel. The test fuel D89E10DEE1CEO15 is found to exhibit better characteristics with respect to performance and emission than the neat diesel. Hence, the blended diesel fuel D89E10DEE1CEO15 may be considered as a promising alternative to the neat diesel, for the single cylinder four-stroke diesel engine.

6 Conclusion

A computerized single cylinder four-stroke diesel engine test rig has been used to investigate for a suitable alternative to the neat diesel. For this, four trial blended diesel fuel samples have been prepared using ethanol, cerium oxide, and diesel. Experiments have been repeated with different loading conditions and fuels. Based on the experimental data, the performance and emission characteristics of test fuels have been compared with diesel and investigated. It has been found that the test fuel D89E10DEE1CEO15 has a 0.05 % increase in volumetric efficiency, 3.2 % increase in brake thermal efficiency, 0.6 % decrease in CO_2, and 252 PPM decrease in NO_x. Economic feasibility study also shows that the average one gallon of diesel and ethanol costs around 3.97$ and 3.41$ (Rashedul et al. 2014), respectively. The nano additives costs may be less than 1$ per kg. If the government takes forward and increase the production of ethanol it may reduce the price with decreasing the sample preparation rate. Based on the above investigation, the test fuel may be considered as a potential alternative for the conventional diesel in the test engine.

References

Aksoy HA, Becerik I, Karaosmanoglu F, Yatmaz HC, Civelekoglu H (1990) Utilization prospects of Turkish raisin seed oil as an alternative engine fuel. Fuel 69(5):600–603. doi:10.1016/0016-2361(90)90144-F

Campos-Fernandez J, Arnal JM, Jose Gomez M, Dorado P (2012) A comparison of performance of higher alcohols/diesel fuel blends in a diesel engine. Appl Energy 95:267–275. doi:10.1016/j.apenergy.2012.02.051

Golimowski W, Pasyniuk P, Berger WA (2013) Common rail diesel tractor engine performance running on pure plant oil. Fuel 103:227–231. doi:10.1016/j.fuel.2012.09.051

Gowen MM (1989) Biofuel v fossil fuel economics in developing countries: how green is the pasture? Energy Policy 17(5):455–470. doi:10.1016/0301-4215(89)90068-2

Huang J, Wanga Y, Li S, Roskilly AP, Hongdong Y, Li H (2009) Experimental investigation on the performance and emissions of a diesel engine fuelled with ethanol–diesel blends. Appl Therm Eng 29:2484–2490. doi:10.1016/j.applthermaleng.2008.12.016

Kannan GR, Karvembu R, Anand R (2011) Effect of metal based additive on performance emission and combustion characteristics of diesel engine fuelled with biodiesel. Appl Energy 88:3694–3703. doi:10.1016/j.apenergy.2011.04.043

Karthikeyan B, Srithar K (2011) Performance characteristics of a glowplug assisted low heat rejection diesel engine using ethanol. Appl Energy 88:323–329. doi:10.1016/j.apenergy.2010.07.011

Kim HG, Choi S-H (2013) The characteristics of biodiesel and oxygenated additives in a compression ignition engine. J Mech Sci Technol 27(5):1539–1543. doi:10.1007/s12206-013-0334-1

Malaya Naik LC, Meher SN, Naik LMD (2008) Production of biodiesel from high free fatty acid Karanja (Pongamia pinnata) oil. Biomass Bioenergy 32(4):354–357. doi:10.1016/j.biombioe.2007.10.006

Parlak A, Ayhan V, Cesur I, Kokkulunk G (2013) Investigation of the effects of steam injection on performance and emissions of a diesel engine fuelled with tobacco seed oil methyl ester. Fuel Process Technol 116:101–109. doi:10.1016/j.fuproc.2013.05.006

Prabhahar and Rajan (2013) Performance and combustion characteristics of a diesel engine with titanium oxide coated piston using Pongamia methyl ester. J Mech Sci Technol 27:1519–1526. doi:10.1007/s12206-013-0332-3

Ramesh Kumar C, Nagarajan (2012) Performance and emission characteristics of a low heat rejection spark ignited engine fuelled with E20. J Mech Sci Technol 26:1241–1250. doi:10.1007/s12206-012-0206-0

Rashedul HK, Masjuki HH, Kalam MA, Ashraful AM, Ashrafur Rahman SM, Shahir SA (2014) The effect of additives on properties, performance and emission of biodiesel fuelled compression ignition engine. Energy Conserv Manage 88:348–364. doi:10.1016/j.enconman.2014.08.034

Sajith V, Sobhan CB, Peterson GP (2010) Experimental investigations on the effects of cerium oxide nanoparticle fuel additives on biodiesel. Adv Mech Eng 1–6. doi:10.1155/2010/581407

Designer Biodiesel: An Optimization of Fuel Quality by Blending Multiple Oils

Anuchaya Devi, Renuka Barman and Dhanapati Deka

Abstract Liquid biofuels such as biodiesel, derived from vegetable oils are gaining worldwide importance as a substitute for diesel fuel owing to their environmental advantages and renewable sources in nature. Since biodiesel has the similar fatty acid composition as the parent oil, its fuel properties are directly affected by the composition of the precursor vegetable oil. Various blends of some locally available vegetable oils such as Jatropha, Nahor, Yellow Oleander, Karanja, and Coconut were prepared by mixing at different volumetric ratios in the range of 1:1–1:10 (*Jatropha*: *Coconut*; *Nahor*: *Coconut*; *Yellow Oleander*: *Coconut*; *Pongamia*: *Coconut*, *Jatropha*: *Yellow Oleander*; *Nahor*: *Yellow Oleander*, *Pongamia*: *Yellow Oleander*, *Pongamia*: *Jatropha, and Pongamia*: *Nahor*). Biodiesel was prepared in laboratory scale from the pure vegetable oils and selected blends through a single step conc. H_2SO_4 catalyzed transesterification process and their fuel properties were analyzed in accordance with the standard test method. Fatty acid composition of the vegetable oils and the blends were elucidated by ^1H NMR. The effect of fatty acid profile/composition on fuel properties such as cetane number, viscosity, calorific value, and iodine number has been measured.

Keywords Biodiesel · Blending · Fatty acid composition · Oxidation stability · ^1H NMR

1 Introduction

The substitution of fossil fuels to safe future energy demand goes on with the first and foremost disquiet over the last decade. With this correlation, biodiesel (Knothe et al. 2005; Mittelbach and Remschmidt 2004) is an option of selection for all sorts of petroleum-based diesel fuel. The outstanding plausible alternate of the usual

---

Anuchaya Devi · Renuka Barman · Dhanapati Deka (✉)
Biomass Conversion Laboratory, Department of Energy, Tezpur University, Napaam, India
e-mail: dhanapati@tezu.ernet.in

S. Kumar et al. (eds.), *Proceedings of the First International Conference on Recent Advances in Bioenergy Research*, Springer Proceedings in Energy,
DOI 10.1007/978-81-322-2773-1_10

diesel fuel known to be biodiesel has generally been constituted with fatty acid methyl esters which are often produced from triglycerides found in vegetable oils via the process of transesterification using methanol/ethanol. Transesterification is an effective way to overcome all these problems associated with biodiesel (Shay 1993). In recent years, biodiesel has recently gained worldwide popularity as a solution of energy crisis and also as domestic, renewable and biodegradable alternative diesel fuel (Singh and Singh 2010). The bright feed stocks for biodiesel production are demonstrated as the vegetable oils which are highly replenished and can be manufactured in industrial level. Biodiesel which is a fatty acid methyl esters (FAME) of several vegetable oils and animal fats have been found appropriately fit for utilizing in diesel engine in the form of a fuel. FAME as biodiesel is environmentally harmless, non-toxic and rapidly deteriorated by atmosphere (Azam et al. 2005). The overall fuel properties including oxidation stability, Cetane number, Calorific value etc. have been robustly affected by the characteristics of each of the fatty acid esters (Knothe and Kenar 2004). The main practical troublesome associated with the biodiesel utilizations in the current hours can be put forward as oxidative stability, low-temperature properties, and a minor enhancement in NO_x exhaust production, even though the last difficulty may become paler over time with the invention of novel exhaust emission control technology. To solve the aforementioned troubles at the same time have demonstrated intricacies on account of the clarification to one of the difficulty always creates disturbances to another ones. The above circumstances can be mainly traced to the belief of fuel features on the structure of fatty acid (Knothe 2005). Hence, it has been revealed that the cetane number which is directly connected to the ignition quality of a diesel fuel lessens with shortening the chain length, extension in branching, and rises in unsaturation in the chain of fatty acid (Harrington 1986). Normally, the greater the cetane number, the superior the ignition quality. Conversely, saturated esters with greater value of cetane numbers have lower cold-flow traits. Fatty acids with unsaturated or polyunsaturated functionalities possessing poorer melting points are enviable for better low-temperature traits and have small value of cetane numbers as well as reduced oxidative stability which are unwanted for the biodiesel. Remedy to one of these complexities always brings about the growing problematic behavior in the other characteristics and have incorporated the application of additives or changing the composition of fatty acid either via physical procedures, for instance winterization, or through genetic adjustments. This has led to the concept of designer biodiesel, with optimized fatty acid compositions for superior fuel properties. To achieve the biodiesel fuel with encouraging features, it is beneficial for the fuel to constitute with merely one major part probably in a large concentration. The characteristics of the different fatty ester components encompassing biodiesel decide the on the whole fuel features of the biodiesel fuel. Thus, the qualities of the several fatty esters are detected by the structural characteristics of the fatty acids and the alcohol functionalities consisting of them (Knothe 2005, 2008). As per our objective is concerned, in this manuscript, different oil seeds from locally available oil bearing trees has been collected, extracted and prepared model feedstock oils by blending of composed vegetable oils at different ratio in order to modify their fatty acid compositions. Additionally, the fuel properties of biodiesel/methyl esters of model oils based on fatty acid compositions has also been predicted and verified the outcomes experimentally. The optimum blending ratio of various feedstock oils for production of biodiesel with superior fuel quality has also been accomplished.

2 Experimental

Methanol (extra pure), Petroleum Ether (40–60 °C), H_2SO_4 (98 % pure), Na_2SO_4 (99 % pure), and $NaHCO_3$ (99 %) were procured from Merck India Ltd, Mumbai. Each of the reagents and chemicals had been made use with no additional purification. Various oil bearing seeds such as Yellow oleander (*Thevetia peruviana*), Nahor (*Mesua ferrea Lin.*), Karanja (*Pongamia pinnata*), and Jatropha (*Jatropha curcas*) were collected from different localities of Assam, India. *Jatropha curcas* and *Pongamia pinnata* seeds were supplied by Kaliabor Nursery, Nagaon (Assam), India. Coconut oil (*Cocos nucifera*; 100 % pure, according to manufacturer) involved in the experiment was obtained from a local supplier in the Tezpur university campus and used without any further purification.

The components of the fatty acids of the vegetable oils and the blends were confirmed by using ^{1}H Nuclear Magnetic Resonance Spectroscopy (^{1}H NMR). ^{1}H NMR spectra were processed using "ACD NMR Processor Academic Edition" (Elyashberg et al. 2008). All ^{1}H NMR analyses were carried out at 25.5 °C with $CDCl_3$ and TMS as solvent and internal standard, respectively. The ^{1}H NMR spectra were carried out on a Jeol JNM-ECS400 spectrometer.

2.1 Mathematical Treatment for Determining Biodiesel Fuel Properties

Based on the fatty acid compositions, important fuel parameters such as cetane number, viscosity, calorific value, iodine value of the vegetable oils, and model oils were determined using the empirical relations as follows (Ramírez-Verduzco et al. 2012).

$$f_b = \sum z_i \times f_i \quad (1)$$

where, z_i is the wt% individual fatty acid/ester determined from ^{1}H NMR and f_i is a function that represents a fuel parameter such as cetane number, viscosity, calorific value, and iodine value (the subscripts b and *i* refer to the biodiesel and the pure *i*th FAME, respectively).

The individual fuel parameters of each FAME component were calculated as follows:

$$\text{Cetane number, CN} = -7.8 + 0.302 \times M_i - 0.20 \times N \tag{2}$$

$$\text{Calorific value (MJ/kg), CV}_i = 46.19 - 1794 \div M_i - 0.21 \times N \tag{3}$$

$$\text{Viscosity (cSt), } \ln(\nu) = -12.503 + 2.496 \times \ln(M_i) - 0.178 \times N \tag{4}$$

$$\text{Iodine value (mg I}_2\text{/g), IV} = \sum \{254 \times (\text{no. of double bonds} \times \text{wt\%})\} \div M_i \tag{5}$$

where, M_i is molecular weight of individual FAME, N is the number of double bonds and η_i is the kinematic viscosity at 40 °C of the ith FAME. For calculation, the entire saturated portion in all vegetable oils and their blends were approximated to be C16 fatty esters except for coconut oil where C12 fatty esters was considered to be the saturated fraction and Nahor oil where C18 fatty esters was considered to be the saturated fraction because ^{1}H NMR method do not allow quantification of individual saturated fatty acids/esters. Oxidation stability as induction period (IP) was predicted based on the oxidation stability (IP) of individual FAME reported in literature and then putting the values in the general expression (1).

2.1.1 Acid Value

The determination of acid value was done by titration of the material in ethanol with aqueous KOH. The calculations of the acid values of the various vegetable oils were done as per ASTM D 664. For this purpose, 0.5 g of sample was dissolved in ethyl alcohol by heating gently on a water bath and then the titration was performed against 0.1 M KOH solution utilizing phenolphthalein as an indicator. The acid value was calculated as follows:

$$\text{Acid value (AV)} = \frac{56.1 \times V \times N}{W}$$

where,

V volume of standard KOH used (mL).
N normality of standard KOH.
W weight of oil (grams).

2.1.2 Iodine Value

Iodine value was determined according to AOAC Official Method 993.20. 0.4 g of the oil was weighed into a conical flask and made miscible in CCl_4 (25 mL). Then W_{ij}'s solution (25 mL) was transferred to it and mixed appropriately. Having done this, concentrated HCl (5 mL) and KI (10 %, 15 mL) were poured into the flask. The flask was then placed in the dark for 30 min and then titration was performed against standard sodium thiosulphate solution by making use of starch as an indicator. The similar method was repeated for the blank test.

The calculation of Iodine Value (IV) is given by:

$$\text{Iodine value (IV)} = 1.269(V_2 - V_1)/W$$

Here,

V_1 volume of thiosulphate utilized for the test (in mL).
V_2 volume of thiosulphate utilized for the blank test (in mL).
W weight of oil under investigation (in g).

2.1.3 Oxidation Stability

The oxidation stability was measured via the induction period (IP) of the prepared sample. The IP value was calculated using Rancimat method EN 14112 on a Metrohm 873 Rancimat instrument. The moisture free biodiesel samples (3 g) were examined underneath a steady flow of air at the rate of 10 L/h into the vessel having double distilled water. At 110 °C, the biodiesel samples were heated and a temperature correction factor ΔT as recommended by the test method had been set to 1.5 °C. To the measuring and recording device the electrode was fixed. The finish point of the induction period was presented as soon as the conductivity begins to rise quickly. The instant, at which the conductivity of these measuring solutions is obtained continually along with achieving an oxidation curve at that point of inflection, is called the induction period which supplies the characteristic numerical value of the oxidation stability for that particular biodiesel sample (Jain and Sharma 2011).

2.1.4 Calorific Value

The calorific value of the vegetable oils and the methyl esters and their blends were measured using an oxygen bomb calorimeter according to ASTM D 2015 standard method.

2.1.5 Kinematic Viscosity

Kinematic viscosities of the prepared samples were determined by falling-ball viscometer (HAAKE, Type C) following the standard method ASTM D1298. The time of falling-ball is directly proportional to its viscosity. The outcomes obtained during the experiment are the falling time for a definite distance which is evaluated by electronic means and changed into the units of viscosity.

The dynamic viscosity 'η' (in mPa s) has been calculated as given below:

$$\eta = K\,(\rho_1 - \rho_2)t$$

Here,

K ball constant (in mPa s cm^3/g s).
ρ_1 density of the ball (in g/cm^3).
ρ_2 density of the liquid to be measured at the measuring temperature (in g/cm^3).
t falling time of the ball (in s).

The dynamic viscosity 'η' and the kinematic viscosity 'v' are related with as shown below:

$$v = \frac{\eta}{\rho}$$

v kinematic viscosity (in mm^2/s) (1 mm^2/s = 1 cSt);
η dynamic viscosity (in mPa s);
ρ density of the liquid sample (in g/cm^3).

2.1.6 Density

Densities of the prepared samples of biodiesel were recorded in accordance with ASTM D 287 standard.

2.1.7 Cloud Point and Pour Point

Cloud point and Pour point were measured by following ASTM D 2700-91 and ASTM D 97-96, respectively. The cloud point (CP) is the temperature at which wax gives a cloudy outward show. To determine the Cloud point, samples were cooled and examined visually until first cloud appeared. Pour point (PP) is the minimum temperature at which the fuel transforms into semi solid and drops its flow properties (Giakoumis 2013).

2.2 Oil Extraction

The seeds were dried in sunlight and in an oven at 110 °C to remove any traces of moisture. The moisture free seeds were dehulled, kernels removed, and grinded using a standard mixture grinder. Oil was extracted by the Soxhlet extraction method using petroleum ether (40–60 °C) as solvent (Dobush et al. 1985) [AOAC 963.15., 1976]. The extracted oil has been dried over Na_2SO_4 (anhydrous) and filtered for removing solid particles and impurities and stored in volumetric flasks.

2.3 Preparation of Model Oils (Mixed Oil Feedstocks)

Various blends were prepared by mixing Jatropha, Nahor, Yellow oleander, Karanja (*Pongamia pinnata*), and Coconut oils in different volumetric proportions, such as by mixing (*Jatropha* with *Coconut*), (*Nahor* with *Coconut*), (*Yellow oleander* with *Coconut*), (*Pongamia* with *Coconut*), (*Jatropha* with *Yellow oleander*), (*Nahor* with *Yellow oleander*), (*Pongamia* with *Yellow oleander*) (*Pongamia* with *Jatropha*) and (*Pongamia* with *Nahor*) oil in volumetric ratios 1:1–1:10. The fatty acid compositions of these model feedstocks were evaluated and fuel properties of their corresponding methyl esters were predicted based on empirical relations. A number of fuel features for instance oxidation stability, iodine number, acid value, calorific value and kinematic viscosity of biodiesel from the mixed feedstocks were determined and compared to predicted values.

2.4 Experimental Methods for Determination of Selected Biodiesel Fuel Properties

For determining Oxidation stability, Acid value, Iodine value, Kinematic viscosity, Calorific value, Density, Cloud point, and Pour point and in order to validate our predicted results (biodiesel fuel parameters of the oils and blends), biodiesel was prepared in laboratory scale from the pure vegetable oils and selected blends through a single step conc. H_2SO_4 catalyzed transesterification process, and their fuel properties were analyzed according to standard test methods. In a typical reaction for this purpose 10 g of preheated oil was added to a 100 mL round bottom flask containing 25 mL MeOH (alcohol) and conc. H_2SO_4 (2 wt% of oil) and stirred vigorously at 600 rpm with a magnetic stirrer. The contents in the flask (reaction mixture) were heated at 100 ± 5 °C for 24 h. As soon as the reaction completed, the entire contents in the flask was moved to the separating funnel and allowed to settle for getting two distinct phases. The layer of the upper phase contained biodiesel and lower phase was having glycerin. The excessive volume of methanol present in the upper phase of the collected layer was evaporated using rotary vacuum evaporator. The methyl ester (biodiesel) present in upper phase was washed with distilled water till neutralizing unutilized H_2SO_4 and the glycerine was removed. Finally, biodiesel is dried over anhydrous Na_2SO_4. Fuel properties like Acid value, Iodine value, Oxidation stability, Calorific value, Density, Kinematic viscosity, Cloud point, and Pour point were recorded according to standard methods discussed above and compared with the predicated values. Calorific values were determined using an oxygen bomb calorimeter.

3 Results and Discussion

3.1 Fatty Acid Profiles/Compositions of Feedstocks

Table 1 shows the fatty acid compositions of single and model oil feedstocks (created by blending) evaluated using ^{1}H NMR spectroscopy.

Samples prepared were characterized using ^{1}H NMR technique. The ^{1}H NMR spectra of Coconut oil, Jatropha oil, Nahor oil, Pongamia oil and Coconut and Jatropha blend (1:1 blending ratio) are shown in Figs. 1, 2, 3, 4, and 5. From Fig. 5, the effect of blending on the patterns of ^{1}H NMR of the oils was seen. The ^{1}H NMR of highly saturated coconut oil shows less number of peaks because the unsaturated fatty acids were not present. The characteristic linolenic acid peak at 0.98–0.96 ppm was absent and peak intensities of characteristic linoleic and oleic acids appearing near ~2.77 and ~2.0 ppm were very low due to the presence of only trace amounts linoleic acid (0.72 %) and oleic acid (4.8 %). However, when blended with unsaturated oil like Jatropha, the pattern of ^{1}H NMR spectrum changed drastically, the peak intensities of unsaturated peaks at 2.77 and 2.0 ppm increased and reached to a maximum. This means that ^{1}H NMR could be used to detect the changes in fatty acid compositions of model oils due to blending. In our study we found that as the percentage of unsaturation was increased, the intensities of peaks at 0.98–0.96, 2.77 and 2.0 ppm were also increased.

From the ^{1}H NMR spectrums of Nahor and Pongamia, we have seen that both are showing oleic, linoleic and linolenic acids characteristic peaks. However, the peak area for linoleic acid is little smaller in Nahor than Pongamia oil. This is because of the presence of more amounts of linoleic acid (22.22 wt%) in Pongamia than Nahor (13.21 wt%). Also, it is seen that Pongamia is more saturated compared to Nahor.

Table 1 Fatty acid composition of single and model oil feedstocks

Sl. No.	Sample's name	Fatty acid profile/composition (wt%)				
		Linolenic (C18:3)	Linoleic (C18:2)	Oleic (C18:1)	Total saturated	OSI (h)
1	C (Coconut)	0	0.72	4.8	94.48	22.81
2	J (Jatropha)	0.5	36.99	53.51	9	4
3	K (Karabi)	1.78	23.81	56.66	17.75	6.07
4	N (Nahor)	3.42	13.21	65.47	17.9	6.25
5	P (*Pongamia pinnata*)	4.51	22.22	51.8	21.47	5.77
6	C+J (1:1)	0	17.23	20.36	62.41	15.71
7	C+J (2:3)	0	20.08	28.42	51.5	13.34
8	C+J (1:2)	0	18.99	26.46	54.55	14
9	C+J (1:3)	0	23.15	33.41	43.44	11.57
10	C+J (1:4)	0	24.29	33.38	42.34	11.32
11	C+J (1:5)	0	24.51	39.99	35.5	9.86
12	C+J (1:8)	0	24.72	39.39	35.9	9.94
13	C+J (1:10)	0	25.88	43.80	30.32	8.73
14	C+N (1:1)	1.81	3.39	26.51	68.29	16.51
15	C+N (1:2)	2.65	4.03	33.29	60.03	15.37
16	C+N (1:3)	3.7	3.22	39.54	53.54	13.98
17	C+N (1:5)	3.94	4.76	44.19	47.13	12.58
18	C+N (1:8)	4.09	5.07	49.08	41.76	11.43
19	C+N (1:10)	4.22	6.32	56.27	33.19	9.59
20	C+K (1:5)	0	17.18	27.54	55.28	14.19
21	C+K (1:10)	0	22.09	40.53	37.38	10.3
22	K+N (1:1)	3.5	15.52	62.67	18.31	13.4
23	K+N (2:3)	3.89	14.07	57.72	24.32	7.57
24	K+N (1:2)	3.68	14.65	62.68	19.01	6.4
25	K+N (1:3)	3.56	14.81	57.61	24.02	7.51
26	K+N (1:8)	5.21	9.53	66.67	18.59	6.41
27	K+J (1:1)	1.29	18.69	64.1	15.92	2.65
28	K+J (2:3)	1.33	30.53	50.66	17.48	5.89
29	K+J (1:2)	0.8	32.32	52.19	14.69	5.28
30	K+J (1:3)	4.9	21.73	56.05	17.32	5.92
31	K+J (2:4)	1.05	29.42	52.19	17.34	5.89
32	C+P (1:4)	2.52	11.3	37.06	49.25	12.96
33	C+P (1:5)	2.41	14.92	33.49	49.18	12.87
34	C+P (1:10)	3.19	19.42	41.8	35.59	9.88
35	P+K (2:1)	3.49	18.51	53.11	24.89	7.62
36	P+J (1:1)	2.51	26.56	51.01	19.92	6.45
37	P+N (6:1)	4.54	20.35	45.84	23.27	7.05
38	P+N+C (2:2:1)	1.32	14.28	38.57	45.83	12.2

The used symbols represent:
J Jatropha (*Jatropha curcas*); *N* Nahor (*Mesua ferrea Lin*); *C* Coconut (*Cocos nucifera*); *K* Karabi (*Thevetia peruviana*); and *P* Karanja (*Pongamia pinnata*)

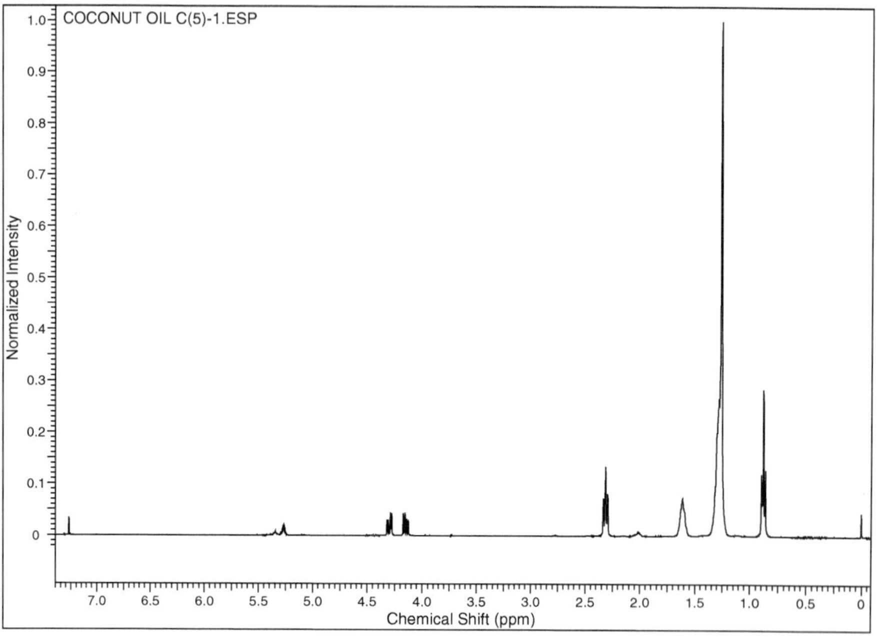

Fig. 1 ^{1}H NMR spectrum of coconut (*Cocos nucifera*) oil

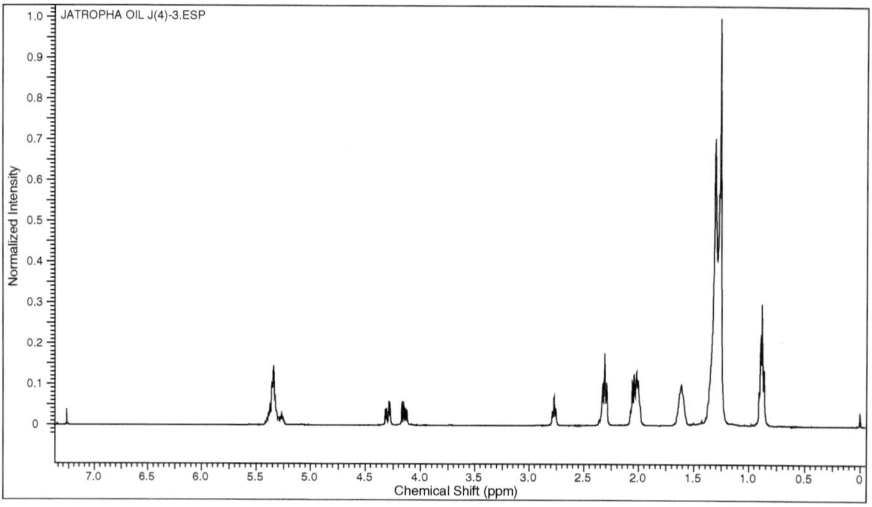

Fig. 2 ^{1}H NMR spectra of Jatropha (*Jatropha curcas*) oil

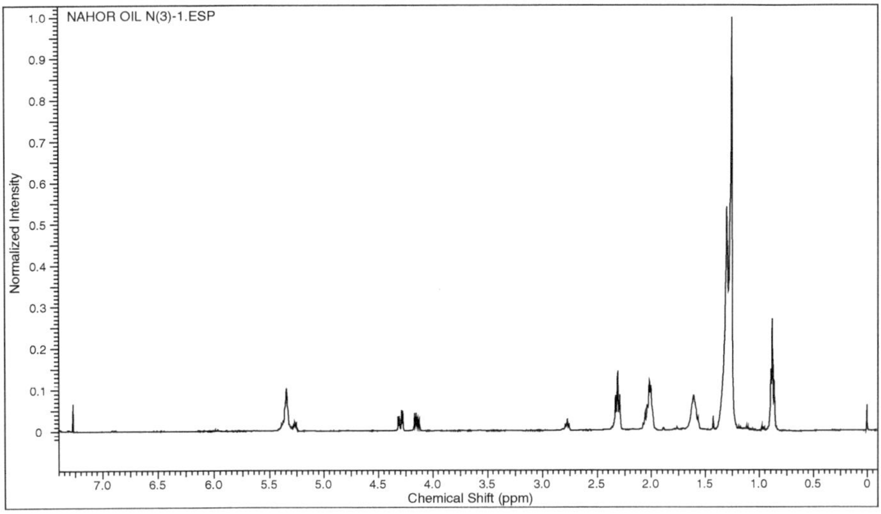

Fig. 3 ^{1}H NMR of *Mesua ferrea* (Nahor) oil

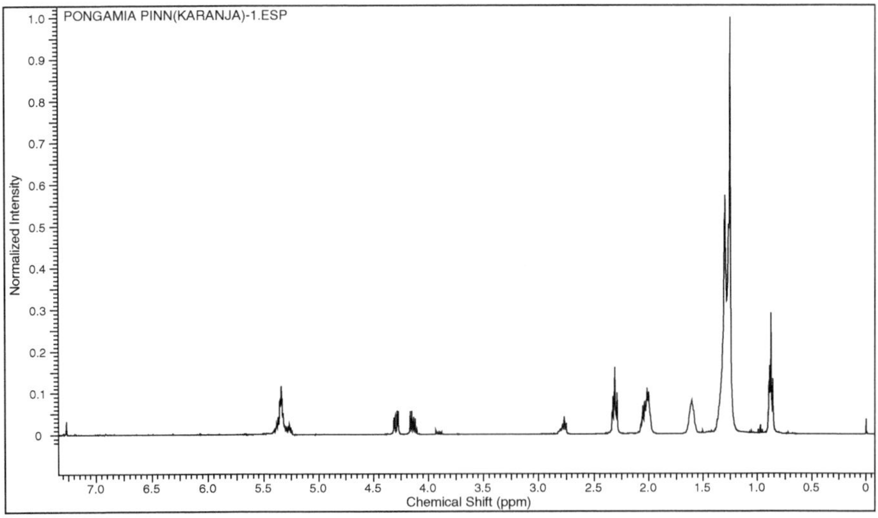

Fig. 4 ^{1}H NMR spectra of Karanja (*Pongamia pinnata*) oil

3.2 Comparison of Experimental and Predicted Properties of Biodiesel

All the predicted and experimental results of cetane number, calorific value, kinematic viscosity, iodine number and oxidation stability that we obtained, are shown in Table 2.

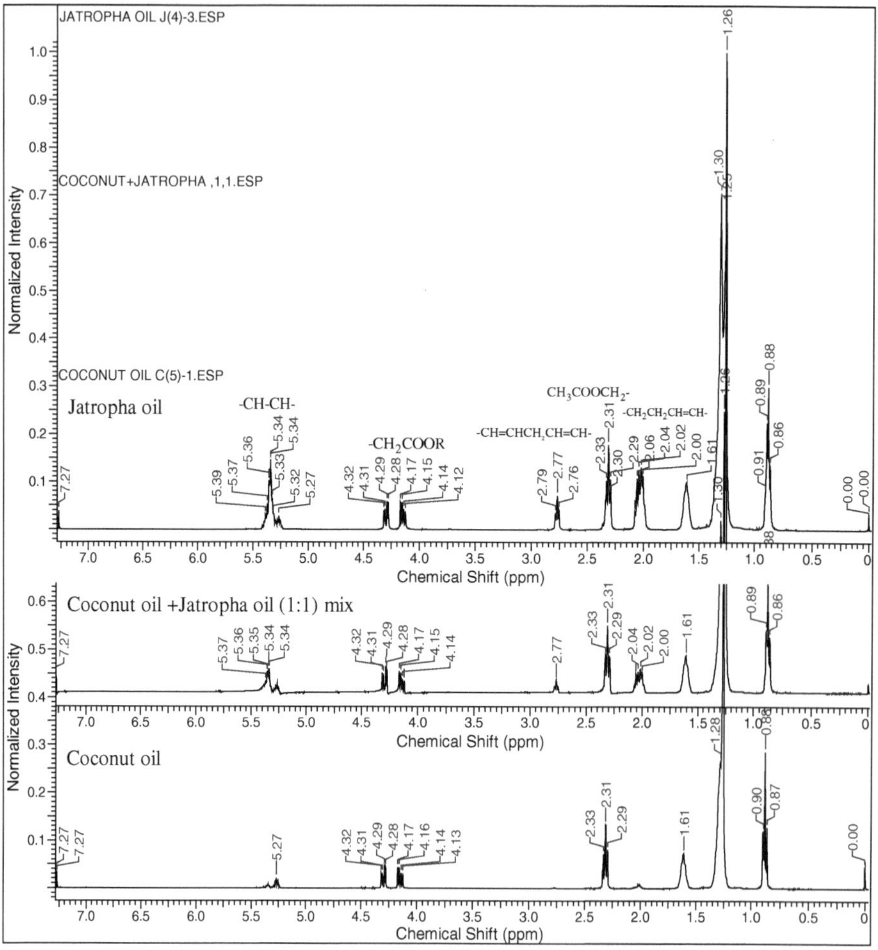

Fig. 5 Effect of fatty acid composition on ^{1}H NMR Spectrum of oils

And it is found that the predicted values that we calculated are nearly equivalent to those obtained experimentally. Thus, it is clear that the method we used is the correct one which can be successfully used for the determination of fuel properties of biodiesel. It is seen from Table 2 that cetane number, kinematic viscosity and oxidation stability increase with the increase in saturation. But unlike these properties, iodine value and density decrease with the increasing degree of saturation. Calorific value increases with increasing chain length. Here, Coconut FAME is the most saturated biodiesel which is illustrated by less viscosity and greater cetane number and also considered as a feedstock that differentiates by a lot from the others.

Table 3 lists the experimental values of some important fuel properties such as density, IV, viscosity, calorific value, OSI, cloud point, and pour point of 15 selected samples. It is seen from Table 3 that, highly saturated coconut biodiesel is having highest oxidation stability >24 h and lowest iodine value and viscosity. When Coconut is blended with highly unsaturated oil such as Jatropha or Nahor or Karabi or Pongamia oil, we are getting significant changes in the fuel properties. Viscosity, iodine value and calorific value were found to be increased with the addition of unsaturated nonedible oils. Overall, the trends for variation of fuel properties were found to be similar to the empirical results presented in Table 2. Additionally, all the characteristics of the biodiesel obtained from the feedstocks were in accordance to the specifications of ASTM D6751 and EN 14214.

3.3 *Effect of Coconut Oil Blending Ratio on Iodine Value and Viscosity*

The effect of blending ratio (percentage of Coconut oil in the blends) on iodine value and viscosity are described in Fig. 6a–c. From the results we could see that as the percentage of unsaturated oils in the blends increased (percentage of Coconut oil in blends decreased) the iodine values also increased and reached its maximum value when the blends contained no added Coconut oil. In all cases the presence of 10 % (1:10 = C:J/P/N) coconut oil in blends was sufficient to reduce the iodine values to <100 mg I_2/g the minimum limit as per the specifications of EN14111.

Figure 6a–c illustrated the interdependence between iodine number (and viscosity) and degree of unsaturation. Degree of unsaturation is found to show a relationship with the iodine number and a statistical connection is established. It is found from the above graph that the iodine value increases with the increasing unsaturation in the blends, as the compositions in the blends becomes richer in unsaturated fatty acids (mainly linoleic and/or linolenic acids). An increase in the concentration of jatropha (or nahor or pongamia) in the blends, which is highly unsaturated, leads to an increase in degree of unsaturation in the blends. Therefore, an extremely strong correlation exists between IV and unsaturation degree substantiating the outcomes of many earlier literature (Sayyed et al. 2013).

Biodiesel is a combination of more than a few fatty esters having each constituents added to the whole value of kinematic viscosity. Hence, the constituents having smaller viscosity can lessen the total kinematic viscosity of biodiesel to a satisfactory range albeit larger value of viscosity may present. Double bonds reduce the kinematic viscosity (Knothe and Steidley 2005). Viscosity of biodiesel enhances with rising degree of saturation and increment in the number of carbon atoms chain. It has been observed that long chain saturated esters (C18:0) have higher viscosity than short chain saturated or long chain unsaturated fatty esters. Because of this combined effect, we can see that coconut methyl ester is the only biodiesel which significantly distinguishes, as observed, in the number of carbon atoms in the chain, obviously demonstrates the least value of viscosity (Giakoumis 2013). In the present study, the viscosities of biodiesel obtained from the blended samples were

Table 2 Comparison of predicted and experimental biodiesel fuel properties

	Cetane number CN	Calorific value (KJ/g) CV		Kinematic viscosity (cSt) ν		Iodine value (IV) (mg I_2 g^{-1})		OSI (h)	
Sample code	*X*	*X*	*Y*	*X*	*Y*	*X*	*Y*	*X*	*Y*
J	52.26	39.66	40.49	4.27	5.05	110.95	109.00	4.00	2.0
K	57.03	39.78	39.79	4.33	5.11	94.25	90.00	6.07	5.82
N	60.95	39.91	39.75	4.61	6.17	88.16	72.54	6.25	3.87
C	61.00	37.65	38.44	2.56	1.99	5.354	8.20	22.81	>24
P	57.89	39.76	40.20	4.30	3.12	94.43	84.33	5.77	3.79
C+J (1:1)	56.42	38.08	38.53	3.51	4.35	47.16	43.61	15.71	13.47
C+J (2:3)	60.35	39.32	38.60	2.70	4.64	58.97	46.83	13.34	11.78
C+J (1:2)	61.22	39.35	–	3.97	–	55.43	–	14.00	–
C+J (1:3)	60.39	39.51	–	4.10	–	77.54	–	11.57	–
C+J (1:4)	60.43	39.56	38.73	4.14	4.85	70.48	54.385	11.32	10.6
C+J (1:5)	60.00	39.62	38.88	4.20	4.97	76.52	49.66	9.86	7.98
C+J (1:8)	60.71	39.65	–	4.24	–	76.38	–	9.94	–
C+J (1:10)	59.60	39.69	38.98	4.26	5	82.15	69.52	8.73	6.73
C+N (1:1)	58.64	38.46	–	3.13	–	33.27	–	16.51	–
C+N (1:2)	67.09	39.57	38.42	4.49	4.24	56.33	51.65	15.37	13.77
C+N (1:3)	67.17	39.71	38.53	4.62	4.41	49.07	43.47	13.98	13.01
C+N (1:5)	66.84	39.82	38.58	4.72	4.59	56.33	55.14	12.58	12.5
C+N (1:8)	63.25	39.88	–	4.76		61.45		11.43	–
C+N(1:10)	64.76	39.89	38.79	4.71	6	70.05	60.12	9.59	10.11
K+C (5:1)	57.33	39.51	–	2.7	–	53.21	–	14.19	–
K+C(10:1)	61.13	39.66	–	4.26	–	72.82	–	10.3	–
K+N (1:1)	59.32	39.85	–	4.61	–	89.58	80.34	13.4	–
K+N (2:3)	61.41	39.86	–	4.5	–	83.83	–	7.57	–
K+N (1:2)	60.58	34.79	–	4.5	–	88.54	–	6.4	–
K+N (1:3)	61.75	39.88	–	4.57	–	84.18	–	7.51	–
K+N (1:8)	61.25	39.9	–	4.6	–	87.13	–	6.41	–
K+J (1:1)	57.61	38.73	–	4.26	–	90.52	–	2.65	–
K+J (2:3)	57.01	39.65	–	4.29	–	99.5	–	5.89	–
K+J (1:2)	56.51	39.77	–	4.29	–	102.54	–	5.28	–
K+J (1:3)	58.8	39.77	–	4.31	–	98.23	86.21	5.92	–
K+J (2:4)	57.33	39.77	–	4.3	–	98.19	–	5.89	–
C+P (1:4)	62.66	39.50	38.62	4.17	4.19	57.81	55.506	12.96	10.5
C+P (1:5)	62.24	39.54	38.98	2.63	3.96	60.71	59.36	12.87	9.1
C+P(1:10)	60.18	39.66	–	4.25	–	77.62	–	9.88	6.2
P+K (2:1)	59.49	39.76	–	4.34	–	86.52	–	7.62	–
P+J (1:1)	57.63	39.78	–	4.3	–	96.06	–	6.45	–
P+N (6:1)	55.05	39.55	–	4.28	–	86.21	–	7.05	–

(continued)

Table 2 (continued)

	Cetane number CN	Calorific value (KJ/g) CV		Kinematic viscosity (cSt) ν		Iodine value (IV) (mg I_2 g^{-1})		OSI (h)	
Sample code	*X*	*X*	*Y*	*X*	*Y*	*X*	*Y*	*X*	*Y*
P+N+C (2:2:1)	63.98	39.66	38.47	4.4	5.41	61.12	53.01	12.2	8.3

The used symbols represent:
X Predicted and *Y* Experimental
J Jatropha (*Jatropha curcas*); *N* Nahor (*Mesua ferrea Lin.*); *C* Coconut (*Cocos nucifera*); *K* Karabi (*Thevetia peruviana*); and *P* Karanja (*Pongamia pinnata*)

Table 3 Experimental determination of fuel properties of 15 selected feedstocks

Samples name	Density @ 15 °C (g/c c)	Viscosity @ 40 °C (cSt)	Calorific value (kJ/g)	Iodine value (mgI_2g^{-1})	OSI (h)	Cloud point (°C)	Pour point (°C)
J	0.878	5.05	40.49	109	2.00	9	−3
K	0.882	5.11	39.79	90	5.82	6	−8
N	0.891	6.17	39.75	72.54	3.87	7	3
C	0.852	1.99	38.44	8.20	>24	4	−6
P	0.884	3.12	40.2	84.33	3.79	15	6
C+J (1:1)	0.867	4.35	38.53	43.61	13.47	8	5
C+J (2:3)	0.872	4.34	38.60	46.83	11.78	0	−5
C+J (1:4)	0.887	4.85	38.73	54.385	10.6	10	−2
C+J (1:5)	0.891	4.97	38.88	49.66	7.98	9	−3
C+J (1:10)	0.895	5.0	38.98	69.52	6.73	–	–
C+N (1:2)	0.877	4.24	38.42	51.65	13.77	4	2
C+N (1:3)	0.889	4.41	38.53	43.466	13.01	9	5
C+N (1:5)	0.890	4.59	38.58	55.14	12.48	12	8
C+N (1:10)	0.893	6.0	38.79	60.12	10.11	16	10
C+P (1:4)	0.810	4.19	38.62	55.506	10.51	12	7
C+P (1:5)	0.887	3.96	38.98	59.36	9.06	19	9
C+P+N (1:2:2)	0.897	5.41	38.47	53.01	8.27	5	1

The used symbols represent
J Jatropha (*Jatropha curcas*); *N* Nahor (*Mesua ferrea Lin.*); *C* Coconut (*Cocos nucifera*); *K* Karabi (*Thevetia peruviana*); and *P* Karanja (*Pongamia pinnata*)

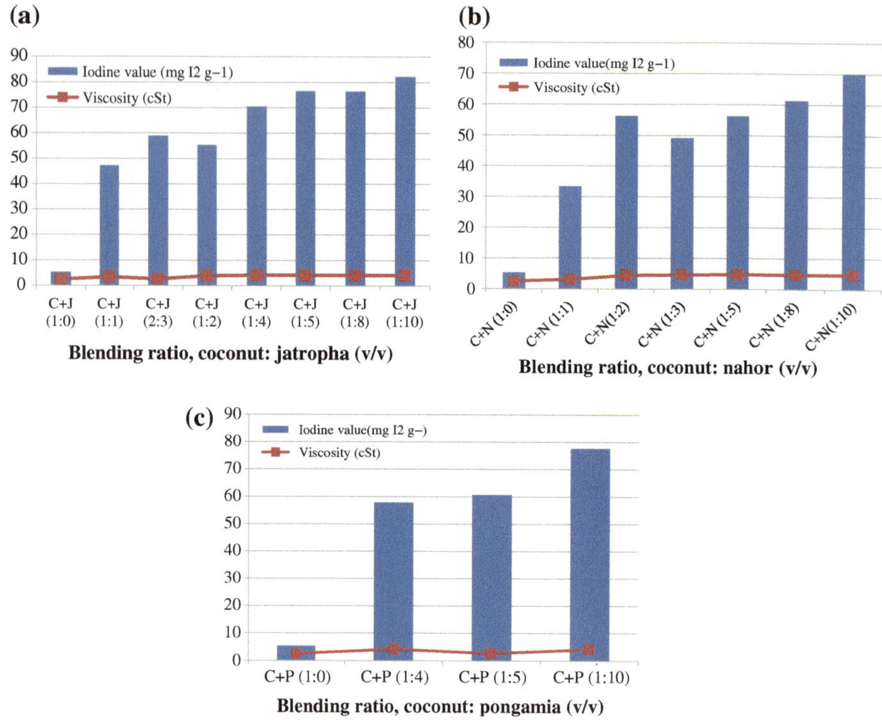

Fig. 6 **a** Effect of blending ratio (C:J) on iodine value and viscosity. **b** Effect of blending ratio (C: N) on iodine value and viscosity. **c** Effect of blending ratio (C:P) on iodine value and viscosity

always found to be lower than the parent oil (such as *Jatropha curcas*, or *Mesua ferrea Lin.*, *Pongamia pinnata* or *Thevetia peruviana*). The viscosities of blends containing 50 % or more coconut oil were found to significantly lower.

3.4 Effect of Blending Ratio on Calorific Value

Figure 7a–c show a correlation between calorific value and different blending ratio of Coconut and Jatropha (C+J), Coconut and Nahor (C+N) and Coconut and Pongamia(C+P). Coconut biodiesel exhibited the lowest calorific value due to the existence of short chain fatty acid methyl esters (C8:C12) in large amounts. It is found that, calorific value increases gradually as the proportion of Coconut oil in the blends were declined. This is due to the fact that, with addition of unsatuarted oil (such as Jatropha, Nahor and Pongamia) amount of oleic acid (C18:1) in the blends also increases. Thus, the amount of long chain methyl esters (C18:1 and C18) in the biodesel obtained from these blends also increases which is responsible for the increased calorific values.

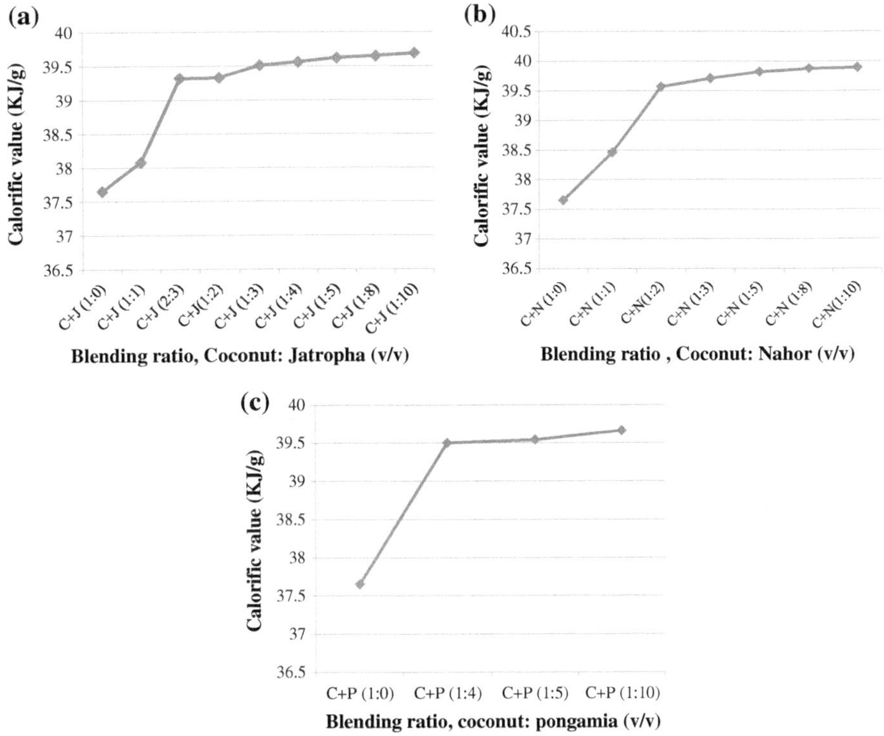

Fig. 7 **a** Effect of blending ratio (C:J) on calorific value. **b** Effect of blending ratio (C:N) on calorific value. **c** Effect of blending ratio (C:P) on calorific value

4 Conclusions

In conclusion, we devised an improved fuel quality of biodiesel from four locally available nonedible vegetable oils namely, *Jatropha curcas* (Jatropha), *Mesua ferrea Lin*. (Nahor), *Thevetia peruviana* (Karabi/Yellow oleander), *Pongamia pinnata* (Karanja) by modifying their fatty acid/ester composition via blending with Coconut oil, a highly saturated short chain oil. Fatty acid/ester composition were modified by blending the different nonedible vegetable oils in different volumetric ratios (1:1–1:10) with Coconut oil. Hence, the mixtures of Jatropha, Nahor, Karabi, Karanja, and Coconut oil can be used as feedstocks for biodiesel production. It has been endowed with that by blending different oil feedstocks prepared using this method can influence magnificent increase in the quality enhancement of the biodiesel which is going to be a breakthrough in domain of industry and human civilization.

References

Azam MM et al (2005) Prospects and potential of fatty acid methyl esters of some non-traditional seed oils for use as biodiesel in India. Biomass Bioenergy 29:293–302

Dobush et al (1985) The effect of apparatus, extraction time, and solvent type on lipid extractions of snow geese Can. J Zool 63:1917–1920

Elyashberg et al (2008) Computer-assisted structure verification and elucidation tools in NMR-based structure elucidation. Prog Nucl Magn Reson Spectrosc 53:1–104

Giakoumis EG (2013) A statistical investigation of biodiesel physical and chemical properties, and their correlation with the degree of unsaturation. Renew Energy 50:858–878

Harrington KJ (1986) Chemical and physical properties of vegetable oil esters and their effect on diesel fuel performance. Biomass 9:1–17

Jain S, Sharma MP (2011) Study of oxidation stability of *Jatropha curcas* biodiesel/diesel blends. Int J Energy Environ 2:533–542

Knothe G (2005) Dependence of biodiesel fuel properties on the structure of fatty acid alkyl esters. Fuel Process Technol 86:1059–1070

Knothe G (2008) Designer biodiesel: optimizing fatty ester composition to improve fuel properties. Energy Fuels 22:1358–1364

Knothe G, Kenar JA (2004) Determination of the fatty acid profile by ^{1}H-NMR spectroscopy. Eur J Lipid Sci Technol 106:88–96

Knothe G, Steidley KR (2005) Kinematic viscosity of biodiesel fuel components and related compounds: influence of compound structure and comparison to petrodiesel fuel components. Fuel 84:1059–1065

Knothe G, Krahl J, Van Gerpen J (eds) (2005) The biodiesel handbook. AOCS Press, Champaign, IL

Mittelbach M, Remschmidt C (2004) Biodiesels—the comprehensive handbook. Mittelbach. Graz, Austria

Ramírez-Verduzco LF et al (2012) Predicting cetane number, kinematic viscosity, density and higher heating value of biodiesel from its fatty acid methyl ester composition. Fuel 91:102–111

Sayyed et al (2013) Effect of acid & iodine value of karanja oil methyl ester (KOME) & it's statistical correlation with gross calorific value. IJRET 02:680–685

Shay EG (1993) Diesel fuel from vegetable oils: status and opportunities. Biomass Bioenergy 4:227–242

Singh SP, Singh D (2010) Biodiesel production through the use of different sources and characterization of oils and their esters as the substitute of diesel: a review. Renew Sustain Energy Rev 14:200–216

Biodiesel Production from *Moringa oleifera* Oil and Its Characteristics as Fuel in a Diesel Engine

Mohd. Zeeshan, Mohit Vasudeva and A.K. Sarma

Abstract Biodiesel is emerging as a promising source for alternative energy, owing to the increasing demand for fuel derived from fossil fuel resources. For developing countries like India, biodiesel can help reduce the dependency on the fossil fuel being imported. Present study is focused on the production of biodiesel from a nonedible oil source (*Moringa oleifera*) along with the engine testing of its blends as fuel. A single-step transesterification process is carried out for the extraction of biodiesel from *M. oleifera* oil with NaOH as catalyst. Properties of biodiesel extracted are tested and met the ASTM D6751 and EN 14214 standards. Engine performance and exhaust emission characteristics of 10, 15, 20 (%vol.) biodiesel blends and diesel are examined on a four-stroke diesel engine at compression ratio of 16. An average increase in brake power of 2.2 % and decrease in brake thermal efficiency of 3.7 % is observed for biodiesel blends compared to diesel. Overall increase in brake specific fuel consumption of 4.1 % for biodiesel blends is observed compared to diesel. On an average carbon monoxide and hydrocarbon emissions of biodiesel blends decreased by 4.9 and 74.4 %, respectively, compared to diesel. Carbon dioxide and NO_x emissions of biodiesel blends increased by 19.5 and 13.9 %, respectively, on an average compared to diesel. *M. oleifera* can be considered as a potential feedstock for biodiesel production and used as diesel engine fuel without making any engine modifications.

Keywords *Moringa oleifera* · Transesterification · Engine performance · Exhaust emissions

Mohd. Zeeshan (✉)
Mechanical Engineering Department, Lovely Professional University, Phagwara, India
e-mail: er.mdzeshan@gmail.com

Mohit Vasudeva
Mechanical Engineering Department, Baddi University of Emerging Sciences and Technology, Baddi, India

A.K. Sarma
R&D-I, Sardar Swaran Singh National Institute of Renewable Energy, Kapurthala, India

S. Kumar et al. (eds.), *Proceedings of the First International Conference on Recent Advances in Bioenergy Research*, Springer Proceedings in Energy,
DOI 10.1007/978-81-322-2773-1_11

1 Introduction

It is the utmost requirement of the time to develop an alternative to the fossil-based petroleum fuels as the dependency on fossil fuels is increasing exponentially with the increase in fossil fuel-based vehicles. The increase in fossil fuel-based vehicles is also contributing to the increase in atmospheric pollutants. Researchers all over the world have been working for decades for an alternative source of fuel that can reduce the overall dependency on petroleum fuels and also help in the reduction of the harmful engine emissions. Use of biodiesel and ethanol as alternative automotive fuel is quite advantageous (Kalam and Masjuki 2008; Ali et al. 2013; Giakoumis et al. 2012), as their blends do not pose any engine problem and reduces the harmful emissions. Use of 5 % ethanol blends with motor spirit is already deployed in different countries (Knothe et al. 2005). Moreover, biofuel development for an alternative and renewable energy source for automobiles has become important in the national effort for maximum self-reliance.

Moringa oleifera is an indigenous oilseed tree of Sub-Himalayan regions of northwest India, Africa, Arabia, and Southeast Asia, the Pacific and Caribbean islands, and South America (Rashid et al. 2008). It is now also distributed in the Philippines, Cambodia, and Central and North America (Rashid et al. 2008). In India, cultivation of Moringa is mainly in the southern states of Tamil Nadu, Karnataka, Kerala, and Andhra Pradesh with Andhra Pradesh leading in production (Rajangam et al. 2001). *M. oleifera* has an oil content of nearly 35 % by weight (Azam et al. 2005). Oil from *M. oleifera* is obtained by cold pressing of seeds in an oil expeller machine. The oil of *M. oleifera* is referred to as "ben oil" due to its behenic acid contents that gives the oil a specific order of behenic acid, which also provides resistance to oxidation for oil. Various studies have been conducted so far to assess the properties of biodiesel derived from various resources such as palm, pongamia pinnata, jatropha curcas, rice bran, rapeseed/canola, sunflower, safflower, etc. and their effects on diesel engines when used fully or partially as fuel with petro-diesel (Mofijur et al. 2014; Kaya et al. 2009; Kalbande et al. 2008; Qiu et al. 2011; Vasudeva et al. 2013; Yuan et al. 2008).

The present study work is focused on extraction of biodiesel from *M. oleifera* oil through transesterification process and evaluating the performance and exhaust emission characteristics of its biodiesel blends with diesel on a stationary four-stroke diesel engine at a compression ratio of 16.

2 Materials and Methods

Crude *M. oleifera* oil, procured from AOS Products Private Ltd., Ghaziabad is used as a feedstock for biodiesel/methyl ester extraction. The physicochemical properties of the oil had been shown in Table 1. All the necessary instruments, chemicals, and

Table 1 Physical and chemical properties of *Moringa oleifera* oil

Properties	
C16:0 Palmitic acid (saturated) (%)	6.5
C18:0 Stearic acid (saturated) (%)	6.0
C18:1 Oleic acid (monosaturated) (%)	72.2
C18:2 Linolic acid (polysaturated) (%)	1.0
C20:0 Arachidic acid (saturated) (%)	4.0
C20:1 Eicosenoic acid (monounsaturated) (%)	2.0
C22:0 Behenic acid (saturated) (%)	7.1
Other fatty acids (%)	1
FFA content (%)	1.66
Density at 15 °C (g/cc)	0.92615
Kinematic viscosity at 40 °C (cSt)	35.2567
Flash point (°C)	153

experimental test setup are generously provided by Sardar Swaran Singh National Institute of Renewable Energy, Kapurthala, Punjab.

Methyl esters of *M. oleifera* oil is extracted by a single-step transesterification process using a Reactor Ready setup RS27 standard. A molar ratio of oil to methanol (1:6) is used for the carrying out the transesterification reaction in the presence of NaOH (1 %wt.) as alkali catalyst. Transesterification reaction is carried out at a temperature of 60° for a time of 2 h with constant stirring at a rate of 500 rpm. After transesterification reaction the mixture is allowed to settle in a separating funnel. After a period of 20 h glycerol is settled at the bottom of the separating funnel separating the biodiesel/methyl ester on top of it. It is then washed with warm distilled water for 3–4 times, then dried in rotary evaporator and filtered using a filter paper. Properties of the *M. oleifera* methyl esters (MOME) after washing and drying are determined and are shown in the Table 2.

Blends of *M. oleifera* methyl esters extracted are further designated as B10 for 10 % by volume of MOME, B15 for 15 % by volume of MOME and B20 for 20 % by volume of MOME in the blend. B0 is designated for neat diesel.

The *M. oleifera* Methyl Ester (MOME obtained after the process was tested for the percentage conversion of oil to biodiesel in a Gas Chromatography System 7890A, manufactured by Agilent technologies, U.S.A. and the system conforms

Table 2 Properties of *M. oleifera* methyl esters (MOME)

Property parameters	MOME	ASTM D6751	EN 14214
Calorific value (MJ/kg)	40.05	–	–
Cetane number	56.3	>47	>51
Density (g/cc)	0.87327	–	0.86–0.9
Flash point (°C)	150.5	>130	>120
Fire point (°C)	155.5	–	–
Kinematic viscosity (in cSt)	4.941	1.9–6	3.5–5

Table 3 Engine specifications

Engine	Make Kirloskar, single cylinder, four-stroke diesel engine
Power rating	3.5 kW
Engine speed	1500 rpm
Cylinder bore	87.50 mm
Stroke length	110.00 mm
Swept volume	661.45 cc
Compression ratio	16
Dynamometer	Type eddy current (water cooled) with loading unit
Load indicator	Digital supply 230 VAC
Rotameter	Engine cooling 40–400 LPH
Software	"EnginesoftLV" engine performance analysis software

with ASTM D6584 for testing. Conversion efficiency of biodiesel was measured in terms of C17:0 and was found to be 68.146 %.

The experimental test setup consists of a computerized Kirloskar-make single cylinder four-stroke diesel engine connected to an eddy current (water cooled) type dynamometer for loading. It is equipped with instruments and sensors for measurement of cylinder pressure, crank angle, fuel flow, airflow, temperature, and load applied. EnginesoftLV software is used for online study of engine performance parameters such as brake power, brake mean effective pressure, brake thermal efficiency, specific fuel consumption. Table 3 shows the engine specifications. An exhaust gas analyzer AVL DiGas 444 manufactured by AVL India Private Ltd., is used to record the exhaust emission characteristics of the engine during testing process.

3 Results and Discussion

Engine performance and exhaust emission characteristics of a four-stroke diesel engine are evaluated at different load conditions with B0, B10, B15, and B20 as fuel.

3.1 Engine Performances Characteristics

Engine performance parameters in terms of brake power, brake thermal efficiency, and brake specific fuel consumption are evaluated with increase in engine load conditions.

Figure 1 shows the brake power output of the engine for different blends with increasing load conditions. Brake power gives the measure of the net power

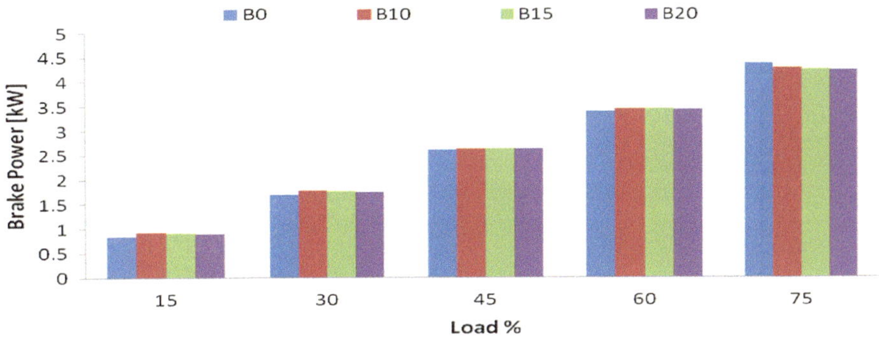

Fig. 1 Variation of brake power with load

delivered by the engine at the crank shaft measured with the help of a dynamometer. Results show an overall 3.2 % increase in brake power for B10, 1.9 % increase for B15 and 1.4 % increase for B20 compared to that of diesel. Figure 2 shows the variation of brake thermal efficiency with load for biodiesel blends and diesel. Brake thermal efficiency is defined as brake power of an engine as a function of the thermal input from the fuel and is used to evaluate the conversion of heat from a fuel to mechanical energy in an engine.

An average decrease of 5.8 % in brake thermal efficiency is observed for B10 compared to diesel. Whereas B15 and B20 have an average decrease of 4.2 and 1.3 % in brake thermal efficiency compared to diesel. Higher brake power along with lower brake thermal efficiency results in higher specific fuel consumption. Results showing the variation in brake specific fuel consumption (BSFC) with load for biodiesel blends and diesel is shown in Fig. 3. Brake specific fuel consumption is the ratio between mass fuel consumption and brake power. It is a type of comparison ratio which looks at an engine's fuel efficiency in terms of amount of fuel the engine consumes versus the power it produces. A nominal 0.25 % increase in BSFC for B10 is observed compared to diesel. Whereas an average increase of 9.7

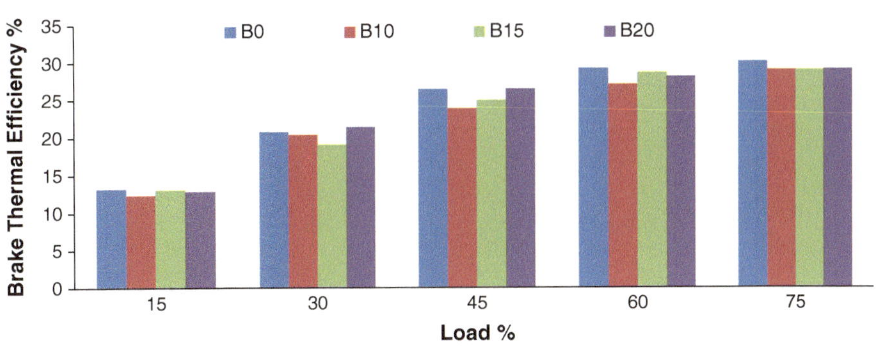

Fig. 2 Variation of brake thermal efficiency with load

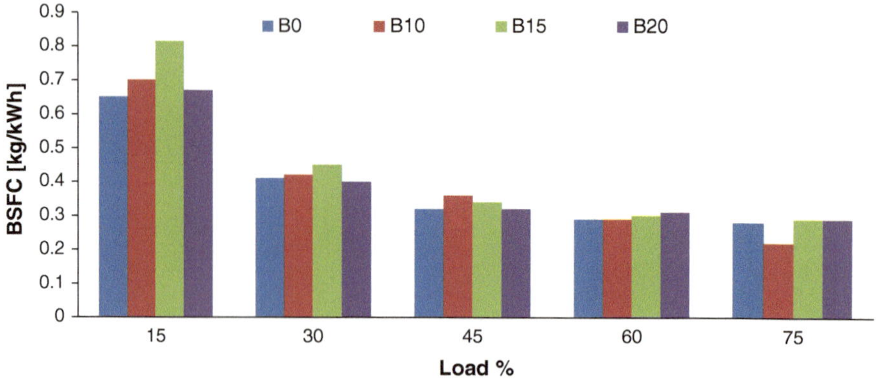

Fig. 3 Variation of brake specific fuel consumption with load

and 2.2 % for B15 and B20, respectively, is observed in comparison to diesel. Higher BSFC and lower brake thermal efficiency are due to the lower energy content of *M. oleifera* biodiesel and its blends. The decrease in brake thermal efficiency can also be attributed to the significantly lower efficiencies of biodiesel especially at lower compression ratios due to a reduction in the calorific value and an increase in fuel consumption as compared to B0.

3.2 Exhaust Emissions Characteristics

Exhaust emission characteristics as carbon monoxide (CO), carbon dioxide (CO_2), hydrocarbons (HC), and nitrogen oxide as NO for biodiesel blends and diesel are evaluated at different load conditions.

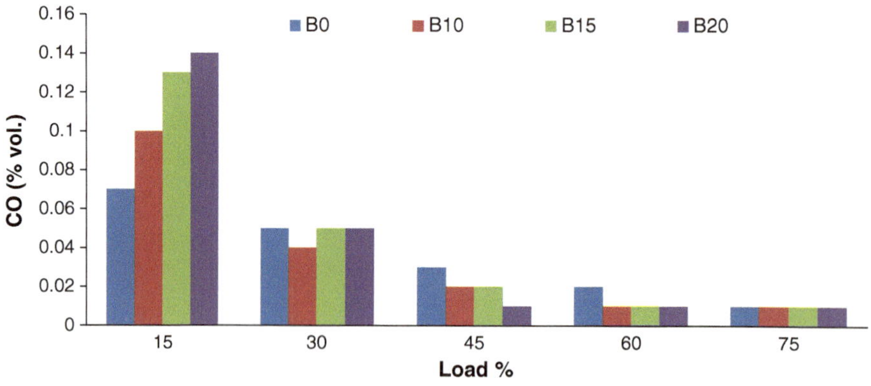

Fig. 4 Variation of carbon monoxide emission with load

Figure 4 shows the variation in carbon monoxide emissions of the engine with load for biodiesel blends and diesel. Results show an overall decrease of 12.1 % in CO emissions for B10, a nominal 0.4 % increase for B15, and 3.3 % decrease for B20 compared to diesel. Reduction in CO emissions may be attributed to the higher oxygen content in biodiesel resulting in better combustion of fuel particles. Also due to the higher cetane number of *M. oleifera* biodiesel, the probability of fuel-rich zones formation decreases, which results in lower CO emissions of biodiesel blends.

Figure 5 shows the variation of the hydrocarbon emissions of engine with load for different biodiesel blends and diesel. Unburnt hydrocarbons in emissions are due to incomplete or improper combustion of fuel in the engine. Results show an overall decrease in HC emissions of 69.7, 76.8, and 76.7 % for B10, B15, and B20, respectively, compared to that of diesel. Improved combustion of biodiesel blends due to the presence of excess oxygen content can be attributed to the reduction in its hydrocarbon emission. Variation in carbon dioxide emissions per unit brake power with load for biodiesel blends and diesel is shown in Fig. 6. An increase in CO_2 emissions for biodiesel blends is observed. CO_2 emissions pose a threat to the environment as it a major constituent for the cause of global warming. But as products of combustion it is more acceptable compared to CO which is a result of incomplete combustion. Results show an overall increase of 16.5 and 39 % in CO_2 emissions per unit brake power for B15 and B20, respectively, compared to diesel. A higher CO_2 emission for biodiesel blends is a result of complete combustion of fuel. Figure 7 shows the variation of emissions of nitrogen oxide as from the engine with load for different biodiesel blends and diesel. Oxide of Nitrogen is highly dependent on the temperature and pressures inside the combustion chamber. Exhaust NO for biodiesel blends increased as compared to that of diesel for all range of loading conditions. There was 5.6 % increase in NO emission for B10, 12.5 % increase for B15, and 23.8 % increase for B20 compared to that of diesel.

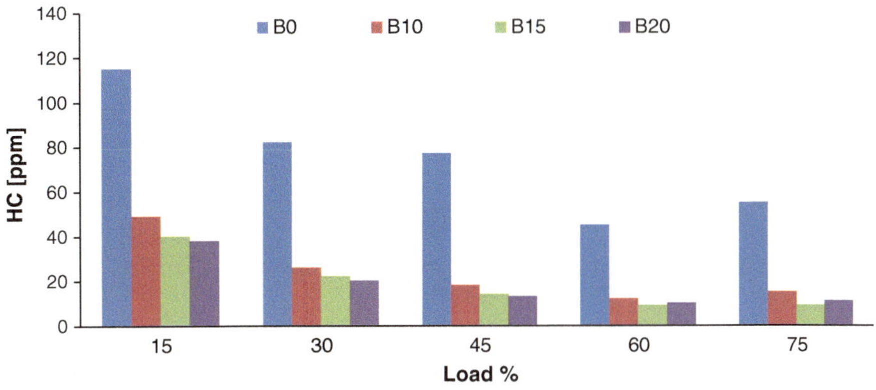

Fig. 5 Variation of hydrocarbon emission with load

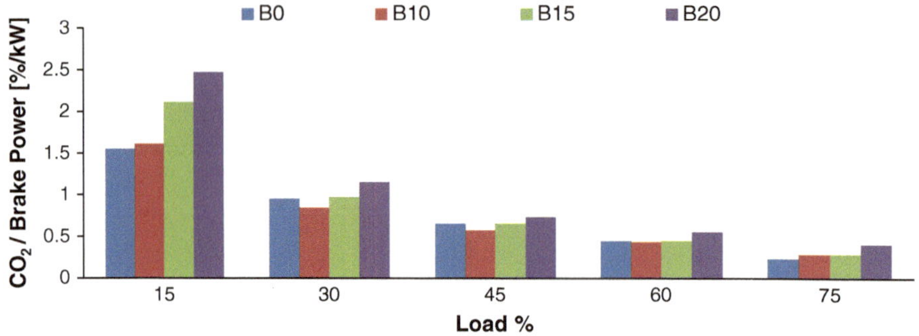

Fig. 6 Variation in carbon dioxide emissions per unit brake power with load

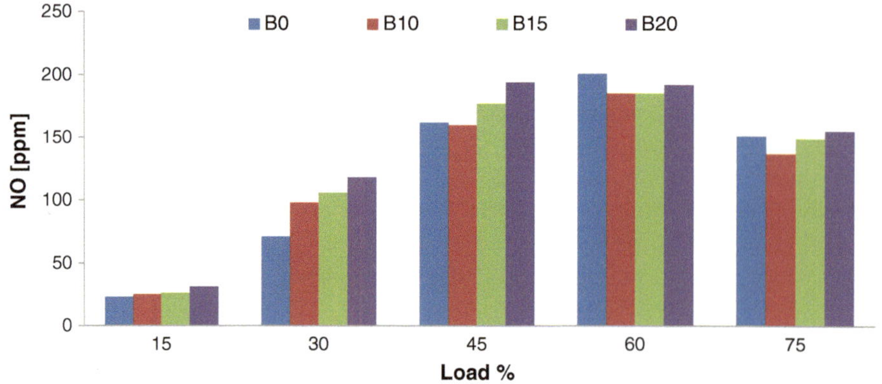

Fig. 7 Variation in emissions as nitrogen oxides as NO with load

4 Conclusions

Following conclusions are drawn from the study:

- In the present study, *M. oleifera* oil shows the potential to be a feedstock for biodiesel production and the performance of blends up to B20 are closely comparable with that of diesel.
- An average increase of 2.2 % in brake power of engine is observed compared to diesel. In addition, average decrease in brake thermal efficiency of 3.7 % is observed for biodiesel blends with an overall increase in brake specific fuel consumption of 4.1 % compared to diesel.
- The use of the biodiesel blends shows effective reduction of 4.9 % in CO emissions and 74.4 % reduction in HC emissions on an average for different loading conditions.

- As a result of improved combustion of biodiesel blends and higher temperature inside the combustion chamber, an increase of 19.5 and 13.9 % is observed for CO_2 and NO emission.

Blends of biodiesel/methyl esters extracted from *M. oleifera* oil can be used as fuel in diesel engine without making any engine modification.

Acknowledgments The authors would like to acknowledge the support of Sardar Swaran Singh National Institute for Renewable Energy for providing resources to conduct the research work. The authors would also like to thank Mr. Balwant Singh and Ms. Neetu Singh for their technical assistance in carrying out the research work.

References

Ali MH, Mashud M, Rubel MR, Ahmad RH (2013) Biodiesel from Neem oil as an alternative fuel for Diesel engine. Procedia Eng 56:625–630

Azam MM, Waris A, Nahar NM (2005) Prospects and potential of fatty acid ethyl esters of some non-traditional seed oils for use as biodiesel in India. Biomass Bioenergy 29:293–302

Giakoumis EG, Rakopoulos CD, Dimaratos AM, Rakopoulos DC (2012) Exhaust emissions of diesel engines operating under transient conditions with biodiesel fuel blends. Prog Energy Combust Sci 38:691–715

Kalam MA, Masjuki HH (2008) Testing palm biodiesel and NPAA additives to control NOx and CO while improving efficiency in diesel engines. Biomass Bioenergy 32:1116–1122

Kalbande SR, More GR, Nadre RG (2008) Biodiesel production from non-edible oils of Jatropha and Karaja for utilization in Electric Generator. Bioenerg Res 1:170–178

Kaya C, Hamamci C, Baysal A, Akba O, Erdogan S, Saydut A (2009) Methyl ester of peanut (Arachis hypogea L.) seed oil as a potential feedstock for biodiesel production. Renew Energy 34:1257–1260

Knothe G, Gerpen JV, Krahl J (2005) The biodiesel handbook. AOCS Press Champaign, Illinois, USA

Mofijur M, Masjuki HH, Kalam MA, Atabani AE, Fattah IMR, Mobarak HM (2014) Comparative evaluation of performance and emission characteristics of *M. oleifera* and Palm oil based biodiesel in a diesel engine. Ind Crops Prod 53:78–84

Qiu F, Li Y, Yang D, Li X, Sun P (2011) Biodiesel production from mixed soybean oil and rapeseed oil. Appl Energy 88:2050–2055

Rajangam J, Azahakia Manavalan RS, Thangaraj T, Vijayakumar A, Muthukrishan N (2001) Status of production and utilisation of moringa in southern India. Development potential for Moringa products, Dar es Salaam, Tanzania

Rashid U, Anwar F, Moser BR, Knothe G (2008) Moringa oleifera oil: a possible source of biodiesel. Bioresour Technol 99:8175–8179

Vasudeva M, Sharma S, Mohapatra SK, Kundu K (2013) Optimizing the compression ratio of compression ignition engine fuelled with esters of crude rice bran oil. Int J Theor Appl Res Mech Eng 2(3):131–134

Yuan X, Liu J, Zeng G, Shi J, Tong J, Huang G (2008) Optimization of conversion of waste rapeseed oil with high FFA to biodiesel using response surface methodology. Renew Energy 33:1678–1684

Growth and Lipid Production from *Scenedesmus* sp. Under Mixotrophic Condition for Bioenergy Application

Monika Prakash Rai and Shivani Gupta

Abstract Microalgae are seen as an alternative and renewable feedstock for biodiesel production. They are effective in producing large amount of oil which can be converted into biodiesel by the process of transesterification. The present work aims to cultivate microalgae under optimized culture condition and lipid was extracted for the production of fatty acid methyl esters (FAMEs). The biomass and lipid content of the species was tried to enhance under photoheterotrophic (mixotrophic) condition by adding glycerol as the organic carbon source. Glycerol is the by-product in transesterification reaction for biodiesel production, its utilization in algae cultivation will support the biorefinery system. The green microalgae, *Scenedesmus* sp. was cultured in Fogg's medium and culture conditions were optimized for best growth. The algae were allowed to grow mixotrophically with glycerol at concentrations ranging from 0–5 % (v/v) and significant improvement was noticed in growth and lipid accumulation. The growth of the *Scenedesmus* sp. was increased up to 1 % glycerol (v/v) and recorded 0.414 g l^{-1} that is approximately 2 times in comparison with photoautotrophic cultivation (0.223 g l^{-1}). The lipid accumulation of 36.47 % was estimated in cultures with 1 % glycerol which is, nearly, 3 times that of autotrophic culture (12.55 %). Maximum lipid accumulation (52.32 %) was achieved under mixotrophic cultivation using 5 % glycerol, but growth is inhibited at very high concentration. Oil from *Scenedesmus* sp. grown under 1 % glycerol was subjected to acid-based transesterification for its proposed application in biofuel. The lipid enhancement in *Scenedesmus* sp. in the presence of glycerol provides a potential route for economic biodiesel production.

Keywords *Scenedesmus* sp. · Mixotrophic condition · Glycerol · FAME

M.P. Rai (✉) · Shivani Gupta
Amity Institute of Biotechnology, Amity University Uttar Pradesh, Sector 125, Gautam Budh Nagar, Noida, India
e-mail: mprai@amity.edu; monika1778@gmail.com

S. Kumar et al. (eds.), *Proceedings of the First International Conference on Recent Advances in Bioenergy Research*, Springer Proceedings in Energy,
DOI 10.1007/978-81-322-2773-1_12

1 Introduction

Energy is the prerequisite for survival, sustainability, and growth of any country, its economy and population. Increasing industrialization creates a burden on the fossil fuels as they are limited owing to their nonrenewable nature. This has created various issues of global concern like increase in crude oil prices and global warming due to burning of fossil fuels in different forms, liquid (oil) and gas (Malcata 2011). This motivates researchers across the globe to quest for alternative sources of energy which are renewable and grown on industrial scale to meet energy demands (Chisti 2007a). Biodiesel, as word signifies, is a diesel from a biological entity that replaces conventional diesel and has gained enormous popularity in recent years (Miao and Wu 2006). Among all candidates, viz., plant sources, oil from animal tissues, and algae, microalgae have been proved to be the most efficient one (Chisti 2007a). They have many advantages as compared to other feedstock like less land requirement and per land usage productivity is more, can make use of arable land, ease of culture and harvest, can feed on waste water, majority of industrial effluent, flue gas, and mitigate extra atmospheric CO_2 to keep up the healthy balance (Li et al. 2014; Smith et al. 2009; Liu et al. 2008; Chisti 2007b). The main challenge in algae biofuel production is to increase cellular lipid accumulation to make this feedstock more economically viable and cost-effective (Bhatnagar et al. 2011). Many researchers have tried different approaches to enhance lipid accumulation like applying variety of culture conditions, CO_2 bubbling in the culture growing photoautotrophically to enhance biomass (Mata et al. 2010), photoperiods (Perez and Izquierdo 2011), and light and salt stress in various strains (Mohan and Devi 2014). According to Marquez et al. (1993), photoheterotrophic growth rate is a combination of autotrophic and heterotrophic growth rates. There have been reports in which some microalgae yield 3–10 times higher biomass when grown mixotrophically (Li et al. 2014; Bhatnagar et al. 2011).

Scenedesmus sp. is known for its high growth rate and lipid content; it can achieve higher lipid accumulation under mixotrophic condition using glycerol, glucose, acetate, and others (Andruleviciute et al. 2014; Dittamart et al. 2014; Makereviciene et al. 2012). The lipid content of *Scenedesmus quadricauda* showed excellent enhancement when grown mixotrophically (33.1 % dry cell weight) compared to grown photoautotrophically and heterotrophically (14–28 %) (Zhao et al. 2012). The use of glycerol increases the usage of resources in the algal oil process more as it is the by-product of transesterification process; so recycling makes process more economically feasible. The present study aims to analyze the most effective concentration of glycerol to obtain high biomass and increased lipid accumulation in *Scenedesmus* sp. for biodiesel application.

2 Materials and Methods

2.1 Culture Maintenance

The axenic culture of *Scenedesmus* sp. was maintained in a batch mode in a 1 l Erlenmeyer flask by taking 300 ml of Fogg's media including 0.2 g $MgSO_4 \cdot 7H_2O$, 0.1 g $CaCl_2 \cdot 2H_2O$, 0.2 g K_2HPO_4, and 0.0745 g Na_2EDTA and trace elements 2.86 mg H_3BO_3, 1.81 mg $MnCl_2 \cdot 4H_2O$, 0.222 mg $ZnSO_4 \cdot 7H_2O$, 0.39 mg $Na_2MoO_4 \cdot 2H_2O$, 0.0494 mg $Co(NO_3)_2 \cdot 6H_2O$, 0.0557 mg $FeSO_4 \cdot 7H_2O$, and 0.079 mg $CuSO_4 \cdot 5H_2O$ in 1000 ml distilled water (Nigam et al. 2011). The microalgae were cultured in a temperature controlled incubator at 25 °C by providing 24 h continuous illumination (40 W, white tube light) of 3000 lx light intensity with intermittent shaking at 120 rpm. Exponentially growing cells were inoculated by keeping the optical density 0.1 O.D. This was referred to as autotrophic control culture.

2.2 Mixotrophic Culture

The *Scenedesmus sp.* was grown in Fogg's media supplemented with different concentrations of technical glycerol, viz., 0, 0.5, 1, 3, and 5 % (v/v). Other media components were kept the same as in autotrophic cultures.

2.3 Growth Curve Analysis

2.3.1 Determination of Dry Weight

The optical density of the *Scenedesmus* cultures was measured spectrophotometrically (UV/Vis Shimadzu) at 660 nm after every 24 h for both autotrophic and photoheterotrophic cultures. After stationary phase, photoheterotrophic cultures with different concentrations of glycerol were harvested by centrifugation (Eppendorf 5810R) at 4000 rpm for 15 min. The pellet was then washed twice with double distilled water and dried in oven at 105 °C to obtain dry cell mass.

2.3.2 Determination of Chlorophyll Content

Chlorophyll content of the microalgae was measured at alternate days to estimate cell growth. 10 ml of homogenized algal suspension was taken and centrifuged at 4000 rpm for 10 min. Chlorophyll was extracted by using the same volume of methanol (95 %) in a water bath at 60 °C for 30 min. Centrifugation was performed to discard the pellets and the absorbance of supernatant was taken at 650 and 665 nm against 95 % methanol in blank (Mackinney 1941).

$$\text{Chlorophyll content}(\mu g \backslash ml) = \{2.55 \times 10^{-2} OD\,(650\,nm) + 0.4 \times 10^{-2} OD\,(665\,nm)\} \times 10^{3}$$

2.4 *Lipid Estimation*

The total lipid was extracted by mixing chloroform: methanol (2:1 v/v) with dried algal samples using Bligh and Dyer method (Bligh and Dyer 1959).

$$\text{Lipid accumulation}(\%\,DCW) = \text{lipid production}\,(gl^{-1}) / (\text{biomass production}\,(gl^{-1}) \times 100$$

2.5 *Production of FAME by Transesterification*

The extracted lipid was transesterified by using methanol and a mixture of acidic catalyst (Rai et al. 2013). The method was used for transesterification of dried lipid extract to obtain the fatty acid methyl esters (FAME). The FAME thus produced was preserved for further analysis.

2.6 *Statistical Analysis*

All the experiments were performed in triplicates for better validation. Also they were repeated in lieu to ensure reproducibility of the results and the average value has been recorded and represented in the results.

3 Result and Discussion

The present work was aimed to explore the efficacy of microalgae *Scenedesmus* sp. for biodiesel production. The work was focused to assess the maximum lipid as well as biomass production of alga under different culture conditions and its efficient conversion to FAME. Apart from lipid, the amount of producible biomass is one of the major factors to determine the potentiality of algae for bioenergy feedstock, and it is necessary to provide growth conditions which favor high biomass content (Mandotra et al. 2014). Cellular growth and lipid productivity of *Scenedesmus* sp. were investigated under various conditions. Biomass and oil production were evaluated under photoautotrophic and photoheterotrophic conditions using Fogg's medium and organic carbon glycerol.

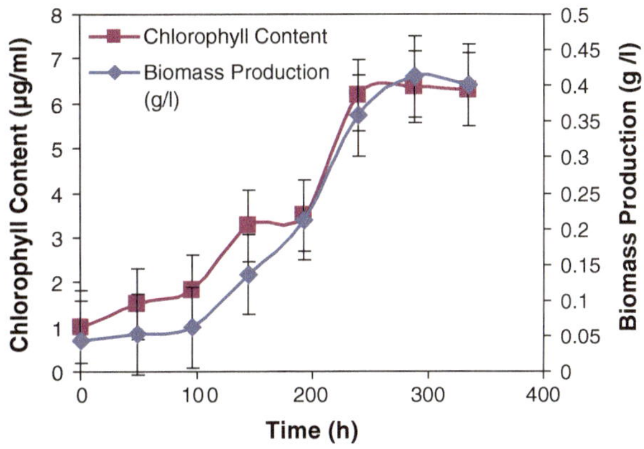

Fig. 1 Growth curve of *Scenedesmus* sp. in Fogg's medium (by estimating dry cell weight and chlorophyll content)

3.1 Growth Curve Analysis

The microalgae were grown under photoautotrophic condition using normal Fogg's medium including major macro- and microelements (Nigam et al. 2011). The *Scenedesmus sp.* growth pattern in Fogg's medium is shown in Fig. 1. During lag phase (up to 72 h) the growth of microalgae is very little and microorganisms adapted to growth conditions. Cellular growth between fourth and sixth days (i.e., by 144 h) increases sharply that strongly express the exponential phase. The maximum concentration was achieved after 264 h (beginning of stationary phase) and reached 0.23 g l^{-1}. The growth curve plotted by taking chlorophyll content of algae is showing parity with the curve plotted by taking dry cell weight (Fig. 1).

3.2 Effect of Glycerol on Algae Growth

To determine the effect of organic carbon source on the growth of *Scenedesmus* sp., glycerol was added at different concentrations 0, 0.5, 1.0, 3.0, and 5.0 % (v/v). The growth curve was plotted in mixotrophic condition at various concentrations of glycerol as given in Fig. 2. During the first six days (up to 144 h), the growth of algae in various concentration of glycerol was very similar (Fig. 2). The fastest growth was obtained during the exponential phase in the culture containing 1 % (v/v) glycerol. The highest concentration obtained after stationary phase (after 288 h) of the media containing 1 % volume fraction of glycerol was 0.42 g l^{-1}. This concentration is almost 2 times higher than the cells obtained in photoautotrophic control, which was 0.23 g l^{-1}. Results show that, under photoheterotrophic conditions (using glycerol), growth of *Scenedesmus* sp. is much better compared to autotrophic control. In mixotrophic cultivation, the microalgae could obtain energy from both photosynthesis and oxidation of organic carbon compounds. Thus, they could utilize some part of the energy for cell division, but the excess energy was

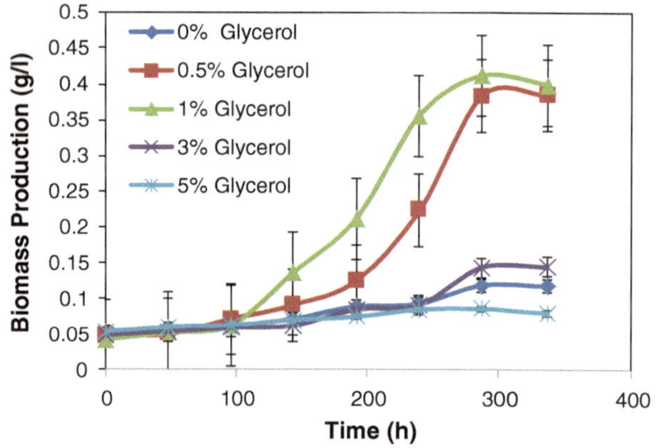

Fig. 2 Effect of concentration of glycerol (0, 0.5, 1.0, 3.0, and 5.0 %) on growth of *Scenedesmus* sp.

stored in the form of lipid granules. Similar kind of results was also obtained by researchers using glycerol to increase the growth of *Scenedesmus* sp. Makarevičienė et al. (2012) have achieved 1.3 times higher growth of alga under mixotrophic condition using 5 g l^{-1} technical glycerol compared to autotrophic growth.

Microalgae growth was not obtained well in all concentrations of glycerol. The algal growth was lower in the growth medium containing more than 1 % volume fraction of the glycerol. High concentration of glycerol, i.e., 3 and 5 % showed an inhibitory effect on biomass production. In our previous research, *Chlorella pyrenoidosa* showed best growth at 0.5 % (v/v) glycerol and was inhibited above this concentration (Rai et al. 2013). Various microalgae have different levels of inhibitory concentration of carbon. After a certain concentration, growth decreases with increasing carbon concentration. Impurities such as free fatty acids, residue of catalyst, soap, etc., could be the possible reason for growth inhibition (Liang et al. 2010).

3.3 *Effect of Glycerol Concentration on Lipid Content of Scenedesmus sp.*

The lipid content obtained under photoautotrophic and with glycerol under photoheterotrophic conditions was estimated. Accumulation of lipid in microalgae cells is improved by applying photoheterotrophic conditions (in the presence of glycerol). The cellular lipid content increases by increasing the amount of glycerol as given in Fig. 3. Under photoheterotrophic cultivation with the use of glycerol, microalgae have increased lipid content approximately 3 times, i.e., from 12.55 to 36.47 % along with double increase in biomass concentration (g l^{-1}) as compared to photoautotrophic condition. The lipid content of 36.47 % is achieved in the growth medium containing 1 % (v/v) glycerol, the lowest lipid concentration (13.02 %) is

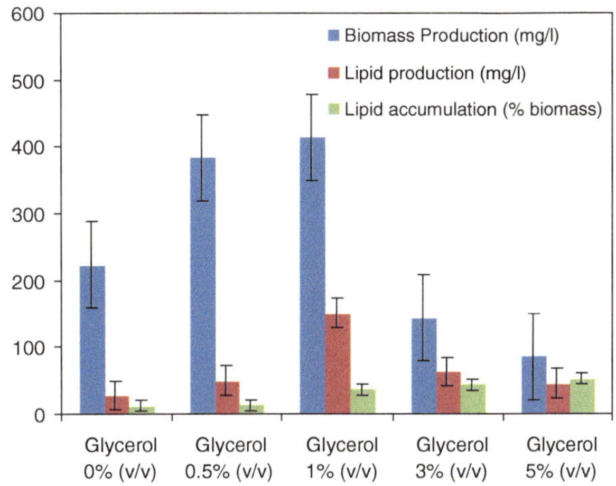

Fig. 3 Effect of glycerol concentration (0, 0.5, 1.0, 3.0, and 5.0 %) on biomass production (mg l^{-1}), lipid production (mg l^{-1}), and lipid accumulation (% biomass) of *Scenedesmus* sp.

estimated in the growth medium containing 0.5 % (v/v) glycerol under photo-heterotrophic cultivation.

The lipid content (% biomass) of microalgae in 3 and 5 % (v/v) glycerol is higher than the content in 1.0 % glycerol, as the lipid accumulation is higher through carbon dependent metabolic pathways; but showed very poor effect on biomass concentration (Fig. 3). Hence, the overall productivity will be the highest at 1 % volume fraction of glycerol. Some early results also showed compatibility of mixotrophic culture condition using glycerol for lipid enhancement in microalgae like *Chlorella* sp., *Scenedesmus* sp., *Chodatella* sp., etc., (Rai et al. 2013; Makarevičienė et al. 2012; Li et al. 2014). Crude glycerol is proved to be an effective carbon source to enhance growth and cellular lipid accumulation in algae and also it provides an effective way for disposal of surplus glycerol in the production of biodiesel.

3.4 *Transesterification of Algal Oil*

The algae were cultivated in 5 l Erlenmeyer flask and 30.127 g wet biomass was collected and dried to produce 4.845 g dry cells. Lipid was extracted by means of solvent extraction using chloroform and methanol (2:1) and 1.406 g of lipid produced from *Scenedesmus* sp. was converted to fatty acid methyl esters (FAME) by the method of transesterification. The total yield of fatty acid from 1.406 g cells was 1.052 g that shows more than 75 % recovery (Table 1).

Table 1 Production of FAME from *Scenedesmus* sp. by transesterification

Wet weight (g)	Biomass production (g)	Lipid production (g)	FAME (g)
30.127	4.845	1.406	1.052

4 Conclusion

Growth and lipid production of freshwater green microalgae *Scenedesmus* sp. was investigated using Fogg's medium for photoautotrophic and glycerol for photo-heterotrophic cultivation. Utilization of glycerol in the cultivation of microalgae, that is a by-product of biodiesel industry will help not only in the removal of waste of the environment, but also makes the biofuel production sustainable. Our results show that the maximum biomass production is obtained under photoheterotrophic conditions by adding 1 % by volume fraction of glycerol into the Fogg's medium. The presence of organic carbon source (glycerol) in the growth medium increases the growth of microalgae but the growth is inhibited at very high concentration of glycerol. Cultivation of *Scenedesmus* sp. under photoheterotrophic condition not only shows better growth but also higher lipid accumulation. The highest lipid accumulation was achieved in the medium containing 5 % glycerol but growth decreases. Hence, the cells grown under 1 % volume fraction (v/v) of glycerol showed highest overall productivity in terms of both biomass and lipid and the process could be utilized for the formation of fatty acid methyl esters. The FAME produced in the present research will be characterized thoroughly for its bioenergy applications.

Acknowledgments The present work is financially supported by a project grant (Ref.No. DST/TSG/AF/2009/101) from Department of Science and Technology, Govt. of India, New Delhi. Authors are grateful for the financial support.

References

Andruleviciute V, Makareviciene V, Skorupskaite V, Gumbyte M (2014) Biomass and oil content of *Chlorella* sp., *Haematococcus* sp., *Nannochloris* sp. and *Scenedesmus* sp. under mixotrophic growth conditions in the presence of technical glycerol. J Appl Phyco 26 (1):83–90

Bhatnagar A, Chinnasamy S, Singh M, Das KC (2011) Renewable biomass production by mixotrophic algae in the presence of various carbon sources and wastewaters. Appl Energ 88:3425–3431

Bligh EG, Dyer WJ (1959) A rapid method of total lipid extraction and purification. Cann J Biochem Phy 37:911–917

Chisti Y (2007a) Biodiesel from microalgae. Biotechnol Adv 25:294–306

Chisti Y (2007b) Biodiesel from microalgae beats bioethanol. Trends Biotechnol 26:126–131

Dittamart D, Pumas C, Pekkoh J, Peerapora pisal Y (2014) Effects of organic carbon source and light-dark period on growth and lipid accumulation of *Scenedesmus* sp. AARL G022. Maejo. Int J Sci Technol 8(02):198–206

Li YR, Tsai WT, Hsu YC, Xie MZ, Chen JJ (2014) Comparison of autotrophic and mixotrophic cultivation of green microalgal for biodiesel production. Energy Procedia 2:371–376

Liang Y, SarkanY N, Cui Y, Blackburn JM (2010) Batch stage study of lipid production from crude glycerol derived from yellow grease or animal fats through microalgal fermentation. Bioresour Technol 101:6745–6750

Liu ZY, Wang GC, Zhou BC (2008) Effect of iron on growth and lipid accumulation in *Chlorella vulgaris*. Bioresour Technol 99:4717–4722

Mackinney G (1941) Absorption of light by chlorophyll solutions. J Biol Chem 140:315–322

Malcata FX (2011) Microalgae and biofuels: a promising partnership. Trends Biotechnol 29:542–549

Mandotra SK, Kumar P, Suseela MR, Ramteke PW (2014) Fresh water green microlaga Scenedesmus abundans: a potential feedstock for high quality biodiesel production. Bioresour Technol 156:42–47

Mata TM, Martins AA, Caeteno NS (2010) Microalgae for biodiesel production and other application: a review. Renew Ener Rev 14:217–232

Miao XL, Wu QY (2006) Biodiesel production from heterotrophic microalgal oil. Bioresour Technol 197:841–846

Makarevciene V, Skorupskaite V, Andruleviciute V (2012) Biomass and oil production of green microalgae *Scenedesmus* sp. using different nutrients and growth, Env Res, Eng Manag, 4 (62):5–13

Mohan SV, Devi MP (2014) Salinity stress induced lipid synthesis to harness biodiesel during dual mode of cultivation of mixotrophic microalgae. Bioresour Technol 165:288–294

Marquez FJ, Sasaki K, Kakizono K, Nishio N, Nagai S (1993) Growth characteristics of *Spirullina platensis* in mixotrophic and heterotrophic conditions. J Ferment Bioeng 76:408–410

Nigam S, Prakash Rai M, Sharma R (2011) Effect of nitrogen on growth and lipid content of *Chlorella pyrenoidosa*. Am J Biochem Biotechnol 7(3):126–31

Perez-pazos JV, Izquierdo F (2011) Synthesis of neutral lipids in *Chlorella sp.* under different light and carbonate conditions. Ciencia Technol Futuro 4:47–58

Rai MP, Nigam S, Sharma R (2013) Response of growth and fatty acid compositions of *Chlorella pyrenoidosa* under mixotrophic cultivation with acetate and glycerol for bioenergy application. Biomass Bioenergy 58:151–57

Smith VH, Strum BS, Denoyelles FJ, Billings SA (2009) The colony of algal biodiesel production. Trends Ecol Evol 25:301–309

Zhao G, Yu J, Jiang F, Zhnag X, Tan T (2012) The effect of different trophic modes on lipid accumulation of *Scenedesmus quadricauda*. Bioresour Technol 114:466–471

Experimental Evaluation of Combustion Parameters of a DI Diesel Engine Operating with Biodiesel Blend at Varying Injection Timings

Abhishek Sharma and S. Murugan

Abstract There is a limited reserve of the petroleum fuels and the world is already facing acute energy crisis. Besides this, the petroleum fuels are the dominant global sources of carbon dioxide emission. This research work provides an overview on the usage of a blend of jatropha methyl ester (JME) with oil obtained by the pyrolysis of waste tyre in diesel engine while varying the injection timing (IT) in order to improve the combustion parameters of the engine. Experiments were conducted in a single cylinder diesel engine at the advanced (24.5 °CA bTDC), standard (23 °CA bTDC), and retarded IT (21.5 °CA bTDC) for the blend and results were compared with diesel fuel and also with JME operation. The combustion parameters of the diesel engine were calculated in terms of cylinder pressure history, cumulative heat release, maximum heat release rate, maximum cylinder pressure, and mass fraction burned. For the blend with advancement of IT, the maximum cylinder pressure and cumulative heat release rate are enhanced.

Keywords Biodiesel · Combustion · Diesel engine · Injection timing · Tyre pyrolysis oil

1 Introduction

Conventional petroleum fuels for example diesel and gasoline are the principal source of energy for the development of any country and their demand is expanding with accelerating rate due to increasing industrialization, overwhelming rise in the automobiles, and growing population. Speedy boom of fuel costs, lack of fossil fuels, and reducing fossil fuel reserves have forced to search for their substitutes which

Abhishek Sharma (✉) · S. Murugan
Internal Combustion Engines Laboratory, Mechanical Engineering Department, National Institute of Technology Rourkela, Rourkela 769008, India
e-mail: drasharma58@gmail.com

S. Kumar et al. (eds.), *Proceedings of the First International Conference on Recent Advances in Bioenergy Research*, Springer Proceedings in Energy,
DOI 10.1007/978-81-322-2773-1_13

might meet the constantly growing claim of the energy. Also, the petroleum fuels are the topmost sponsors to pollute air and main contributors of greenhouse gases, and are reflected to be the leading reason behind the climate change. Automobiles significantly contaminate the environment by emissions of pollutants like carbon monoxide (CO), carbon dioxide (CO_2), nitric oxides (NO_x), sulfur dioxide (SO_2), unburnt or partially burnt hydrocarbons, and particulates (Sher 1998). These circumstances have forced researchers and expertise to look for alternatives to fossil fuels that would preferably be economical and environmental friendly.

Biodiesel has gained a promising extender for diesel and is known as the methyl or ethyl esters of fatty acid prepared from renewable sources such as virgin and used vegetable oils (edible and non-edible) and animal fats (Knothe 2005). It is considered as clean burning and renewable fuel, and various advantages of biodiesel are nontoxic, biodegradable, which can mitigate exhaust emissions of a diesel engine. Engine fueled with biodiesel results in considerable decrease of hydrocarbons (HC), CO, and particulate matters. It is considered to have almost no sulfur, no aromatics and improves the combustion quality of the fuel due to their high cetane number. As a renewable source of supply, biodiesel imparts a promising influence on the surroundings, and it can replace the diesel fuel in compression ignition (CI) engines. Extensive consumption of biodiesel will cut back the relay on fossil fuels, which is presently the key source of energy for the world. For countries without or restricted petroleum reserves, the usage of biodiesel will cut back imports of fossil fuels, thus resulting in substantial foreign exchange savings. In India, the biodiesel production from renewable sources such as edible and nonedible vegetable oil has been established in the last three decades. Use of edible oils like soybean, sunflower oil, and coconut oil to produce biodiesel is not practical as these oils are being consumed as food in India and there is an enormous gap in their demand and supply in the country. Several researchers in India have tried different non-edible feed stocks such as Jatropha curcas (Sharma and Murugan 2014, 2015b), Pongamia pinnata (Srivastava and Verma 2008), Madhuca indica (Ghadge and Raheman 2005), Linseed caster, jojoba, etc., for the production of biodiesel. Thus in Indian conditions, solely non-edible oil seed plants can be considered for the biodiesel production.

Several studies have been carried out to investigate the potential application of tyre pyrolysis oil (TPO) as a replacement of base fuel in both petrol engines and diesel engines. The TPO shows longer ignition delay compared to diesel fuel owing to higher viscosity of the test fuel (Murugan et al. 2008). A Kirloskar diesel engine was used to examine the influence of an ignition improver (Di ethyl ether) when it was admitted at three different flow rates (65,130 and 170 g/h) (Hariharan et al. 2012). Beside this, ignition delay of TPO can be upgraded by blending it with high cetane fuels.

In an earlier investigation, reported by the authors, experiments were conducted in a diesel engine with different blends of jatropha methyl ester (JME) and TPO.

The TPO was blended with JME in the blend from 10 to 50 % in interval of 10 % on a volume basis. Experimental investigation revealed that, the combustion parameters of the engine fueled with JMETPO20 (80 % JME + 20 % TPO) blend were better than other blends considered in that study but poorer than that of diesel operation (Sharma and Murugan 2013). The inferior combustion behavior of the blend was achieved due to the relatively higher density and poor volatility of the blend. Injection timing (IT) plays a vibrant role on the behavior of a CI engine, as the pressure and temperature change significantly when piston approaches near top dead center. In the recent past, research works have been documented study of the influence of IT on the behavior of a CI engine running with different alternative fuels (Sayin and Gumus 2011; Xu et al. 2014; Sayin and Canakci 2009). This test is carried out to estimate the influence of IT on the combustion characteristics of a CI engine running on the optimum blend, i.e., JMETPO20.

2 Materials and Methods

For this experiment, the JME was collected from a commercial biodiesel plant in Raipur, India. The TPO was collected from a pilot pyrolysis plant located at Rourkela, India. The TPO was blended with the JME on a 20/80 % volume basis and the blend was kept under observation for 30 days, to ensure its stability. The details of preparation of JME and TPO have been described by the authors elsewhere (Sharma and Dhakal 2013). It was noticed that the TPO was not separated from the JME in the blend. Table 1 gives properties of diesel, JME, TPO, and the JMETPO20 blend.

Table 1 Properties of diesel, JME, TPO, and the JMETPO20 blend

Properties	ASTM test method	Diesel	JME	TPO	JMETPO20
Specific gravity	D 4052	0.830	0.881	0.913	0.887
Viscosity (cSt)	D 445	2.6	5.6	3.35	5.2
Heating value (MJ/kg)	D 4809	43.8	39.4	38.1	38.82
Flash point (°C)	D 93	50	156	49	132
Fire point (°C)	D 93	56	171	58	145
Cetane number	D 613	50	55	28	52
Carbon (%)	D 3178	86.2	77.1	86.92	79.26
Hydrogen (%)	D 3178	13.2	11.81	10.46	11.31
Nitrogen (%)	D 3179	Nil	0.119	0.65	0.23
Sulfur (%)	D 3177	0.3	0.001	0.95	0.18
Oxygen by difference (%)	E 385	Nil	10.97	1.02	9.02

3 Experimental Section

The combustion tests were carried out in a Kirloskar make single cylinder, four stroke, direct injection, diesel engine, with a rated power of 4.4 kW. The layout of the test setup is shown in Fig. 1. The specifications of the test are given Table 2.

It was observed during the stability test that, the TPO is miscible with the JME and also found that the blend was stable during the test. First experiment was started with base fuel diesel and after that it was moved over to JME, and then to the blend at the standard IT of 23 °CA bTDC for obtaining the base data. Further, the tests were done at varied IT for the blend. The standard IT was advanced and retarded by adjusting the number of shims fitted under the plunger in the pump. Addition or removal of a single shim of thickness 0.25 mm changes the IT by about 1.5 °CA. The experiments were carried out using the blend at three different injection timings, of 24.5, 23, and 21.5 °CA bTDC, respectively. Three shims were used in the fuel pump while engine runs at standard IT of 23 °CA bTDC.

The study was carried out with 1.5 °CA advanced and retarded IT for the blend, and the findings were compared with reference fuel diesel and JME. The Kistler type piezoelectric pressure sensor was attached on the cylinder head which measures the cylinder pressure of the engine. A top dead center encoder was used to detect the crank angle (CA). The engine setup was attached with a control panel, which is connected with the pressure sensor, and to convert the signal from the pressure sensor to the analog voltage signal, which was fed to the computerized data acquisition system (DAS).

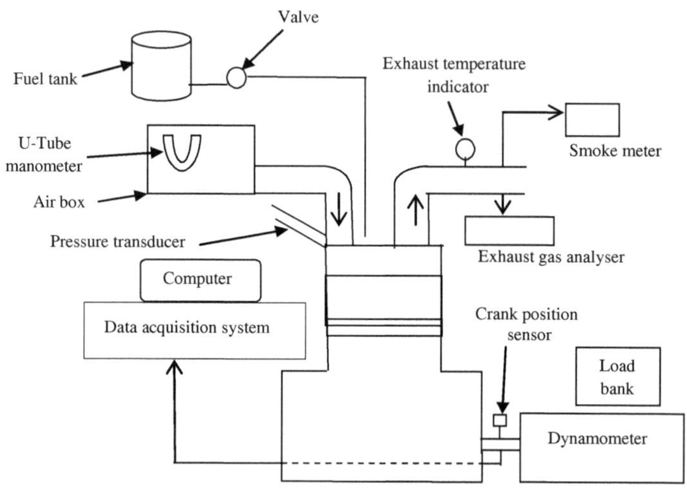

Fig. 1 Schematic diagram of the test setup

Table 2 Test engine specifications

Model	Kirloskar TAF 1
Engine type	Single cylinder, four stroke, constant speed, direct injection diesel engine
Rated power	4.4 kW
Speed	1500 rpm (constant)
Bore	87.5 mm
Stroke	110 mm
Displacement volume	661 cm^3
Compression ratio	17.5
Nozzle opening pressure (bar)	200 bar
Start of fuel injection	23 °CA bTDC
Dynamometer	Eddy current
Injection type	3-Hole pump-line-nozzle injection system

4 Results and Discussion

4.1 Cylinder Pressure History

Figure 2 compares the cylinder pressure variation at full load with crank angle for diesel, JME, and the JMETPO20 blend and at different injection timings. The highest cylinder pressure for diesel, JME, and blend, are approximately 80 bar at 371.6 °CA, 81 bar at 370.3 °CA, and 80 bar at 371 °CA, respectively, with the standard IT.

At full load and standard IT, the start of combustion process occurs by about 348.51, 347.15, and 348.36 °CA in the case of base fuel diesel, JME, and blend, respectively. For the blend with advancement and retardation of IT by 1.5 °CA,

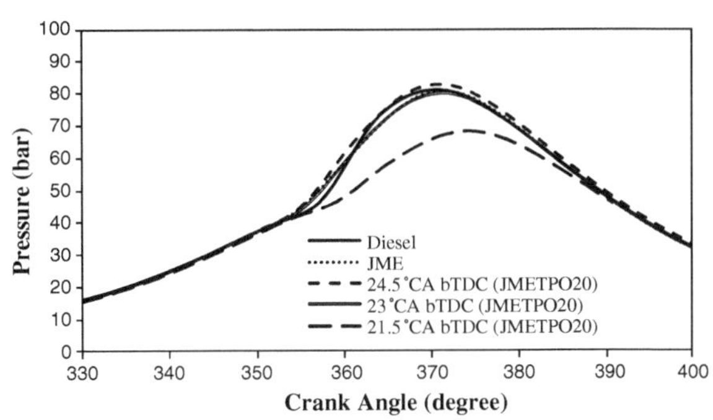

Fig. 2 Variation of the cylinder pressure with crank angle at different injection timings

the start of combustion occurs at about 346.21 and 356.14 °CA, respectively. It can be observed for blend that at the advanced IT and full load, combustion starts earlier and causes increased cylinder pressure in comparison with standard IT. This is a result of faster burn at premixed combustion because of longer ignition delay (Ganapathy et al. 2011).

4.2 *Cumulative Heat Release (CHR)*

The CHR rate pattern at full load, with respect to the crank angle for diesel, JME, and JMETPO20 blend at different injection timings is shown in Fig. 3. The rate of heat release and CHR is determined to evaluate the rate of pressure rise in the cylinder and combustion characteristics for different operating conditions. The CHR value is higher for diesel compared to the JME and blend owing to higher calorific value of diesel.

The CHR values of 2210, 2124, and 2018 kJ are obtained for the blend at full load and injection timings of 24.5, 23, and 21.5 °CA bTDC, respectively. The maximum value of the CHR for blend is attained at CA of 400° and an IT of 1.5 °CA advancement, because mixing process of fuel with the air is superior due to the availability of more time with advanced IT.

4.3 *Maximum Heat Release Rate (MHRR)*

Figure 4 demonstrates the variation of the MHRR for base fuel diesel, JME, and JMETPO20 blend with brake power at different injection timings. It can be observed that with the growth in the brake power, the MHRR rises throughout the combustion period for all the test fuels in this investigation. The MHRR values for diesel, JME, and blend are 56.4, 52.43, and 50.36 J/°CA, respectively, at full load and original IT.

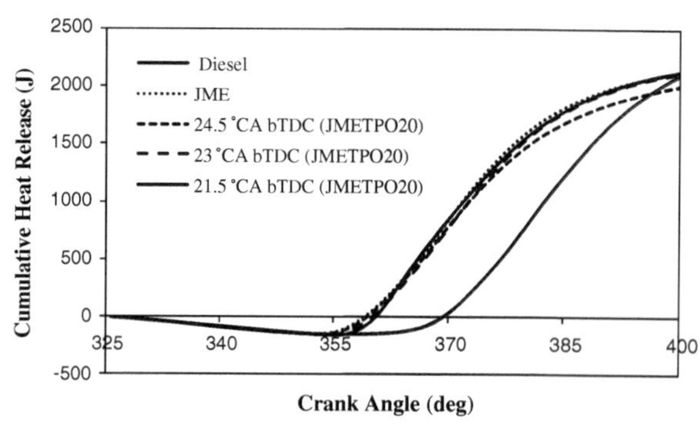

Fig. 3 Variation of the CHR rate with crank angle at different injection timings

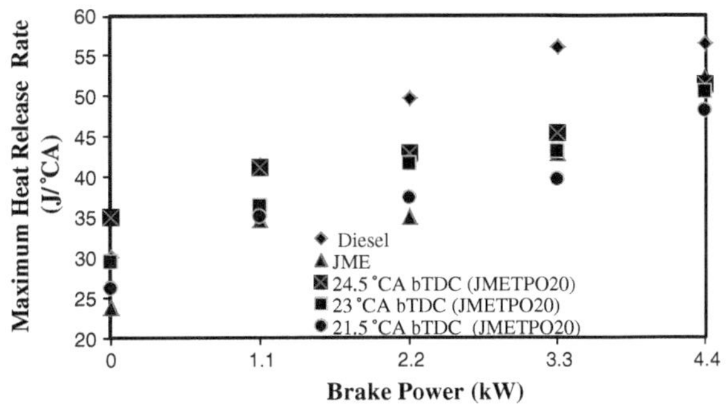

Fig. 4 Variation of the MHRR with brake power at different injection timings

The MHRR for blend is found to be higher in comparison with standard IT. This is owing to increase in the volume of fuel delivered throughout the delay period because of the longer ignition delay (Sharma and Murugan 2015a).

4.4 Maximum Cylinder Pressure (P_{max})

The effects of injection timings on $P_{\max}$ are compared between diesel, JME, and the JMETPO20 blend and are shown in Fig. 5. The $P_{\max}$ values for diesel, JME, and blend at the standard IT and full load are found to be about 80.96, 80.6, and 79.95 bar, respectively.

It is apparent from the figure that advancing the IT, rises the $P_{\max}$ for the blend. It is due to early starting of fuel injection that results in a more amount of mixture accumulated that undertakes combustion process, which results in an increase in $P_{\max}$ (Ganesan 2010). While retarding the IT, a lower $P_{\max}$ is obtained as a result of less fuel being accumulated in the delay period for blend. For the blend, at full load and an advanced and retarded IT of 24.5 and 21.5 °CA bTDC, respectively, the $P_{\max}$ values are found to be 83.66 and 78.32 bar.

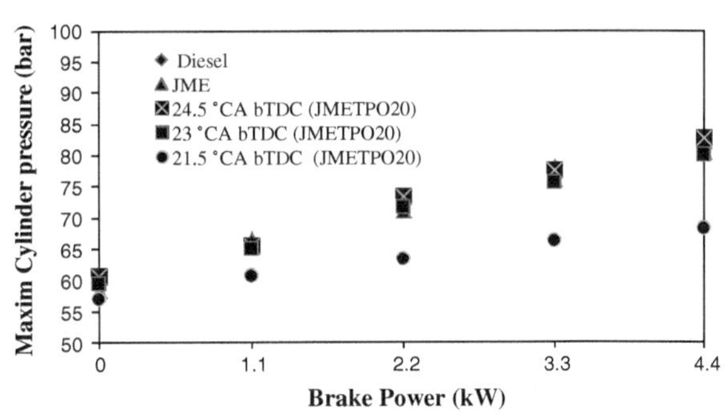

Fig. 5 Variation of the $P_{\max}$ with brake power at different injection timings

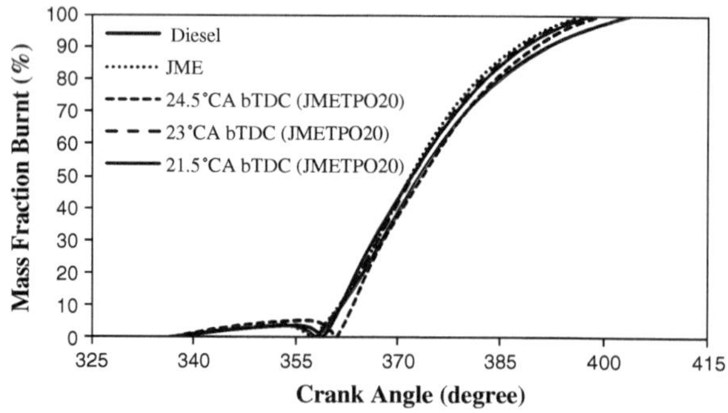

Fig. 6 Variation of the MFB with crank angle at different injection timings

4.5 *Mass Fraction Burnt (MFB)*

Figure 6 shows the MFB of the fuels tested in full load condition with respect to the crank angle for diesel, JME, and the JMETPO20 blend at different injection timings. The MFB provides an information about the percentage of fuel that has burned at any given engine crank angle during the combustion process. Numerous key MFB parameters are explored to record the rate and nature of combustion phenomenon. The study of the MFB duration provides a better outlook on the rate of burning during the combustion process.

The 10–50 % MFB duration indicates the rate of the early part of combustion while the 10–90 % MFB duration provides a sense of the overall combustion duration (Jacobs 2011). It can be observed from the figure that at original IT, the MFB duration for the JME is shorter in comparison with diesel fuel due to the oxygen present in the JME, which supports for a faster combustion. The lowest value of MFB duration for the blend is recorded at advanced IT, because of rapid burning of more fuel accumulated owing to longer delay period.

5 Conclusions

The aim of this investigation was to study the combustion process of a CI engine fueled with biodiesel blend, when the experiments were carried out at three different injection timings: 24.5, 23, and 21.5 °CA bTDC. Based on the test results obtained for the JMETPO20 blend, the following specific conclusions are given:

- The rise in the P_{max} and CHR was found at advanced IT, while the reverse trend was noticed at retarding IT.
- Advancing the IT by 1.5 °CA compared to standard IT, the maximum cylinder pressure was found to be higher by 2.7 bar at full load.

- The MHRR and CHR were found to be higher with advancing the IT by 1.5 °CA in comparison with standard IT. The MHRR and CHR at advanced IT of 24.5 °CA bTDC were noticed 51.5 J/°CA and 2210 J, which were higher by about 1.2 J/°CA and 86 J than standard IT.

References

Ganapathy T, Gakkhar RP, Murugesan K (2011) Influence of injection timing on performance, combustion and emission characteristics of jatropha biodiesel engine. Appl Energy 88:4376–4386

Ganesan V (2010) Internal combustion engines. 3rd edn. Tata McGraw-Hill, New Delhi

Ghadge SV, Raheman H (2005) Biodiesel production from mahua (MadhucaIndica) oil having high free fatty acids. Biomass Bioenerg 28:601–605

Hariharan S, Murugan S, Nagarajan G (2012) Effect of diethyl ether on Tyre pyrolysis oil fueled diesel engine. Fuel

Jacobs TJ (2011) Summary and basic technical analysis of Pulstar plugs engine data. A technical report; September 26

Knothe G (2005) Dependence of biodiesel fuel properties on the structure of fatty acid alkyl esters. Fuel Process Technol 86(10):1059–1070

Murugan S, Ramaswamy MC, Nagarajan G (2008) The use of tyre pyrolysis oil in diesel engines. Waste Manag 28:2743–2749

Sayin C, Canakci M (2009) Effects of injection timing on the engine performance and exhaust emissions of a dual-fuel diesel engine. Energy Convers Manag 50:203–213

Sayin C, Gumus M (2011) Impact of compression ratio and injection parameters on the performance and emissions of a DI diesel engine fueled with biodiesel-blended diesel fuel. Appl Thermal Eng 31:3182–3188

Sharma A, Murugan S (2013) Investigation on the behaviour of a DI diesel engine fueled with Jatropha Methyl Ester (JME) and Tyre Pyrolysis Oil (TPO) blends. Fuel 108:699–708

Sharma A, Dhakal B (2013) Performance and emission studies of a diesel engine using biodiesel tyre pyrolysis oil blends. SAE Technical Paper, 2013-01-1150

Sharma A, Murugan S (2014) Impact of fuel injection pressure on performance and emission characteristics of a diesel engine fueled with jatropha methyl ester tyre pyrolysis blend. SAE Technical Paper, 2014-01-2650

Sharma A, Murugan S (2015a) Combustion, performance and emission characteristics of a DI diesel engine fuelled with non-petroleum fuel: a study on the role of fuel injection timing. J Energy Inst 88:364–375

Sharma A, Murugan S (2015b) Potential for using a tyre pyrolysis oil-biodiesel blend in a diesel engine at different compression ratios. Energy Convers Manag 93:289–297

Sher E (1998) Handbook of air pollution from internal combustion engines: pollutant formation and control. Academic Press, San Diego

Srivastava PK, Verma M (2008) Methyl ester of Karanja oil as alternative renewable source energy. Fuel 87:1673–1677

Xu Z, Li X, Guan C, Huang Z (2014) Effects of injection timing on exhaust particle size and nanostructure on a diesel engine at different Loads. J Aerosol Sci 76:28–38

Performance and Emission Characteristics of CI Engine Fueled with Alternative Fuel with Special Reference to Modification for Combustion: A Literature Review

Himansh Kumar, Y.K. Yadav and A.K. Sarma

Abstract The future of alternative fuels for compression-ignition engine has become imperative due to exhaustion of petroleum products and its major contribution in pollution. As Dr. Rudolf Christian Karl Diesel had invented the compression-ignition engines and the first exhibition of compression-ignition engines was carried out at World's fair in Paris fueled with peanut oil (Alternative Fuel) in 1911. However, due to the ample availability of petrodiesel and the unmatched properties of alternative fuel, researchers ignored it. But, after 1970 due to the hike in price and increased demand-supply gap, researchers had investigated the new alternative fuel source to overcome the dependency on crude oil and foreign exchange. This review paper is about the various types of alternative fuels and the modification needed in compression-ignition engines to run effectively fueled with biofuels. All the modifications are considered on single cylinder four-stroke CI engine with rated power of 5.2 kW at 2000 rpm. Six types of modifications in compression-ignition engines are detailed and almost every performance and emission test results were described, i.e., BSFC, BTE, hydrocarbon emission, oxide of nitrogen emission, and carbon monoxide emission in this review paper. The outcomes of this survey shows that each and every modification have their own advantages and disadvantages but for modification of guide vanes at the inlet passage increased the breathing capacity of the CI engine and due to this brake thermal efficiency was increased and emissions were decreased.

Keywords Compression-ignition engine · Modification · Alternative fuel · Emissions · Performance · Biodiesel

List of symbols

BSFC	Brake specific fuel consumption
CI	Compression ignition
BTDC	Before top dead center

Himansh Kumar (✉) · Y.K. Yadav · A.K. Sarma
Sardar Swaran Singh National Institute of Renewable Energy, Kapurthala 144601, India
e-mail: himansh.rmu@gmail.com

S. Kumar et al. (eds.), *Proceedings of the First International Conference on Recent Advances in Bioenergy Research*, Springer Proceedings in Energy,
DOI 10.1007/978-81-322-2773-1_14

BTE	Brake thermal efficiency
NO_2	Nitrogen dioxide
NO_x	Nitrogen oxides
CFPP	Cold filter plugging point
BP	Brake power
CO	Carbon monoxide
ppm	Parts per million
CO_2	Carbon dioxide
rpm	Revolutions per minute
HC	Hydrocarbon
wt	Weight
CNG	Compressed natural gas
D-base	Engine without modification
Md1	First type of modification
Md2	Second type of modification
Md3	Third type of modification
Md4	Fourth type of modification
Md5	Fifth type of modification
Md6	Sixth type of modification

1 Introduction

The past scenario shows that the production of petroleum oil was no longer than the supply and demand. Therefore, different techniques were used to prepare the alternative fuels which can take petroleum fuel place. As alternative fuels are defined as any material or substance, other than petroleum used to provide energy to power for power producing device such as engine. So the term biodiesel was referred to any petrodiesel equivalent processed fuel, prepared from biological sources (Jaichandar and Annamalai 2011; Subramaniam et al. 2013; Kumar et al. 2010). It is a processed fuel, which was ready to be used in common compression-ignition engines vehicles and it was distinguished as SVO or WVO (Xue et al. 2011; Pandey et al. 2012). As biofuel contained no petroleum source, it can be blended up to any percentage with petrodiesel to create a biodiesel–diesel blend. Some of the commonly used alternative fuels are shown in Fig. 1.

The prime objective of this review paper is to discuss about CI engine modifications required for fruitful run of bioliquid fuels in conventional compression ignition. Almost every modern CI engines would run on biodiesel quite happily provided that the biodiesel has enough high quality, moreover biodiesel required much less engine modification but for low class engines there were some basic problems such as failure of rubber seal, cold starting, fuel injector failure due to high viscosity, poor atomization, less lubrication, etc. (Heywood 1998; Bari et al. 2002).

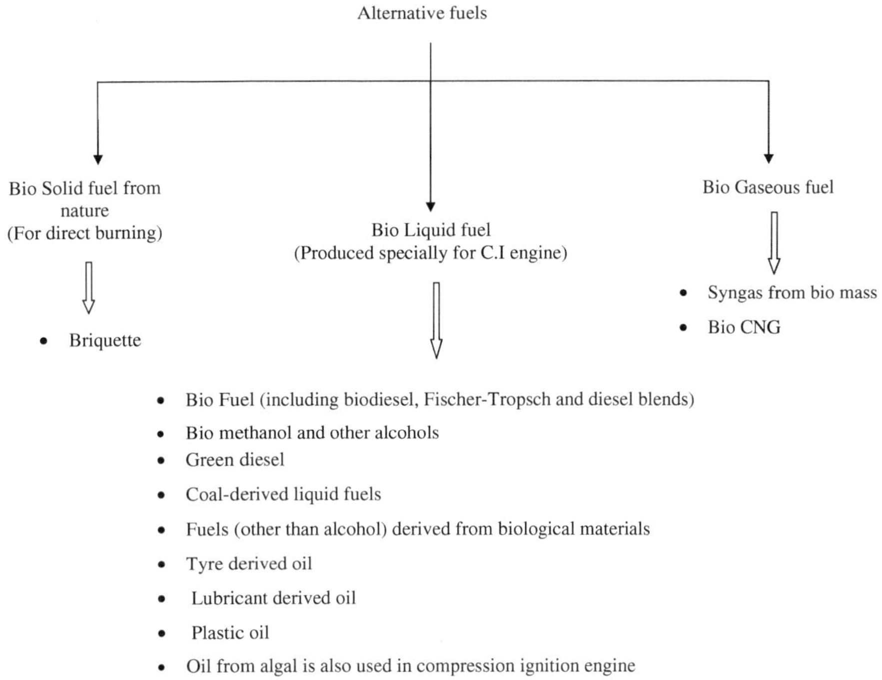

Fig. 1 Classification of alternative fuels focused on bio liquid fuel

Engine modification is used to overcome all of these difficulties. In CI engine modification, like preheating of alternative fuel and intake of air through exhaust gasses, was also a good method for using alternative fuel efficiently. Some little changes like use of separator or additional fuel filter to protect the CI engine from water and foreign waste that had missed during prefiltering of biodiesel was also a modification in CI engine (Prasad et al. 2011). Use of nonrubber products like synthetic rubber (Viton) for rubber seal and special grade lubricant oil for biodiesel is also come in the category of modification (Pulkrabek 2003; Agarwal 2006).

The design methodology for modification in compression-ignition engines, performance and emission analysis were discussed in this review paper. The outcome of this study was indicated that the following modifications allowed the CI engine to efficiently utilize the biodiesel as a petrodiesel substitute (Papagiannakis 2013; Pandiyarajan et al. 2011).

2 Types and Designs Available for CI Engine

CI engine is a heat engine in which combustion takes place when the air–fuel mixture self-ignites due to increased suction air temperature in the combustion chamber caused by high compression. Various designs are available for CI engine on the basis of injection of fuel in the combustion chambers like direct, indirect, and common rail injection and on the basis of geometry of combustion chamber like open, divided combustion chamber (Ganeshan 2012).

CI engine researches aimed to improve the brake thermal efficiency, BSFC and to reduce emissions, noise pollution, and vehicle vibration. With different types of designs, BTE increases up to 7 % to values as high as total 40 % (Pulkrabek 2003). In present scenario CI engine emission control requirement is one of the major considerations in design and operation.

3 Modification in CI Engine with Respect to Alternative Fuel

The physicochemical properties of alternative fuels are closed to petrodiesel but due to their inherently high kinematic viscosity and less volatility were created poor atomization, evaporation, and air–fuel mixing formation. This is also creates improper combustion, lower BTE, and higher pollutant emission. The increased kinematic viscosity of alternative fuel created operational problems like poor engine starting, missing in ignition, and decrement in brake thermal efficiency. In long-term operation, failure problems like coking in nozzle, carbon deposition on fuel injector and valve seat of CI engine, and lubricating oil dilution was reported (Çetinkaya et al. 2005). So these problems are created due to direct use of biofuels in CI engines without any modifications. Almost every CI engine runs on biofuel and it blends quite happily, but still it needs some modification for proper combustion and increased its efficiency. In this review paper, six types of modifications are discussed and these are given below (Sects. 3.1–3.6).

3.1 Using Guide Vanes at Air Inlet Passage (Md1)

The guide vanes at inlet passage were installed to increase the breathing capacity of CI engine. Guide vanes were used to increase the turbulent flow, kinematic velocity, and scrolling strength of suction air (Brasz 1996). Simulation technique was used to evaluate the accurate inlet guide vane angle by analyzing the turbulent flow, kinematic velocity, vorticity, and scrolled air strength created by the inlet guide vanes (Bari and Saad 2015). It was reported that due to inlet guide vanes CI engine performance was increased and exhaust emissions decreased compared to any other engine that runs with biodiesel without using inlet guide vanes (Payri et al. 2004). Five inlet guide vane models ranging from of 25° to 45° were used and tested on a generator CI engine fueled with biodiesel. Brake specific fuel consumption was increased in between 0.69 and 1.77 % with use of inlet guide vanes compare to without modification and 35° inlet guide vane angle indicated the maximum performance (Saad et al. 2013; Naser and Gosman 1995). In terms of brake thermal efficiency the guide vanes were improved in the range of 0.7–1.82 % compare to without modification, also 35° inlet guide vane angle showed the highest performance (Bari and Saad 2015). There was an increased of NO_x emission in the ranges of 0.48–3.79 % with use of guide vanes compared to without guide vanes. Carbon monoxide was also reduced due to inlet guide vanes at air inlet passage in the ranges of 1–8.85 % to the highest reduction by 40° inlet guide vane angle redirected by 35° inlet guide vane angle (Bari and Saad 2015; Naser and Gosman 1995). With different inlet guide vanes, the alternative fuel emission was reduced in the variation of 1.19–7.49 % and 35° inlet guide vane angle showed the highest reduction (Bari and Saad 2015).

3.2 Using Preheated Alternative Fuel (Md2)

Higher kinematic viscosity of alternative fuel was one of the major problems of carbon deposition on fuel injectors and engine valve seat also poor fuel atomization results incomplete combustion (Kalam and Masjuki 2004; Masjuki et al. 2001). Preheating and blending with petrodiesel fuel was the possible solution for decreasing the kinematic viscosity of biodiesel (Utlu and Koc 2008; Pugazhvadivu and Jeyachandran 2005). It was reported that at 135 °C, the kinematic viscosity of biodiesel was same as petrodiesel at 30 °C. The BTE was improved and the CO and smoke emissions were reduced with preheated biodiesel (Pugazhvadivu and Jeyachandran 2005). This waste heat recovery technique was used while the exhaust gas temperature would ultimately reduce the overall energy requirement and create healthy environment (Bari et al. 2002; Yilmaz and Morton 2011).

3.3 Using Thermal Barrier Coating on CI Engine Parts (Md3)

The BTE of every range of CI engine was in the range of 28–36 % generally. It was meant that the 70 and 64 % of the fuel energy content was lost in the form of losses and radiation of heat. Experimentally 30 % of energy was wasted in the exhaust gasses and the remaining was removed by the cooling water in the environment (Aydın and Sayın 2014). Coating of ceramic materials was used to reduce the heat loss from CI engine combustion chamber wall that results in energy saving which was useful to increase the BTE of CI engine. The use of ceramic materials created low thermal conductivity and high thermal expansion coefficient and this was useful for insulation for CI engine radiated components (Shrirao and Pawar 2011). For the daily application of biofuels, it was necessary to use CI engine with little modification because its properties are not as same as that of petrodiesel (Hüseyin 2013). Al_2O_3, $3Al_2O_3 \cdot 2SiO_2$ (mullite), ZrO_2, and AE_2O_4 (spinel), where A is divalent and E are trivalent, are the common ceramics for coating on internal surface and parts of combustion chamber (Hazar and Oztürk 2010). It was reported that the BTE increased with decrease of BSFC and increased in NO_x with different loads

Table 1 Coating materials with their properties

Materials/properties	ZrO_2	Al_2O_3	$3Al_2O_3 \cdot 2SiO_2$ (Mullite)	AE_2O_4 (Spinel)
Melting point	High	High	High	High
Thermal conductivity	Low	High	Low	High
Coefficient of thermal expansion	High	Low	Low	Low
Phase	Stable	Unstable	Unstable	Stable
Oxidation resistance	Low	High	Low	High
Corrosion resistance	Low	Moderate to high	High	High

after using thermal barrier coating (Hazar and Oztürk 2010). Table 1 shows some coating materials and their properties.

3.4 *Using Dual Fuel Mode in the Compression-Ignition Engine (Md4)*

Conservation of fossil fuels is necessitated due to limited resources of stocks in nature. Dual fuel CI engine was fueled with both liquid and gaseous fuels simultaneously. Minor modification was needed to convert a CI engines for dual fuel operation (Tarabet et al. 2014). Gaseous fuel was used as a primary fuel to reduce the consumption of petrodiesel for power generation in CI engine. Premixed combustion was also used to increase the BTE and BSFC and decrease the emissions as compared to CI engine without modification (Mohsin et al. 2014; Yoon and Lee 2011). It was reported that the use of compressed natural gas in a dual fuel CI engine reduced the engine noise, BSFC consumption, and NO_x emissions (Karabektas 2009; Makareviciene et al. 2013). Also the researchers were reported that with the use of biogas–petrodiesel and biogas–biodiesel in a dual fuel mode, BSFC increased and BTE decreased, respectively, compared to without modification in CI engine (Duc and Wattanavichien 2007; Tippayawong et al. 2007). However, NO_x and CO_2 emission of biogas-biodiesel was higher than biogas–petroldiesel dual fuel CI engine; whereas CO and HC emission was lower (Bedoya et al. 2009; Banapurmatha et al. 2008). Dual fuel operation results improved BTE, BSFC, and lower emissions also less in noise pollution. Also it was a cheap technique for power generation (Luijten and Kerkhof 2011; Mustafi et al. 2013).

3.5 *Using Varying Combustion Chamber Geometry in Compression-Ignition Engine (Md5)*

The impact of combustion chamber geometry variation and use of different shapes of piston crowns were increased the BTE and BSFC of compression-ignition engines fueled with biodiesel also improved emissions and combustion. Hemispherical combustion chamber, toroidal combustion chamber, and shallow depth combustion chamber were reported to increase the efficiencies of CI engine. The test results showed that the BTE for toroidal combustion chamber was highest compared to other reported combustion chambers (Jaichandar and Annamalai 2012). Also the reduction in particulate matters, CO, and UHC was reported for toroidal combustion chamber with others (Subhash et al. 2013). An increment in NO_x was reported for the toroidal combustion chamber (Subhash et al. 2013; Hribernik and Kegl 2007). The combustion characteristics showed that the results were improved for toroidal combustion chamber with respect to without modification with CI engine (Venkateswaran and Nagarajan 2010). Also due to proper air–fuel mixing in toroidal combustion chamber results maximum cylinder pressure compared to two other engine combustion chambers fueled with biodiesel (Venkateswaran and Nagarajan 2010; Prasad et al. 2011).

3.6 *Using Intake Air Preheating in Compression-Ignition Engine (Md6)*

In slow speed engines, preheating of intake air to the heat exchangers by the use of hot water of engine jacket and again heated by engine exhaust gasses with another heat exchanger is useful for increasing the volumetric efficiency of compression-ignition engines (Kumar and Balamurugan 2014). Thermocouples were used to calculate and maintain the temperature at various locations in the slow speed compression-ignition engines (Papagiannakis 2013). It is reported that the volumetric efficiency and other performance parameters increases with increase in both preheated suction air and its temperature (Pandiyarajan et al. 2011). It is very useful for complete combustion in CI engine, in the case of biofuel (Papagiannakis 2013). There was a drawback for inlet air preheating without using intercooler because it created incomplete combustion in compression-ignition engines for high speed engine. Also it was very useful for full load conditions because of increased cylinder pressure (Kumar and Balamurugan 2014).

4 Discussion

This review paper consisted of mainly six types of techniques for modification in CI engine with respect to alternative fuel. Although all the techniques were useful for modification in CI engine, but there are some advantages and disadvantages. The following Table 2 is showing the techniques used and the changes happened due to the modification.

Table 2 Modification in CI engine related to different researchers

S. No.	Techniques used	Related researchers	Changes in CI engine due to modification
1	Guide vane	S. Bari, Idris Saad, J.J. Brasz	Additional guide vane arrangement attached in the air intake passage
2	Preheating the alternative fuel	M. Pugazhvadivu, Nadir Yilmaz	Introduce a heat exchanger to preheat the intake fuel
3	Thermal barrier coating on engine parts	Selman Aydın, Selman Aydın, P.N. Shrirao, Hüseyin Aydin	Coating of ceramic material on the combustion chamber inner wall, valves seat, etc.
4	Dual fuel technique for CI engine	L. Tarabet, R. Mohsin, N.R. Banapurmatha	Additional injection system to inject the two fuels simultaneously
5	Varying piston bowl shapes	S. Jaichandara, G.V. Subhash	Different types of piston crown are used
6	Intake air preheating	R.G. Papagiannakis, K. Sureshkumar	Use of turbocharger or heat exchanger before the air inlet passage

4.1 *Performance and Emissions Analysis of CI Engine with Modification*

In this review paper performance and emission analysis of CI engine such as BSFC, BTE, UHC, CO, and NO_x were compared for different modification and determined. For understanding the effects properly some coding was taken. For guide vanes at intake air passage type modification symbol Md1 was taken likewise for second modification, i.e., preheating of the alternative fuel Md2 was taken for third modification, i.e., use of thermal barrier coating Md3 was taken, for fourth modification, i.e., dual fuel technique Md3 was taken, for fifth modification, i.e., varying piston bowl shapes Md5 was taken, for sixth modification, i.e., intake air preheating Md6 was taken, and finally standard engine D-base was taken. The following graphs show the performance and emissions results of concerned research papers and these results were compared to standard engine with respect to alternative fuel.

4.1.1 Brake Specific Fuel Consumption (BSFC)

BSFC is basically the ratio of mass fraction of fuel burned to brake power (Singh et al. 2015). The following review paper shows the brake specific fuel consumption with increased engine load with respect to different modification in CI engine. The graph shows that the BSFC for Md1 type modification was almost similar with D-base engine and fuel consumption was increased in all the remaining modifications because of low calorific value of alternative fuel, but in Md1 modification complete combustion occurs due to the turbulent flow of intake air. Figure 2 shows the BSFC of biodiesel with different modification with respect to load.

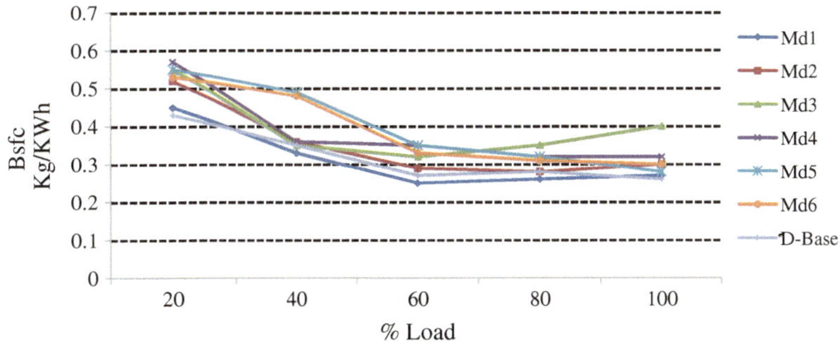

Fig. 2 Variations in brake specific fuel consumption for different modifications with respect to loads

4.1.2 Brake Thermal Efficiency (BTE)

Brake thermal efficiency is an important efficiency for CI engine, it means how much thermal energy of fuel input was converted into shaft output (Singh et al. 2015). Figure 3 shows the comparison of BTE for CI engine without modification to all modified engine fueled with alternative fuel. Generally 31.7 % efficiency is reported on the papers on full load condition and after Md1 modifications become 32.3 %, hence, there is an increment of 0.6 % of the efficiency (Bari and Saad 2015). Although, all the remaining modifications have similar efficiency as that of base fuel, except Md4 modification because bio gas is not sufficient to work like biofuels comparatively. A better BTE in Md1 modification was possible due to the proper mixture formation of alternative fuel with air. Better swirling of air in combustion chamber leads to better combustion of alternative fuel and results increased in brake thermal efficiency.

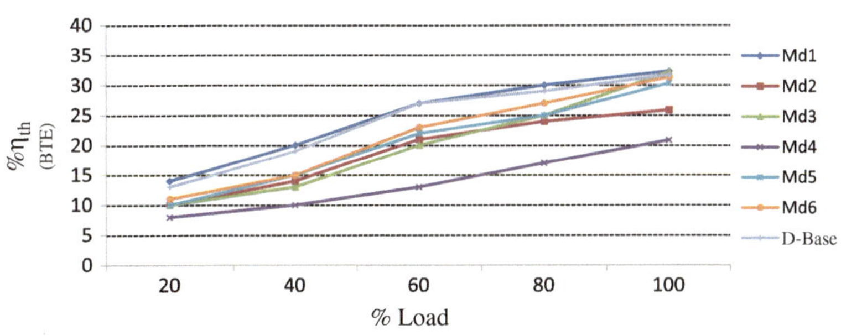

Fig. 3 Variation of BTE for different modifications in CI engine with respect to load

4.2 Emission Analysis

USA was the first country to start the control of air pollution by vehicle in 1964 (Karabektas 2009). After that, all countries had started a numbers of emission regulations in form of emission standards for CI engine. Euro 6 was the latest engine emission standards used in almost every country in the world. In this review paper, the harmful emissions like carbon monooxide (CO), unburnt hydrocarbon (UHC), and NO_x were studied. It is reported in the books that the combustion of a petroldiesel produces carbon dioxide and water after complete combustion but in practice it can not be possible, so this study is to find out the best technique to produce minimum pollutants by CI engine in the exhaust process (Heywood 1998).

4.2.1 Unburnt Hydrocarbon Emission (UHC)

Unburned hydrocarbon is produced due to the incomplete combustion in the combustion phase because it is the unburned part of fuel which was exhausted in atmosphere during exhaust stroke of CI engine (Singh et al. 2015). The unburnt hydrocarbon emissions for Md1, Md2, Md3, Md4, Md5, and Md6 were compared to petroldiesel and shown in Fig. 4. The UHC emissions were reduced from no load to full load with modification of preheating the biodiesel (Md2) also for modification of air inlet preheating (Md6). This was because of complete combustion of alternative fuel with proper mixing of air and fuel, due to increase in intake of air temperature or fuel. There was a reduction of 30 % UHC emissions in Md2 with respect to petroldiesel and almost similar result for Md6 also for all entire range of load with diesel fuel. For modification Md4 the UHC emission is high due to incomplete combustion.

4.2.2 Carbon Monoxide Emission (CO)

Carbon monoxide emission is mainly because of incomplete combustion, slow burning of fuel during final phase of combustion, and due to lower air–fuel ratio. In

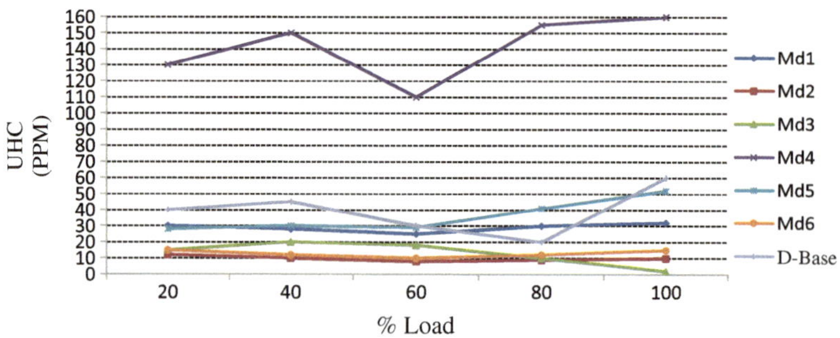

Fig. 4 Variation of unburned hydrocarbons for different modifications with respect to load

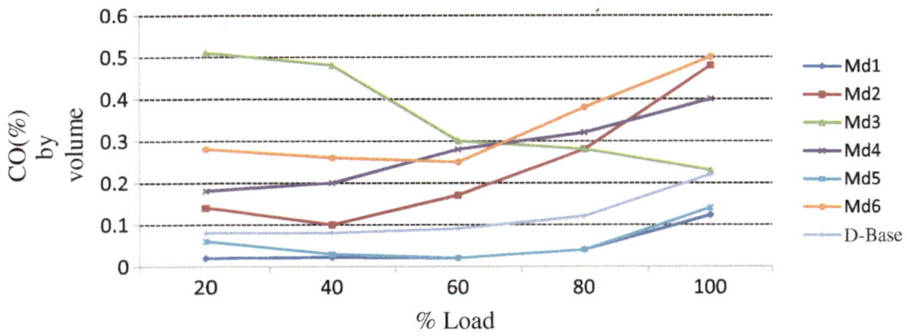

Fig. 5 Variation of carbon monooxide emissions for different modification with respect to load

case of complete combustion CO was converted into CO_2 and if the combustion was incomplete because of less inducted air and temperature, CO_2 turned in CO (Singh et al. 2015). High emission of CO was due to rich air–fuel mixture as well as when combustion occurs with lean mixture CO turned into CO_2. Figure 5 shows the comparison of CO emissions for different modification in CI engine with respect to CI engine without modification. In every engine load conditions CO formation for modifications Md1 and Md5 were lower to the standard CI engine but for remaining modifications, CO was more than that of standard diesel engine. However, on full load condition, CO formation was decreased significantly when compared to standard CI engine. In modification Md1 and Md5, CO formation was hugely reduced due to higher air swirl results better combustion. For remaining modifications, like Md2, Md3, Md4, and Md6, it increased to the standard CI engine.

4.2.3 Oxides of Nitrogen Emission (NO_x)

NO_x is formed because of chain reactions in air with the presence of nitrogen and oxygen. These reactions are depended on reaction temperature. NO_x emission was induced after 1200 °C reaction temperature (Singh et al. 2015). Figure 6 shows the

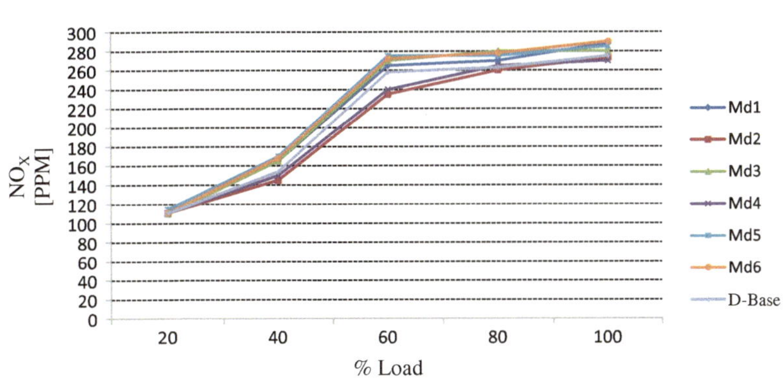

Fig. 6 Variation of oxides of nitrogen emission for different modifications with respect to load

variations in NO_x formation for standard CI engine with different modifications. The NO_x emissions have a little bit decrement for Md2 and Md4 to the standard engine is due to use of alternative fuel in huge quantity. Increased in NO_x emission was due to high combustion product temperature results complete combustion in case of CI engine without modification. For standard CI engine, increase in NO_x emission was due to the high part of the combustion that was completed before BTDC due to lower ignition delay of fuel.

5 Conclusion

From the above indicated modifications it is concluded that the use of alternative fuel or its blends with petroldiesel in modified CI engine can create a good impact on the efficiencies and emissions of the CI engines. It is very useful in developing countries like India because India has great potential for alternative fuels. The compression-ignition engines easily adopt the characteristics of alternative fuels like higher flash point, kinematic viscosity, API density, and lower calorific value due to these modifications. Stationary slow speed compression-ignition enginess were suitable for fueled with straight vegetable oil and their blends without much environmental impact and these types of modifications were very good for stationary engines for increasing their brake thermal efficiency. The main conclusions derived from present investigation are summarized below.

- Modification like air intake in a combustion chamber through the guide vanes (Md1) is very helpful for proper air–fuel mixing and proper combustion due to increased in volumetric efficiency and turbulent nature of intake air.
- Alternative fuel preheating through exhaust gasses was also an optional solution to reduce the kinematic viscosity of liquid alternative fuel because it created the major problems like failure of rubber seals, injector coking, cold shut in winter time, etc.
- Thermal barrier coating is also a useful modification technique for increasing the efficiency and decreasing the emission, but due to insulation, inside temperature gets increased and NO_x emission gets increased, but by adding additives in the alternative fuel it can be reduced.
- Dual fuel CI engine is also a good technique for using gaseous fuel, but the negative point is that the efficiency gets decreased and emissions gets increased, but the positive point is that the fuel cost gets decreased.
- Varying piston bowls type modification is also useful for proper mixing of the air–fuel mixture and complete combustion and toroidal combustion chamber type piston bowl is reported for good results.
- Intake air preheating by exhaust gasses is also a good modification technique for increasing the BTE, BSFC and less emissions, but for stationary engine it was not good because the breathing capacity decreased because of addition of heat exchanger in the intake passage, but for automobile industries, by the use of

turbocharger and supercharger, the breathing capacity was also increased and an intercooler was added to control the temperature of intake air.

Although all the modifications have some advantages and disadvantages, but the modification Md1 (guide vanes at air inlet passage) creates more breathing capacity of the engine and due to this thermal efficiency gets increased and emissions decrease. Also this arrangement does not need much attention for modification in compression ignition engine. So this type of modification is best suited for CI engine in case of alternative fuel. This review paper concludes that alternative fuel and its blends can hold a good future as a fuel for compression-ignition engines with little modifications; further research is needed for elaboration of modification in CI engine.

References

Agarwal A (2006) Biofuels application as fuel for IC engines. J Prog Energy Combust Sci 33:233–271

Aydın S, Sayın C (2014) Impact of thermal barrier coating application on the combustion, performance and emissions of a diesel engine fueled with waste cooking oil biodiesel–diesel blends. Fuel 136:334–340

Banapurmatha NR, Tewaria PG, Hosmath RS (2008) Experimental investigations of a four-stroke single cylinder direct injection diesel engine operated on dual fuel mode with producer gas as inducted fuel and Honge oil and its methyl ester (HOME) as injected fuels. Renew Energy 33:2007–2018

Bari S, Saad I (2015) Performance and emissions of a compression ignition (C.I) engine run with biodiesel using guide vanes at varied vane angles. Fuel 143:217–228

Bari S, Lim TH, Yu CW (2002) Effect of preheating crude palm oil on injection system, performance and emissions of a diesel engine. Renew Energy77:339–351

Bedoya ID, Arrieta AA, Cadavid FJ (2009) Effects of mixing system and pilot fuel quality on diesel—biogas dual fuel engine performance. Bioresour Technol 100(24):6624–6629

Brasz JJ (1996) Aerodynamics of rotatable inlet guide vanes for centrifugal compressors international compressor engineering conference. Paper 1196

Çetinkaya M, Ulusoy Y, Tekin Y, Kapaosmanoğlu F (2005) Engine and winter road test performances of used cooking oil originated biodiesel.Energy Convers Manage 1279–1291

Duc PM, Wattanavichien K (2007) Study on biogas premixed charge diesel dual fuelled engine. Energy Convers Manage 48(8):2286–2308

Ganeshan V (2012) Internal combustion engines, 4th edn. Tata McGraw Hill Education Private Limited, New Delhi

Hazar H, Oztürk U (2010) The effects of Al_2O_3-TiO_2 coating in a diesel engine on performance and emission of corn oil methyl ester. Renew Energy 35:2211–2216

Heywood JB (1998) Internal combustion engines fundamentals. McGraw Hill, Newyork

Hribernik A, Kegl B (2007) Influence of biodiesel fuel on the combustion and emission formation in a direct injection (DI) diesel engine. Energy Fuels 21:1760–1767

Hüseyin A (2013) Combined effects of thermal barrier coating and blending with diesel fuel on usability of vegetable oils in diesel engines. Appl Thermal Eng 51:623–629

Jaichandar S, Annamalai K (2011) The status of biodiesel as an alternative fuel for diesel engine—an overview. J Sustain Energy Environ 2:71–75

Jaichandar S, Annamalai K (2012) Effects of open combustion chamber geometries on the performance of pongamia biodiesel in a DI diesel engine. Fuel 98:272–279

Kalam MA, Masjuki HH (2004) Emissions and deposit characteristics of a small diesel engine when operated on preheated crude palm oil. Biomass Bioenergy 27:289e97

Karabektas M (2009) The effects of turbocharger on the performance and exhaust emissions of a diesel engine fuelled with biodiesel. Renew Energy 34:989–993

Kumar KS, Balamurugan M (2014) Effects of intake air preheat and fuel blend ratio on a diesel engine performance characteristics operating on bio diesel and its blends. ISSN 2348 – 8034

Kumar N, Varun, Kumar A (2010) Biodiesel as an alternative fuel for CI engines: environmental effect. Indian J Sci Techn 3(5):602–606

Luijten CCM, Kerkhof E (2011) Jatropha oil and biogas in a dual fuel CI engine for rural electrification. Energy Convers Manage 52(2):1426–1438

Maizonnasse M, Plante JS, Oha D, Laflamm CB (2013) Investigation of the degradation of a low-cost untreated biogas engine using preheated biogas with phase separation for electric power generation. Renew Energy 55:501–513

Makareviciene V, Sendzikiene E, Pukalskas S, Rimkusb A, Vegneris R (2013) Performance and emission characteristics of biogas used in diesel engine operation. Energy Convers Manage 75:224–233

Masjuki HH, Kalam MA, Maleque MA, Kubo A, Nonaka T (2001) Performance, emissions and wear characteristics of an indirect injection diesel engine using coconut oil blended fuel. IMechE, Part D 215:393–404

Mohsin R, Majid ZA, Shihnan AH, Nasri NS, Sharer Z (2014) Effect of biodiesel blends on engine performance and exhaust emission for diesel dual fuel engine. Energy Convers Manag 88:821–828

Mustafi NN, Raine RR, Verhelst S (2013) Combustion and emissions characteristics of a dual fuel engine operated on alternative gaseous fuels. Fuel 109:669–678

Naser J, Gosman A (1995) Prediction of compressible subsonic flow through an axisymmetric exhaust valve-port assembly. Proc Inst Mech Eng Part D: J Automob Eng 209:289–295

Pandey RK, Rehman A, Sarviya RM (2012) Impact of alternative fuel properties on fuel spray behavior and atomization. Renew Sustain Energy Rev 16:1762–1778

Pandiyarajan V, Chinna Pandian M, Malan E, Velraj R, Seeniraj RV (2011) Experimental investigation on heat recovery from diesel engine exhaust using finned shell and tube heat exchanger and thermal storage system. Appl Energy 88:77–87

Papagiannakis RG (2013) Study of air inlet preheating and EGR impacts for improving the operation of compression ignition engine running under dual fuel mode. Energy Convers Manag 68:40–53

Payri F, Benajes J, Margot X, Gil A (2004) CFD modeling of the in-cylinder flow indirect-injection diesel engines. Comput Fluids 33:995–1021

Prasad BVVSU, Sharma CS, Anand TNC, Ravikrishna RV (2011) High swirl-inducing piston bowls in small diesel engines for emission reduction. Appl Energy 88(7):2355–2367

Pugazhvadivu M, Jeyachandran K (2005) Investigations on the performance and exhaust emissions of a diesel engine using preheated waste frying oil as fuel. Renew Energy 30:2189–2202

Pulkrabek WW (2003) Engineering fundamentals of the internal combustion engine. TBS Limited, New Jersey

Saad I, Baria S, Hossain SN (2013) In-cylinder air flow characteristics generated by guide vane swirl and tumble device to improve air-fuel mixing in diesel engine using biodiesel. Proc Eng 56:363–368

Shrirao PN, Pawar AN (2011) Evaluation of performance and emission characteristics of turbocharged diesel engine with mullite as thermal barrier coating. Int J Eng Technol 3(3):256–262

Singh N, Kumar H, Jha MK, Sarma AK (2015) Complete heat balance, performance, and emission evaluation of a CI engine fueled with Mesua ferrea methyl and ethyl ester's blends with petroldiesel. J Therm Anal Calorim 2015. doi:10.1007/s10973-015-4777-8

Subhash GV, Vara Prasad V, Pasha SkM (2013) An experimental study on a CI engine fuelled with soapnut oil-diesel blend with different piston bowl geometries. (IJERT), ISSN: 2278–0181

Subramaniam D, Murugesan A, Avinash A, Kumaravel A (2013) Biodiesel production and its engine characteristics-an expatiate view. Renew Sustain Energy Rev 22:361–370

Tarabet L, Loubar K, Lounici MS, Khiari K, Belmrabet T, Tazerout M (2014) Experimental investigation of DI diesel engine operating with eucalyptus biodiesel/natural gas under dual fuel mode. Fuel 133:129–138

Tippayawong N, Promwungkwa A, Rerkkriangkrai P (2007) Long-term operation of a small biogas/diesel dual-fuel engine for on-farm electricity generation. Biosyst Eng 98(1):26–32

Utlu Z, Koc¸ AkMS (2008) The effect of biodiesel fuel obtained from waste frying oil on direct injection diesel engine performance and exhaust emissions. Renew Energ 33:1936–41

Venkateswaran SP, Nagarajan G (2010) Effects of the re-entrant bowl geometry on a DI turbocharged diesel engine performance and emissions—a CFD approach. J Eng Gas Turb Power 132(12):122803–122813

Xue J, Grift Tony E, Hansen Alan C (2011) Effect of biodiesel on engine performances and emissions. Renew Sustain Energy Rev 15:1098–1116

Yilmaz N, Morton B (2011) Effects of preheating vegetable oils on performance and emission characteristics of two diesel engines. Biomass Bioenergy 35:2028–2033

Yoon SH, Lee CS (2011) Experimental investigation on the combustion and exhaust emission characteristics of biogas—biodiesel dual-fuel combustion in a CI engine. Fuel Process Technol 92(5):992–1000

Part III
Thermochemical Conversion

Upgradation of Bio-oil Derived from Lignocellulose Biomass—A Numerical Approach

Anjani R.K. Gollakota, Malladi D. Subramanyam, Nanda Kishore and Sai Gu

Abstract The envisaged upgrading of lignocellulose biomass derived feed stock to partially replace crude oil is receiving much attention due to increasing demand for renewable and CO_2 neutral energy sources. One prospective method is to convert the biomass into bio-oil through pyrolysis and upgrading the pyrolytic oil into biofuel. The bio-oil has a high content of oxygen and therefore lowers the stability over the time and lowers the heating value as well. Hence, upgrading is desirable to remove the oxygen and enhance the properties of the bio-oil resembling as a second generation fuel which is an alternative to conventional fossil fuels. Hydrodeoxygenation (HDO) process is found to be very promising route to upgrade pyrolysis bio-oil for producing liquid transportation fuels. The present study deals with the CFD modeling of hydrodynamics and reaction kinetics in a fluidized bed reactor using ANSYS Fluent 14.5. By integrating a lumped kinetic model with Eulerian multiphase fluid flow, a comprehensive model for the hydrodynamics and reaction kinetics for the upgradation of bio-oil is developed. CFD analyses are performed and influence of operating parameters are evaluated for the weight hourly space velocity (WHSV) of 2 h^{-1} and a constant pressure of 8720 kPa at temperature in the range 623–673 K.

Keywords Bio-oil · Upgradation · Fluidized bed reactor · Hydrodeoxygeneation · Hydrodynamics · Reaction kinetics

---

A.R.K. Gollakota · M.D. Subramanyam · Nanda Kishore (✉)
Department of Chemical Engineering, Indian Institute of Technology Guwahati, Guwahati 781039, Assam, India
e-mail: nkishore@iitg.ernet.in; mail2nkishore@gmail.com

Sai Gu
Centre for Biofuel, School of Energy, Environmental and Agrifood, Cranfield University, Cranfield, UK

S. Kumar et al. (eds.), *Proceedings of the First International Conference on Recent Advances in Bioenergy Research*, Springer Proceedings in Energy,
DOI 10.1007/978-81-322-2773-1_15

List of symbols

$\overrightarrow{F_q}$	External forces (N)
$\overline{\overline{\tau_q}}$	Stress-strain tensor (Pa)
$\emptyset_{ls}$	Energy exchange (kJ)
e_{ss}	Coefficient of restitution for particle collisions
$\vec{g}$	Gravitational force (m s^{-2})
$\gamma\theta_s$	Collision dissipation energy (kg s^{-3})
σ_k	Turbulent Prandtl numbers
A	Arrhenius constant (min^{-1})
Al_2O_3	Alumina
b	Buoyancy
C	Model constant
C_d	Drag coefficient
CO	Carbon-monoxide
CO_2	Carbon-dioxide
Co-Mo	Cobalt-molybdenum oxide
D	Diffusion coefficient of species (m^2 s^{-1})
D	Diameter (m)
E	Activation energy (kJ mol^{-1})
Fr	Friction
G	Generation of energy (kJ)
H	Specific enthalpy (kJ)
HDO	Hydrodeoxygenation
HNV	Heavy non volatiles
i,j	Components
K	Interface exchange coefficient
k	Mean velocity gradients
l	Liquid phase
LNV	Light non volatiles
Max	Maximum value
Ni-Mo	Nickel molybdenum oxide
Nu	Nusselt number
P	Pressure (Pa)
p	Secondary phase
Pr	Prandtl number
Pt	Platinum
q	Heat flux (J s^{-1})
r	Rate of reaction (mol l^{-1} s^{-1})
Re	Reynolds number
s	Solid phase
S	Source term
T	Temperature (K)
WHSV	Weight hourly space velocity (h^{-1})

Y	Local mass fraction of species
α	Volume fraction
ε	Rate of dissipation ($m^2\ s^{-3}$)
ρ	Density (kg m^{-3})
v	Velocity (m s^{-1})
μ	Viscosity (Pa s)

1 Introduction

The modern day energy consumption of the world reached to threshold in comparison to the earlier times owing to the present lifestyle and the increased population. The transportation sector consumption constitutes one fifth of the total energy produced according to (van Ruijven and van Vuuren 2009). With the population growth and the convenience in the transportation, there is a definite need in search of alternate source of energy to meet the current demands. This poses major challenge of the near future as the present fuel reserves are depleting drastically. Materialistic research work is being carried out in the fields of energy to find the replacement fuels of gasoline and diesel. The alternative fuel that is equivalent to the conventional fuels should be compatible, sustainable, and reduce the CO_2 emissions thereby decreasing the adverse environmental effects. Biomass derived fuels seems to be the future perspective in accomplishing the aforementioned criteria, and produced within relatively short cycles to reduce the environmental hazards. Unlike first generation, bio-fuels (based on food grains) have been implemented till now creating a problem of food scarcity particularly food grains and the energy efficiency per unit land of the crop is relatively low. To overcome these difficulties, advanced bio-fuels are produced from a wide range of lignocellulosic biomass feed stocks, including waste, and their production does not necessarily compete with food or feed production. The possible option available to produce advanced or second generation bio-fuels is the coprocessing of upgraded pyrolysis oil (also known as bio-oil). Bio-oils are complex natured with a mixture of several compounds with an appreciable proportion of water from both the original moisture and reaction product including char content. These second generation bio-fuels from pyrolysis process are incompatible with the conventional fuels mainly due to the higher oxygen content, high solids content, high viscosity, high moisture content, and chemically unstable, along with higher contents of char formation. Hence, upgradation of the bio-oil from the pyrolysis process is highly advisable. Among the various upgradation processes hydrodeoxygenation (HDO) appears to be the promising technique carried out in the presence of a conventional catalyst using H_2 gas. This process is also known as catalytic upgradation of pyrolytic oil. Thus in this work, numerical investigation of upgradation of pine pyrolytic bio-oil in the presence Pt/Al_2O_3 catalyst using H_2 gas has been carried out within the range of operating conditions as: temperature, $T = 623$–673 K, weight hourly space velocity, WHSV = 2 h^{-1}, and a constant pressure of 8720 kPa.

2 Literature Review

Substantial research work on HDO process is mainly initiated by Weisser and Landa (1973) followed by Furimsky (1978, 1979, 1983) based on various feed stocks and different reaction kinetics mechanisms of furan and tetrahydrofuran. Further, Ternan and Brown (1982) studies on HDO of coal distillate liquids with high oxygen content observed higher conversion rates of reactants to products by the reaction of smaller oxygenated compounds (phenols). Later, Elliott and Baker (1984) performed HDO studies on the products of the liquefaction process in the presence of CO at a pressure of 138 bar and at temperatures ranging between 350–450 °C in a batch reactor. Johnson et al. (1988) investigated the conversion of lignin into phenols and hydrocarbons by mild HDO process. Train and Klein (1986) developed Monte-Carlo simulations of HDO of lignin using the kinetics obtained from the experimental results. Elliott (1988) performed two-stage catalytic hydro-treatment in the presence of Co/Mo supported alumina catalyst. The reaction results are significant and further confirmed by Baker and Elliott (1988). The two stages are combined to a single-stage non-isothermal process with a yield of 33 % of lighter hydrocarbons and high gas formation. Sheu et al. (1988) performed experimental studies on the upgradation of pine pyrolytic oil by HDO process in a trickle bed reactor. The authors developed two models, one for the overall oxygen removal and the other for the compositional changes in hydro-treated oil. Piskorz et al. (1988) confirmed the findings of Elliott results particularly the necessity of high hydrogen gas pressure. Baldauf et al. (1994) performed hydroprocessing of flash pyrolysis oil using the conventional Co–Mo, Ni–Mo catalysts in which the major yield of about 60 % is the moisture content with high deoxygenation rates. However, bio-oil contains all other kinds of oxygenated chemical groups like aldehydes, ketones, carboxylic acids, esters, aliphatic and aromatic alcohols (Horne and Williams 1996). Oasmaa and Czernik (1999) performed catalytic hydro-treatment of peat pyrolysate oils in a batch wise reactor at a higher pressure of 21.5 MPa for two hours at 390 °C with 10 wt% catalyst, 4 % CoO, 15.5 % MoO_3, on alumina support. Oxygen content is reduced from 22 to 3 % in the process. Elliott (2007) reviewed the insights and achievements of HDO process in the past 25 years. The important aspect in HDO process is the hydrogen consumption. Recent results of De Miguel Mercader et al. (2010), and Venderbosch et al. (2010) indicate that stabilization of pyrolysis oil by low severity HDO is sufficient for upgradation process and reduces the hydrogen consumption. Xu et al. (2013) proposed two-step catalytic hydrodeoxygenation of fast pyrolysis oil to hydrocarbon liquid fuels. In the first step, organic solvents were employed to reduce the coke formation and promote higher rates of HDO process. In the second step, conventional reaction mechanism of reactants to products in the presence of catalyst along with hydrogen gas is performed to obtain hydrocarbon fuel. Recently, Parapati et al. (2014) mentioned that the customary practice of using two-stage hydro-processing units at moderate temperatures also yields approximately similar to the single-stage reactor. Hence, in order to increase the throughput and decrease the extra financial commitments, a single-stage hydro-processing unit is sufficient. Finally, on the basis of aforementioned detailed literature review, it is clear that the HDO process has been extensively used for the upgradation process of bio-oils; however, optimized operating conditions are not available. Therefore, in this work, a numerical investigation has been carried out to optimize the range of temperature for the upgradation of pine pyrolytic bio-oil using hydrodeoxygenation in the presence of Pt/Al_2O_3 catalyst with WHSV = 2 h^{-1} and P = 8720 kPa at T = 623–673 K.

3 CFD Modeling

The numerical modeling and simulations of reactive multiphase flows in fluidized bed reactors are extremely complex because of the turbulence of flow of gas and catalyst particles. Thus, to make the model reliable turbulent particle flow phenomena, particle turbulence and the turbulence interaction between the fluid and particle phases are included in the present simulation studies. In the present study, Eulerian multiphase fluid flow approach has been chosen to see the particle interaction alongside pursuing the chemical reaction kinetics using a lumped parameter model adopted/marked as benchmark experimental studies of Sheu et al. (1988). To perform these reaction kinetic studies numerically, the chemical species transport of five lumped parameter model is chosen in a commercial computational fluid dynamic-based solver ANSYS Fluent 14.5. By integrating the five lumped kinetic model (Sheu et al. 1988), with the turbulent flow model, a multiphase flow and reaction model is established. The governing equations are described as follows:

3.1 Mass Conservation Equation

The continuity equation for the gas, oil, and particle phases is written as

$$\frac{\partial}{\partial t}\left(\alpha_q \rho_q\right) + \nabla \cdot \left(\alpha_q \rho_q \vec{v}_q\right) = 0 \qquad (1)$$

The species equation of the gas phase is expressed as follows which represents the coupling of hydrodynamics and reaction kinetics

$$\frac{\partial}{\partial t}(\rho Y_i) + \frac{\partial}{\partial x_j}(\rho u_j Y_i) = \frac{\partial}{\partial x_j}\left(\rho D_i \frac{\partial Y_i}{\partial x_j}\right) + R_i + S_i \tag{2}$$

3.2 Fluid–Fluid Momentum Equation

$$\begin{aligned}\frac{\partial}{\partial t}\left(\alpha_q \rho_q \vec{v_q}\right) + \nabla . \left(\alpha_q \rho_q \vec{v_q} \vec{v_q}\right) &= \alpha_q \rho_q \vec{g} + \nabla . \overline{\overline{\tau_q}} - \alpha_q \nabla p \\ &+ \sum_{p=1}^{n} \left(K_{pq}\left(\vec{v_p} - \vec{v_q}\right)\right) + \vec{F_q}\end{aligned} \tag{3}$$

The first term in the left hand of the equation represents the unsteady state acceleration; second term represents convective acceleration of the respective fluids inside the reactor. The right hand side terms denote the summation of different forces acted upon the fluid–fluid interaction.

3.3 Fluid–Solid Momentum Equation

$$\begin{aligned}\frac{\partial}{\partial t}\left(\alpha_s \rho_s \vec{v_s}\right) + \nabla \cdot \left(\alpha_s \rho_s \vec{v_s} \vec{v_s}\right) &= \alpha_s \rho_s \vec{g} + \nabla \cdot \overline{\overline{\tau_s}} - \alpha_s \nabla p \\ &- \nabla p_s + \sum_{l=1}^{n} \left(K_{ls}\left(\vec{v_l} - \vec{v_s}\right)\right)\end{aligned} \tag{4}$$

This equation is quite similar to the Eq. 3 with an additional term on the right side ∇p_s represents the solids pressure due to solids collision which is modeled using kinetic theory of granular flow. In this equation K_{ls} denotes interphase/momentum exchange coefficients between liquid and solid phases as given by Gidaspow et al. (1992) model which is a combination of Wen and Yu (1966) and Ergun and Orning (1952), and is given as:

$$K_{ls} = C_d \frac{3}{4} \rho_l \frac{\alpha_l \alpha_s}{d_s} \left|\vec{u_s} - \vec{u_l}\right| \alpha_l^{-2.65} + \frac{150 \alpha_s (1 - \alpha_l) \mu_l}{\alpha_l d_s^2} + \frac{1.75 \alpha_s \rho_l \left(\vec{u_s} - \vec{u_l}\right)}{d_s} \tag{5}$$

and K_{pq} is between the continuous phases and is given by Schiller and Naumann (1933)

$$K_{pq} = \frac{\alpha_{\mathrm{p}}\rho_{\mathrm{p}}\left(1 + 0.15\mathrm{Re}^{0.687}\right)}{\tau_{\mathrm{p}}} \tag{6}$$

3.4 Enthalpy Conservation Equation

$$\begin{aligned} \frac{\partial}{\partial t}\left(\alpha_{\mathrm{q}}\rho_{\mathrm{q}}h_{\mathrm{q}}\right) + \nabla \cdot \left(\alpha_{\mathrm{q}}\rho_{\mathrm{q}}h_{\mathrm{q}}\overrightarrow{u_{\mathrm{q}}}\right) &= -\alpha_{\mathrm{q}}\frac{\partial_{pq}}{\partial t} + \overline{\overline{\tau}}_{\mathrm{q}} : \nabla\overrightarrow{u_{\mathrm{q}}} - \nabla \cdot \overrightarrow{q_{\mathrm{q}}} \\ &+ S_{\mathrm{q}} + \sum_{p=1}^{n} \frac{\left(6c_{\mathrm{q}}\alpha_{\mathrm{q}}\alpha_{\mathrm{p}}Nu\left(T_{\mathrm{p}} - T_{\mathrm{q}}\right)\right)}{d_{\mathrm{p}}^{2}} \end{aligned} \tag{7}$$

Here, h_{q} is the specific enthalpy of qth (primary) phase, $\overrightarrow{q_{\mathrm{q}}}$ is the heat flux, S_{q} is the source term that includes the sources of enthalpy (e.g., due to chemical reaction or radiation), Q_{pq} is the intensity of heat exchange between pth (secondary phase) and qth phases, and h_{pq} is the interphase enthalpy. Ranz and Marshall (1952) correlation is applied for fluid–fluid multiphase interaction as follows

$$Nu_{\mathrm{p}} = 2.0 + 0.6\mathrm{Re}_{\mathrm{p}}^{1/2}\,\overset{1/3}{\mathrm{Pr}} \tag{8}$$

and for fluid–solid multiphase interaction, Gunn (1978) correlation is applied and given as

$$Nu_{\mathrm{s}} = \left(7 - 10\alpha_{\mathrm{f}} + 5\alpha_{\mathrm{f}}^{2}\right)\left(1 + 0.7\mathrm{Re}_{\mathrm{s}}^{0.2}\,\overset{1/3}{\mathrm{Pr}}\right) + \left(1.33 - 2.4\alpha_{\mathrm{f}} + 1.2\alpha_{\mathrm{f}}^{2}\right)\mathrm{Re}_{\mathrm{s}}^{0.7} \tag{9}$$

3.5 k − ε Turbulence Model

The turbulent kinetic energy (k) equation for the multiphase is defined as

$$\frac{\partial}{\partial t}(\rho k) + \frac{\partial}{\partial x_i}(\rho k u_i) = \frac{\partial}{\partial x_j}\left[\left(\mu + \frac{\mu_{\mathrm{t}}}{\sigma_{\mathrm{k}}}\right)\frac{\partial k}{\partial x_j}\right] + G_{\mathrm{k}} + G_{\mathrm{b}} - \rho\epsilon - Y_{\mathrm{M}} + S_{\mathrm{k}} \tag{10}$$

The dissipation rate equation (ε) of the turbulent kinetic energy for all the phases is given by

$$\frac{\partial}{\partial t}(\rho\epsilon)+\frac{\partial}{\partial x_i}(\rho\epsilon u_i)=\frac{\partial}{\partial x_j}\left[\left(\mu+\frac{\mu_t}{\sigma_\epsilon}\right)\frac{\partial\epsilon}{\partial x_j}\right]+C_{1\epsilon}(G_k+C_{3\epsilon}G_b)-C_{2\epsilon\rho}-\frac{\epsilon^2}{k}+S_\epsilon \tag{11}$$

3.6 Kinetic Theory of Granular Flow

The particle fluctuation is modeled by kinetic theory of granular flow. The equation for the turbulent fluctuating energy of the particle phase is expressed by Gidaspow and Ding (1990)

$$\frac{3}{2}\left[\frac{\partial}{\partial t}(\rho_s\alpha_s\theta_s)+\nabla\cdot\left(\rho_s\alpha_s\theta_s\vec{v_s}\right)\right]=\left(-p_s\bar{\bar{I}}+\bar{\bar{T}}\right):\nabla\vec{v_s}+\nabla\cdot\left(K_{\theta_s}\nabla\theta_s\right)-\gamma\theta_s+\emptyset_{ls} \tag{12}$$

where the first and second terms in the right hand side of Eq. (12) represent the generation of energy by the solid stress tensor, and the diffusion of energy, where K_{θ_s} is the diffusion coefficient given by Gidaspow et al. (1992)

$$K_{\theta_s}=\frac{150\rho_s d_s\sqrt{\theta\pi}}{384(1+e_{ss})g_{0,ss}}\left[1+\frac{6}{5}\alpha_s g_{0,ss}(1+e_s)\right]^2+2\rho_s\alpha_s^2 d_s(1+e_{ss})g_{0,ss}\sqrt{\frac{\theta_s}{\pi}} \tag{13}$$

and $\gamma\theta_s$ is the collision dissipation of energy, $\emptyset_{ls}$ is the energy exchange between the fluid and solid phase. Other model equations can be found in Fluent users guide (Fluent Inc. 2006).

3.7 Lumped Kinetics Model

The reaction pathway (Fig. 1) and kinetic parameters for the present simulation study is adopted from experimental study of Sheu et al. (1988).

Fig. 1 Reaction pathway of hydro-processing of pine pyrolytic oil (Sheu et al. 1988)

The kinetic parameters (Sheu et al. 1988) that are used in the present simulation studies are listed below:

$$r_1 = -k_1\rho_1 \quad \left(E_1 = 74\,\mathrm{kJmol^{-1}},\, A_1 = 3860\,\mathrm{min^{-1}}\right) \tag{14}$$

$$r_2 = k_1\rho_1 - k_2\rho_2 - k_3\rho_2 \quad \left(E_2 = 91.8\,\mathrm{kJmol^{-1}},\, A_2 = 75400\,\mathrm{min^{-1}}\right) \tag{15}$$

$$r_3 = k_3\rho_2 - k_4\rho_3 \quad \left(E_3 = 80.6\,\mathrm{kJmol^{-1}},\, A_3 = 8300\,\mathrm{min^{-1}}\right) \tag{16}$$

$$r_4 = k_2\rho_2 + k_4\rho_3 - k_5\rho_4 \quad \left(E_4 = 62.3\,\mathrm{kJmol^{-1}},\, A_4 = 950\,\mathrm{min^{-1}}\right) \tag{17}$$

$$r_5 = k_5\rho_4 \quad \left(E_5 = 69.6\,\mathrm{kJmol^{-1}},\, A_5 = 4000\,\mathrm{min^{-1}}\right) \tag{18}$$

4 Simulation System and Numerical Procedure

4.1 Simulation System

A schematic representation of the fluidized bed reactor used in the present simulation study is shown in Fig. 2. The dimensions of the reactor are chosen exactly same as in the experimental study due to Sheu et al. (1988). The height of the reactor is chosen to be 0.813 m and the width of the reactor is chosen to be 0.01564 m. The reactor is packed with Pt/Al_2O_3 catalyst volume fraction of 0.0286 with no oil and gas volume fractions at the inlet at time t = 0 s. Pine pyrolytic oil

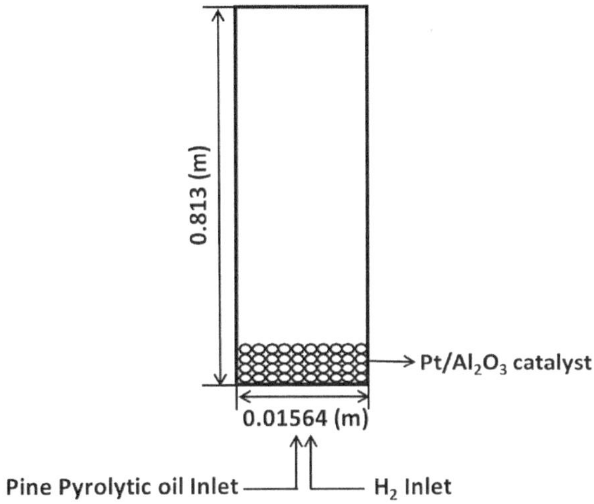

Fig. 2 Schematic representation of fluidized bed reactor model for upgradation of bio-oil

Table 1 Properties of three phases (oil, gas and catalyst) in present simulations

Phase	Compound	Density (kg m^{-3})	Viscosity (Pa s)	Specific heat (Jkg^{-1} K^{-1})	Thermal conductivity (Wm^{-1} K^{-1})
Pine pyrolytic oil phase	HNV	841.15	0.00092	1833.817	0.127
	LNV	679.5	0.00040	2223.19	0.14
	Phenol	1030	0.1842	1430	0.19
	Aromatics	880	0.00081	1699.84	0.131
	Alkane	0.669	0.000018	2222	0.033
Gas phase	H_2 gas	0.8189	0.000008	14283	0.167
	Water vapor	0.5542	0.000013	2014	0.0261
Catalyst phase	Pt/Al_2O_3	21450	0.000017	130	71.6
	Coke	375	1.206	850	0.2

consists of various lumping groups including heavy nonvolatile (HNV), light non-volatile (LNV), phenols, alkanes aromatics mixture, and H_2 gas is introduced from the bottom of the reactor to pass through the catalyst bed. The properties of the three phases used in the present simulation studies are listed in Table 1.

4.2 Numerical Procedure

The governing differential equations (as presented in the previous section) are solved using commercial computational fluid dynamics-based software ANSYS Fluent 14.5 and all simulations are carried out in two-dimensional computational domain. The grids are generated using Gambit 2.2.30 and are exported to Fluent 14.5 software. The structured hexahedral grid is used with a total mesh number of 6000 cells, 12170 faces with 7000 node points. The bottom inlet of the reactor is designated as "velocity inlet" boundary in Fluent software. At the inlet, the velocities, mass fractions, volume fractions, temperatures of the three phases are specified according to the desired operating conditions. The outlet at the top of the reactor is set as "pressure outlet" boundary; and for the other two walls, the boundary conditions are specified as "wall" boundaries which are defined as no-slip boundaries for all the phases.

4.3 Model Validation

The numerical solver used in the present study is validated by comparing the present values of the mass fractions of the individual species of the oil phase with the existing experimental literature of Sheu et al. (1988). As the research related to the upgradation of bio-oil using HDO process through a numerical approach is very scarce, and hence to the best of the knowledge only Sheu et al. (1988) reported their experimental results in terms of numeric for the HDO process. The present results are in close proximity with the existing literature values which boosted us the confidence to proceed further in order to check the effect of temperature at a constant pressure and at a constant catalyst loading values.

5 Results and Discussions

5.1 Hydrodynamics

Figure 3 shows the volume fraction distribution of all phases at different operating temperatures ranging between 623 K < T < 673 K at a constant WHSV = 2 h^{-1} with a bio-oil flow rate of 0.033 × 10^{-6} m^3 s^{-1} and H_2 gas flow rate of 100 cc $min^{-1}g^{-1}$, whereas the velocity of the gas feed is used in the calculation of total velocities of feed to the reactor at a constant pressure P = 8720 kPa. Figures 4, 5, and 6 illustrate the prototype contours of volume fraction representing the flow behavior of catalyst, pine oil, and hydrogen gas in the reactor for different time steps ranging between t = 0 and t = 150 s at T = 673 K. At t = 150 s the bed has reached pseudo steady state with no further deviations with respect to bed characteristics and temperature, the averaged results are plotted and analyzed. From Fig. 3, it is seen that the volume fraction of the catalyst is slightly decreased with increase in the temperature. This denotes the utilization of the catalyst in continuous formation of different product fractions leads to catalyst deactivation. Further, the formation of the aromatics enables the coke formation which is deposited on the surface of the

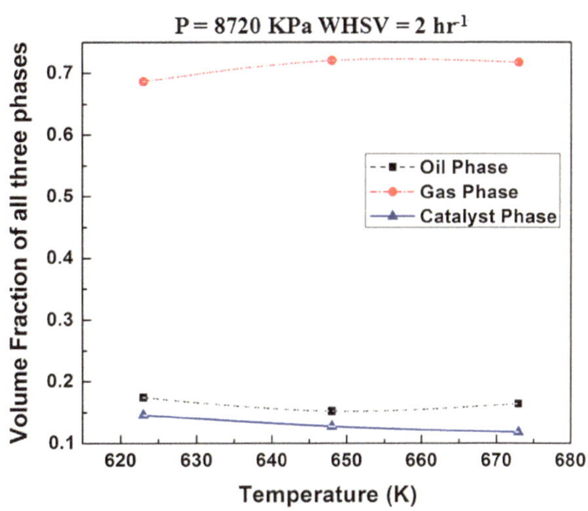

Fig. 3 Volume fraction of all phases varying with respect to temperature

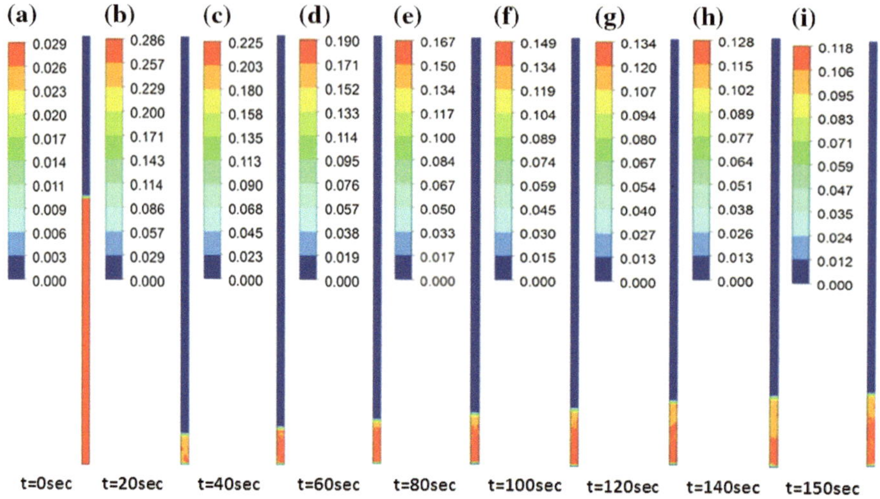

Fig. 4 Catalyst volume fraction images of a fluidized bed reactor at T = 673 K and 8720 kPa

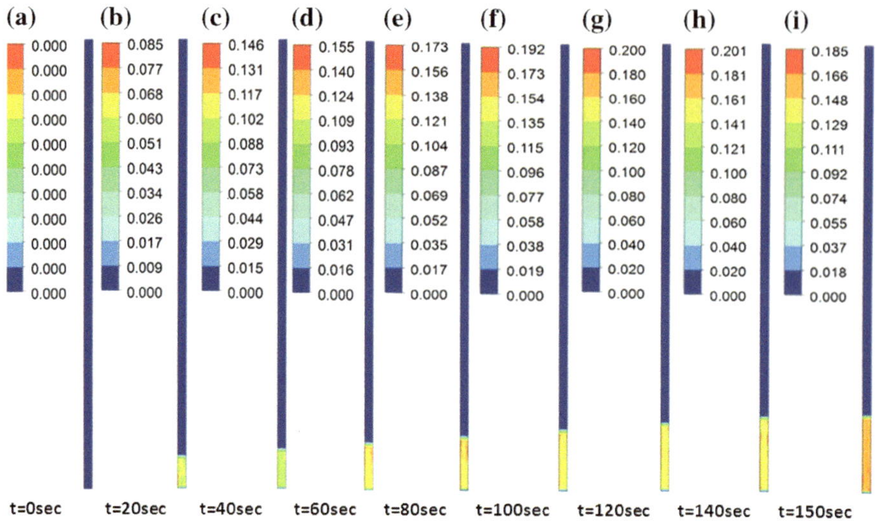

Fig. 5 Pine oil volume fraction images of a fluidized bed reactor at T = 673 K and 8720 kPa

catalyst leads to catalyst deactivation. Hence, proper measures are to be taken to maintain the optimum physiochemical properties.

The volume fraction of oil is slightly decreasing till T = 648 K and further there is no deviation observed even with the increase in the temperature as seen in Fig. 3. This denotes that the increase in the temperature enhances the thermodynamic equilibrium which increases the cracking rate to convert HNV, LNV, phenol (oil

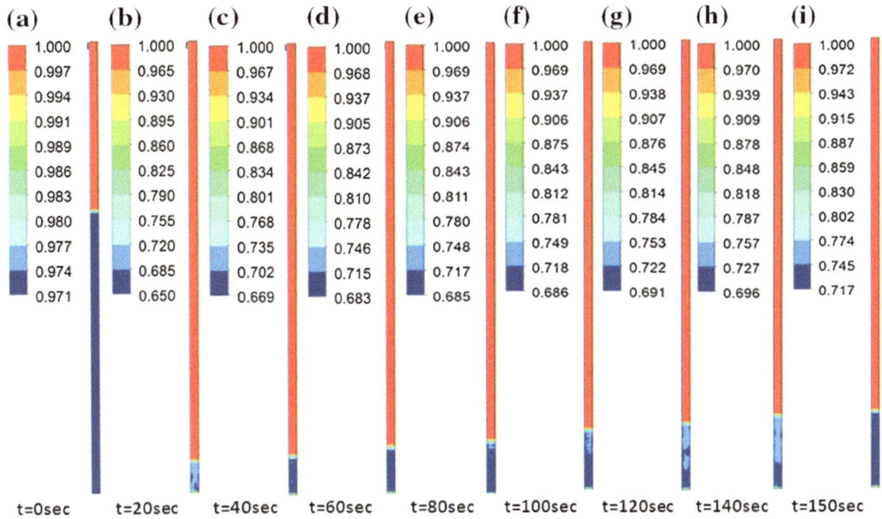

Fig. 6 H_2 gas volume fraction images of a fluidized bed reactor at $T = 673$ K and 8720 kPa

phase nonvolatiles) to desired product streams (gas phase), thus decreasing the volume fraction of the oil phase. Finally, the volume fraction of the gas increases gradually till $T = 648$ K and further shows a constant trend. As mentioned earlier, the converted oil phase reactants are transformed to gas phase products, thereby enhancing the volume fraction of the gasses. In summary, it is observed that the change in the volume fraction is significant at an intermediate temperature of $T = 648$ K in oil and gas phases and further no significant changes is observed.

5.2 *Reaction Kinetics*

The bio-oil mixture containing many organic compounds undergoes several hydro-treating processes simultaneously, for removing the oxygen in the form of water and hydrocracking step for reducing heavy components to middle components. Figure 7 depicts the prototype of the mass fraction contours of three species (oil, gas, catalyst) of upgraded bio-oil in a fluidized bed reactor at WHSV = 2 h^{-1}, pressure $P = 8720$ kPa, and temperature $T = 673$ K. The initial and upgraded mass fractions of three phases (oil, gas, and catalyst) in fluidized bed resulted through HDO are presented in Table 2 and the corresponding values are in close proximity with the experimental (literature) values. Table 3 denotes the product mass fraction species of the lumped bio-oil after HDO at different temperatures and at constant pressure, WHSV. These values are obtained through a post-processing tool of CFD-POST which is an integral part of ANSYS Fluent 14.5. The respective reaction pathway shown in Fig. 1 favors the formation of alkanes and aromatics,

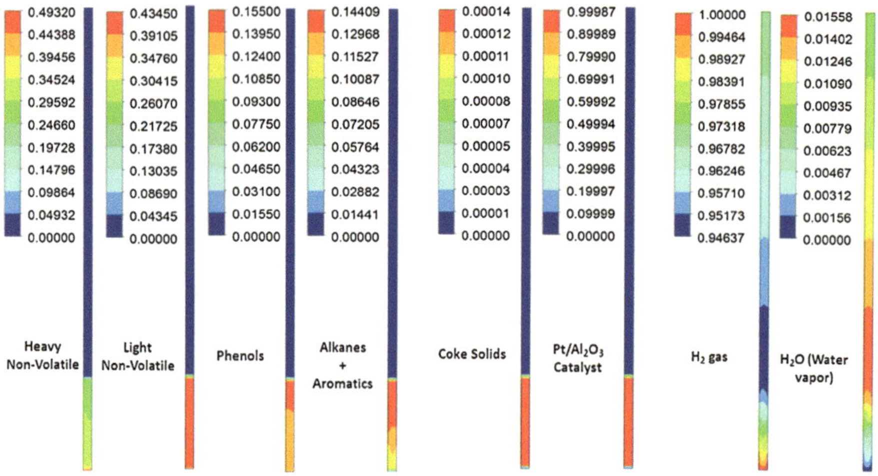

Fig. 7 Upgraded pyrolytic oil species mass fractions of (oil, gas and solid catalyst) for temperature T = 673 K and pressure P = 8720 kPa

Table 2 Mass fraction of lumped product species in the fluidized bed reactor at T = 673 K, P = 8720 kPa and WHSV = $2h^{-1}$

Lumped Fraction	Pine pyrolytic oil (wt%) (Sheu et al. 1988)	Hydrotreated pyrolytic oil (wt%)	
		Experimental (Sheu et al. 1988)	Present
HNV	0.4932	0.2457	0.2102
LNV	0.3690	0.2941	0.4083
Phenol	0.1232	0.1063	0.1552
Alkane + Aromatic	0.0146	0.1952	0.1962
Coke + H_2O + gasses	0	0.1587	0.000169

Table 3 Mass fraction of lumped compounds (wt%) in the fluidized bed reactor at different temperatures and constant pressure P = 8720 kPa

Temperature	Heavy non volatiles	Light non volatiles	Phenol	Alkane + Aromatic
T = 623 K	0.2651	0.4389	0.1486	0.1538
T = 648 K	0.2414	0.4363	0.1456	0.1765
T = 673 K	0.2102	0.4083	0.1552	0.1962

which are replaceable to conventional gasoline and middle distillates from non-volatile, volatile, and phenol compounds of unprocessed bio-oil. It is observed that the mass fraction of heavy nonvolatiles are decreased drastically from 0.4962 wt% to the range of 0.21–0.25 wt% signify the conversion of heavy nonvolatiles to alkanes and aromatics. The massive formation of undesirable LNV attributes to many secondary reactions occurred during hydro-treating and hydrocracking steps.

Very small quantities of LNV are converted to phenol which is further reduced to a combination of alkanes and aromatics. This clearly shows that the phenol reaction rates of formation and utilization are equal.

In summary, there is no much deviation in the mass fractions of product species formed from HDO of unprocessed bio-oil at different temperature ranges. The formation of alkane and aromatics is highly desirable at higher temperatures as shown in Table 3 along with higher quantities of deleterious species (lighter non-volatiles) needs to be processed further.

6 Conclusions

A CFD prototype for fluidized bed reactor for upgradation of pine pyrolytic bio-oil using hydrodeoxygenation using Pt/Al_2O_3 catalyst is successfully modeled in ANSYS Fluent 14.5 incorporating all the continuity, momentum, energy equations, drag behavior exhibited by the fluid and solid particles, turbulence motion, and lumped species reactions. Simulations are conducted at different operating temperatures to examine the variation in mass fractions of product compounds. As observed, there is a significant change in mass fractions of alkane and aromatic mixture species as temperature increases. It is also noted that the higher nonvolatile formation is due to the secondary reactions during hydrocracking mechanism. Significant conversion of heavy nonvolatiles to alkane and aromatics is observed through HDO process.

References

Baker EG, Elliott DC (1988) Catalytic upgrading of biomass pyrolysis oils. In: Bridgwater AV, Kuester JL (ed) Research in thermochemical biomass conversion, pp. 883–895. Elsevier Science Publishers, England

Baldauf W, Balfanz U, Ruppa M (1994) Upgrading of flash pyrolysis oil and utilization in refineries. Biomass Bioenergy 7:237–244

De Miguel Mercader F, Groeneveld MJ, Kersten SRA, Schavaerien NWJ, Hogendoorn JA (2010) Production of advanced bio-fuels: Co-processing of upgraded pyrolysis oil in standard refinery units. Appl Catal B 96:57–66

Ding J, Gidaspow D (1990) A bubbling fluidization model using kinetic theory of granular flow. AIChE J 36:523

Elliott D (1988) Relation of reaction time and temperature to chemical composition of pyrolysis oils from biomass: Producing, analyzing, and upgrading. In: Soltes EJ, Milne TA (ed) ACS Symp. Ser 376:55–65

Elliott DC (2007) Historic developments in hydroprocessing bio-oils. Energy Fuels 21:1792–1815

Elliott DC, Baker EG (1984) Upgrading biomass liquefaction products through hydrodeoxygenation. Biotechnol Bioeng Symp 14:159–174

Ergun S, Orning AA (1952) Fluid flow through packed columns. Chem Eng Prog 48:89–94

FLUENT Inc (2006) FLUENT User's Guide. FLUENT Inc

Furimsky E (1978) Catalytic deoxygenation of heavy gas oil. Fuel 57:494–496

Furimsky E (1979) Catalytic removal of sulfur, nitrogen, and oxygen from heavy gas oil. AIChE J 25:306–311

Furimsky E (1983) Chemistry of catalytic hydrodeoxygenation. Catal Rev: Sci Eng 25:421–428

Gidaspow D, Bezburuah R, Ding J (1992) Hydrodynamics of circulating fluidized beds, Kinetic theory approach. Fluid VII. In: 7th proceedings of the engineering foundation conference fluid, pp 75–82

Gunn DJ (1978) Transfer of heat or mass to particles in fixed and fluidized beds. Int J Heat Mass Transfer 21:467–476

Horne PA, Williams PT (1996) Reactions of oxygenated biomass pyrolysis model compounds over a ZSM-5 catalyst. Renew Energy 7:131–140

Johnson DK, Ratcliff MA, Posey FL, Maholland MA, Cowley SW, Chum HL (1988) Hydrodeoxygenation of a lignin model compound. In: Bridgwater AV, Kuester JL (ed) Research in thermochemical biomass conversion, pp. 941–955. Elsevier Science Publishers, England

Oasmaa A, Czernik S (1999) Fuel oil quality of biomass pyrolysis oils-state of art for end users. Energy Fuels 13:914–921

Parapati DR, Vamshi KG, Venkata Penamatsa K, Phillip SH, Satish Tanneru K (2014) Single stage hydroprocessing of pyrolysis oil in a continuous packed bed reactor. AIChE J 33:726–731

Piskorz J, Scott DS, Radlein D (1988) Composition of oils obtained by fast pyrolysis in different woods, in pyrolysis oils from biomass: Producing, analyzing and upgrading, Soltes EJ, Milne TA (eds) ACS Symp Ser 376:167–178

Ranz WE, Marshall WR (1952) Evaporation of drops, Part 1. Chem Eng Prog 48:141–146

Schiller L, Naumann A (1933) Uber die grundegenden berch-nungen bei der schwerkraftaufbereitung. Z Ver Dtsch Ing 77:318–320

Sheu EYH, Rayford GA, Soltes EDJ (1988) Kinetic studies of upgrading pine pyrolytic oil by hydrotreatment. Fuel Process Technol 19:31–50

Ternan M, Brown JR (1982) Hydrotreating a distillate liquid derived from sub-bituminous coal using a sulfide $CoO\text{-}MoO_3/Al_2O_3$ catalyst. Fuel 61:110–118

Train PM, Klein MT (1986) Chemical and stochastic modeling of lignin hydrodeoxygenation. ACS Div Fuel Chem 32:240

van Ruijven B, van Vuuren DP (2009) Oil and natural gas prices and greenhouse gas emission mitigation. Energy Policy 37(11):4797–4808

Venderbosch RH, Ardityanti AR, Wildschut J, Oasmaa A, Heeres HJ (2010) Stabilization of biomass derived pyrolysis oils. J Chem Technol Biotechnol 85:674–686

Weisser O, Landa S (1973) Sulphide catalysts: Their properties and applications. Academia publishing house, Pergamon, Prague

Wen CY, Yu YH (1966) Mechanics of fluidization. Chem Eng Prog Symp Ser 62:100–111

Xu X, Zhang C, Liu Y (2013) Two–step catalytic hydrodeoxygenation of fast pyrolysis oil to hydrocarbon liquid fuels. Chemosphere 93:652–660

Sustainability Assessment of Biomass Gasification Based Distributed Power Generation in India

Amit Kumar Singh Parihar, Virendra Sethi and Rangan Banerjee

Abstract Access to energy is a precursor to sustainable development. In India, at present, the number of people without access to electricity remains high, despite several efforts by the government. Non-feasibility of grid extension is one of the major reasons for this poor energy access. Even in the areas with grid connectivity, reliable electricity supply still remains a challenge. Biomass gasification based renewable distributed power generation sources may provide electricity access by generating it at the point of consumption using locally available resources. Most of the prior studies have evaluated biomass gasification system based on cost of electricity generation neglecting its impact on environment and society. Instead, sustainability aspects including social, technical, environmental, and resource should be an integral part of this practice along with economic aspect and compared in transparent way. This not only necessitates identification of appropriate indicators to reflect relevant sustainability dimensions, but also requires development of general framework comparing different qualitative and quantitative indicators. Framework should bring out various trade-offs and bounce associated with indicators. This paper analyses different indicators and assessment methodologies including single-criterion and multi-criteria decision methods discussed in the literature for sustainability assessment of power generation systems followed with specific focus on sustainability assessment methodologies suggested for biomass gasification systems. Research gaps with scope for conceptual contributions in the prevailing sustainability assessment process from selection of appropriate indicator to assessment methodologies are discussed. Based on the literature review, this article identifies appropriate criteria and indicators to be considered for assessing the sustainability of biomass gasification based power generation systems. A total of 18 indicators representing economic, technical, environment, resource, and social aspect of sustainability along with their relevance and importance for sustainability assessment are discussed. The set of indicators identified may be used for comparing sustainability of possible distributed power generation systems in Indian context.

A.K.S. Parihar (✉) · Virendra Sethi · Rangan Banerjee
Department of Energy Science & Engineering, Indian Institute of Technology Bombay, Bombay, India
e-mail: aksparihar@iitb.ac.in

S. Kumar et al. (eds.), *Proceedings of the First International Conference on Recent Advances in Bioenergy Research*, Springer Proceedings in Energy,
DOI 10.1007/978-81-322-2773-1_16

Keywords Sustainability assessment framework · Criteria and indicators · Life cycle assessment · Multi-criteria decision-making

1 Introduction

Access to energy is a precursor to sustainable development (Demirtas 2013). In India, around 70 % of the total population lives in remote areas where grid electricity is either not available or suffers with poor supply if available (Castellanos et al. 2015). While percentage village electrification has improved due to various rural electrification schemes including RGGVY, percentage of household electrification still remains low (Harish et al. 2014). Further, electrification through conventional fuel-based power generation leads to environmental pollution due to GHGs emission. Renewable energy based distributed power generation systems are being looked upon as potential source to solve this problem of providing electricity without causing damage to the environment. Due to local availability of fuel and carbon neutral characteristics, biomass gasification based distributed power generation systems may be one potential option among various renewable based distributed power generation systems. However, field level success of this technology has not been able to provide desired success due to reasons including combination of cost and non-cost factors. Therefore, there is a need to examine impact of non-cost parameters along with cost parameters on long-term sustainable operation of these systems. Most of the prior relevant studies have focused on estimating the cost of electricity generation (Banerjee 2006; Nouni et al. 2007; Palit et al. 2011) neglecting other parameters related to technical, environment, resource, and social aspect.

This paper analyses different indicators and assessment methodologies suggested for sustainability assessment of distributed power generation systems with focus on small scale biomass gasification based power generation system in India. The typical scale being targeted in the present study ranges from few kW to 150 kW, as this range is sufficient to meet electrical needs of a typical village or clusters of villages in off-grid mode. Gaps in present methodologies are identified. Finally, present study proposes 18 indicators belonging to economic, technical, environment, resource, and social aspect of sustainability along with discussion about their relevance and importance for developing a general and transparent sustainability assessment framework.

2 Sustainability Assessment

The Brundtland Report initiated the concept of "sustainability" defined as "to satisfy our needs without compromising the ability of future generations to meet their own needs" (Brundtland 1987). The process of evaluating most suitable technological option considering multiple parameters may be termed as sustainability assessment.

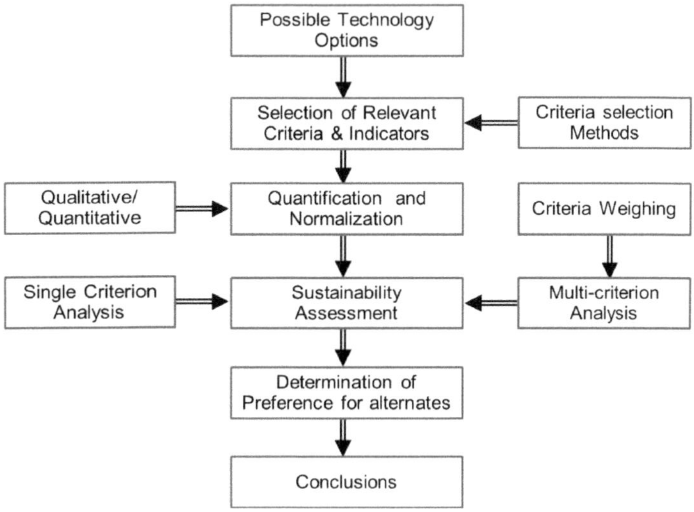

Fig. 1 Flowchart for general sustainability assessment methodology

While economic, environment, and social are the three major dimensions suggested for sustainability assessment, technical and resource dimensions are also reported for sustainability assessment of renewable energy systems (Frangopoulos and Keramioti 2010; Afgan et al. 2000; Wang et al. 2009).

Figure 1 depicts the flow chart of general sustainability assessment process as reported in the literature. The process starts with identification of probable alternatives followed by identification of relevant criteria and indicators using which the various alternates are to be compared. Wang et al. (2009) have discussed guiding principle as well as elementary methods for selection of suitable criteria and indicators for sustainability evaluation. Selected indicators are scored for each alternate option and normalized on a common scale to facilitate comparison. Thereafter, technology alternates are ranked using either single- or multi-criteria assessment methodology. In case of multi-criteria assessment, selected indicators and criteria are assigned weights to reflect relative importance. Finally these weightage and indicator scores are aggregated using multi-criteria decision-making methods to rank various alternate options (Wang et al. 2009; Frangopoulos and Keramioti 2010).

3 Sustainability Criteria and Indicators

In the sustainability assessment process, evaluation of alternatives is carried out considering different criteria comprising of relevant indicators (NEEDS 2009). As technology alternatives are being ranked using sustainability criteria and indicators, selection of relevant indicators becomes important for sustainability assessment.

Several studies provide indicator selection methodologies (Wang et al. 2009; Musango et al. 2011; Dale et al. 2013). Selected indicators should be simple, quantifiable, and understandable. For sustainability assessment of renewable power generation systems technical, economic, environmental, material requirement, and social are some of the main criteria used (Wang et al. 2009; Afgan et al. 2000). Some of the selected criteria and indicators reported in the literature for sustainability assessment of power generation systems are depicted in Fig. 2. There is no clear demarcation for few indicators as they have been grouped under different criteria. For example, efficiency has been considered under economic criteria, whereas it may be more appropriate under technical criteria. Some of commonly used indicators in sustainability assessment studies are capital cost (Rs/kW), cost of electricity generation (Rs/kWh), GHG emissions (kg CO_2/kWh), number of jobs created (Man-hr/kWh), etc. (Varun et al. 2009; Streimikiene 2010; Wang et al. 2009; Dale et al. 2013; Evans et al. 2010). In case of sustainability assessment of renewable energy technologies capacity factor, land and water requirement are also used for evaluation (Chatzimouratidis and Pilavachi 2009; Evans et al. 2009; Manish et al. 2006).

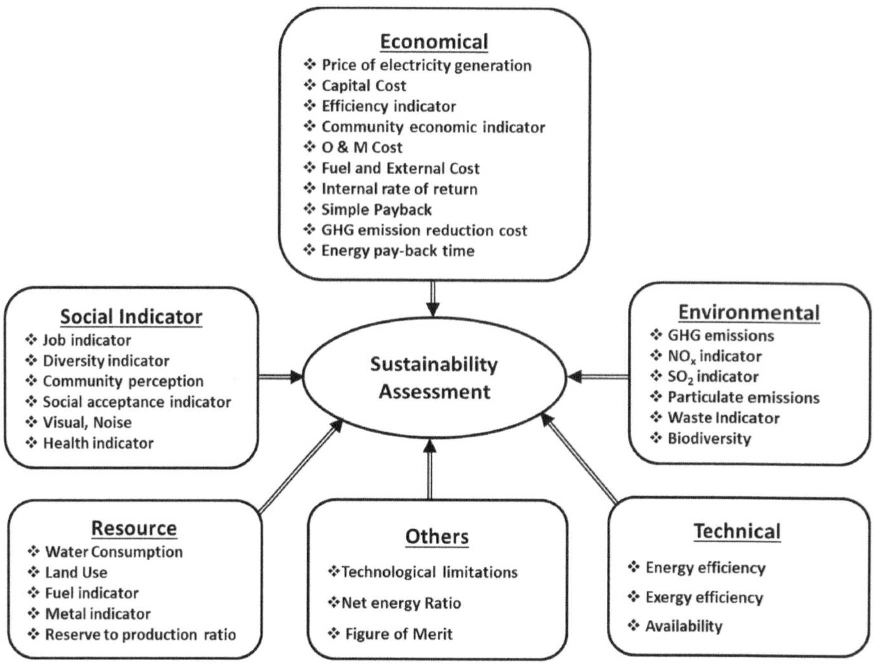

Fig. 2 Different categories of indicators and sub-indicators for sustainability assessment (compiled from Evans et al. 2009; Varun et al. 2009; Onat and Bayar 2010; Afgan et al. 2000, 2002; Dale et al. 2013; Evans et al. 2010; Manish et al. 2006; Buytaert et al. 2011; Dimakis et al. 2011; Streimikiene 2010; Wang et al. 2009)

In addition to the conventional indicators, some specific indicators including net energy ratio (Manish et al. 2006), energy payback time (Varun et al. 2010) have been deployed to assess sustainability of renewable energy technologies. The concept of net energy ratio defined as ratio of total energy output of the system to the total non-renewable energy input is important for assessing sustainability of renewable energy systems as it reflects total non-renewable energy consumed as against its renewable energy generation over its complete life cycle. Its less than one value (<1) will imply that plant consumes more non-renewable energy during its construction, operation, and decommissioning than it generates during its lifetime and therefore may not be considered as renewable and vice versa. Similarly, energy payback time reported by Varun et al. (2010) is defined as ratio of total primary energy required throughout its life cycle in (GJ) to annual primary energy generation by the system in (GJ/year). These indicators are more relevant and useful in comparing sustainability of renewable power generation systems.

4 Sustainability Assessment Methodology

Single-criterion analysis and multi-criteria analysis are the two major assessment methodologies reported in the literature for sustainability evaluation of power generation systems (Afgan et al. 2000; Bhattacharyya 2012). In single-criterion analysis, probable technology options are compared with respect to a single indicator at a time (Evans et al. 2010; Streimikiene 2010). Mostly, researchers have used cost of electricity generation to compare different technological options (Nouni et al. 2007; Banerjee 2006). Nouni et al. (2009) have made economic comparison considering cost of grid extension as baseline. Using single-criterion analysis may be simple but requires high data accuracy to avoid any misinterpretation (Afgan et al. 2000). Further, single-criterion analysis neglects other relevant aspect including impact to environment and society which becomes imperative to be considered in sustainability evaluation studies (Bhattacharyya 2012). Multi-criteria analysis on other hand, overcome such limitations by comparing probable alternatives using different set of criteria and indicators. However, indicators need to be assigned weights to quantify their relative importance towards assessing the sustainability of alternatives. The selected indicators may be assigned the weights by either using equal weighing methods or by rank order methods. Though, assigning weightage using equal weighing to all indicators is a simple process, it has certain limitations of ignoring relative importance among criteria. In rank order weighing methods, indicators are given different weights using different subjective, objective, or mix weighing methods. In subjective weighing methods, preference of decision makers dominates whereas in objective weighing methods, weights are decided mathematically using available data. Mix weighing method incorporates both to avoid limitation of both the methods. A detailed description of these methods is provided in Wang et al. (2009).

Once the weights are assigned, next step in the process is the evaluation of different technical options using multi-criteria decision-making methods. The literature provides detailed description of several multi-criteria decision-making methods along with their pros and cons (Pohekar and Ramachandran 2004; Wang et al. 2009; Bhattacharyya 2012). In evaluation of different renewable energy systems, weighted sum method (WSM) and analytical hierarchy process (AHP) are the most widely used methods. In weighted sum methods, the product of score and the weight of each indicator are added to obtain the overall score of individual alternative. Among different multi-criteria decision-making methods, AHP is found to be the most popular method as it facilitates better comparison of indicators through pairwise comparison of evaluation criteria and alternative (Bhattacharyya 2012; Stein 2013; Wang et al. 2009). Although inclusion of weights to indicator may be helpful in making multi-criteria analysis simple, there is always scope of inherent subjectivity and biasness. This could be avoided to some extent by adopting participatory approaches in deciding weights to individual indicators.

Assessing sustainability of technologies using life cycle methodology has also been discussed in the literature (Finkbeiner et al. 2010). Pairwise comparison is the other method of sustainability assessment in which two indicators are compared simultaneously (Manish et al. 2006). Using this approach, Manish et al. (2006) have assessed the sustainability of power generation systems comparing life cycle GHG emissions and life cycle cost against net energy ratio.

5 Sustainability Assessment of Biomass Gasification Systems

The need for sustainability assessment of bioenergy systems is very well contextualized in prior studies. Most of these studies focused on biofuel generation from biomass rather than power generation as evident from Table 1 summarizing bioenergy sustainability assessment studies.

Evans et al. (2010) have assessed sustainability of biomass based power generation systems using data reported in the literature. However, technologies considered in the study include biomass based power generation using combined cycle operation mostly at MW scale. The indicators considered in the study included cost of electricity generation, efficiency, greenhouse gas emissions, availability, limitations, land use, water use, and social impacts. The study found biomass based power generation systems unfavorable on land, water requirement, and social impacts. This is due to the fact that the study has assessed sustainability over complete cycle of biomass. However, this may not be true if agriculture residues are available as byproduct. Further, score of some of the indicator may be different for small scale biomass gasification based power generation using reciprocating engine. For example, for smaller scale power generation, biomass gasification has better efficiency than biomass combustion based power generation technologies (Mukunda 2011).

Table 1 Summary of studies on sustainability assessment of bioenergy systems

Reference	Focus of the study	Remarks
Dale et al. (2013)	Selection of Indicators for socio-economic sustainability of bioenergy systems	• More focus on indicator selection as compared to assessment methodology development • Over complete supply chain, more focus is on fuel supply side than conversion technology
McBride et al. (2011)	Identification of 19 measurable indicators to assess environmental sustainability of bioenergy system	
Buchholz et al. (2009)	Identification and ranking of sustainability criterion for bioenergy systems based on expert feedback	
Myllyviita et al. (2013)	Framework development for sustainability assessment of wood-based bioenergy systems	• Discussion on economic, ecological, social, and cultural dimensions of sustainability • No discussion on technical and resource aspect • Discussion mostly around biomass combustion based technologies
Sadamichi et al. (2012)	Development of sustainability assessment methodology for bioenergy utilization in eastern Asia	• Sustainability assessment methodology development considering GHG Savings, Total Value Added and Human Development Index • Methodology used as pilot for biomass to fuel generation case study
Bürgi (2003)	To compare biomass gasification with other CDM projects	• Carry out sustainability assessment using Multi-Attributive Utility Theory (MAUT) Methodology to make comparison with other CDM projects • Technology discussed is biomass gasification in dual fuel operation mode • No indicators selected for net energy analysis • Grid supply and diesel generators are considered as baseline • Assessment methodology included equal weighing of criteria
Evans et al. (2010)	Sustainability assessment of biomass power generation systems	• For MW scale biomass based power generation technology using turbine as prime-mover • Found technology favorable on price, efficiency, emissions, availability • Unfavorable on high land, water usage and social impacts

Myllyviita et al. (2013) developed framework for sustainability assessment of wood-based bioenergy systems considering economic, ecological, social, and cultural dimensions of sustainability. The framework suggested in the study incorporates weighing of criteria to calculate aggregate sustainability index to compare sustainability of local heat production plant, a combined heat and power production plant, and a wood pellet processing plant. However, study does not include technical and resource criteria to assess sustainability which may also be critical in deciding long-term sustainability of biomass gasification based power generation system.

Burgi (2003) have applied multi-attributive utility theory (MAUT) to assess sustainability of biomass gasification system operating in dual fuel mode to compare it with other CDM projects. In the proposed methodology, values of selected indicators are converted to a single utility value between −1 and 1 depending upon its performance against baseline value. Electricity generation from diesel generator and grid electricity supply are considered as base case scenario. The proposed methodology considered equal weighing of indicators.

6 Gaps in the Literature

Based on the literature review, key gaps identified in the prevailing sustainability assessment methodologies for power generation systems are briefly summarized in Fig. 3 and discussed in following section.

- As sustainability assessment includes evaluation of technology options against multiple contrasting criteria, assessment framework should be able to bring out various trade-offs in transparent manner. Most of the earlier studies on sustainability assessment appear to be limiting on this front. Saunders and Pope (2013) in their study about conceptualizing and managing trade-offs in sustainability assessment have discussed the concept of negotiable and non-negotiable trade-offs.
- Most of the prior studies lack to provide general framework for sustainability assessment without incorporating location effect. To further explain, several

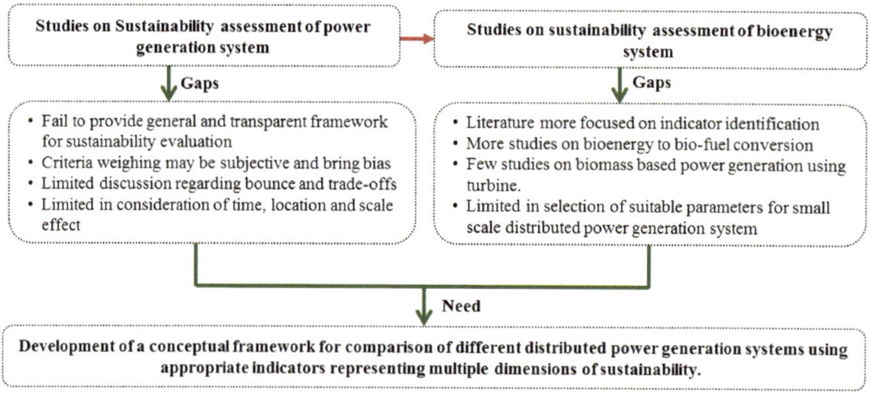

Fig. 3 Key gaps identified in the prevailing sustainability assessment methodologies

studies comparing different renewable energy systems reports wind based power generation as most sustainable solution which may only be true for location having favorable wind speeds.

- Almost all the multi-criteria decision-making methods used for sustainability assessment offers weightage to assign relative importance to different criteria. This process of criteria weighing may be subjective and therefore may introduce bias. Further, some of the indicators used for sustainability assessment fail to justify weightage against their importance. For example, significant weightage given to energy efficiency may only be justified in the case of scarce input resources. Further, there is very limited clarity regarding consideration of time and scale effect on sustainability assessment methodologies being reported in the literature.
- Sustainability assessment methodologies should consider impact of one indicator on other by establishing interrelationship between identified indicators. Prevailing studies lack in establishing this interlinkage in the assessment methodologies. For example, part load performance characteristics of technology would affect overall efficiency thereby cost of electricity generation.
- Most of the studies carried out in bioenergy domain are more focused on sustainability assessment of bioenergy to biofuel conversion with almost negligible focus on biomass gasification based distributed power generation technologies. The studies carried out on biomass gasification based power generation systems are mostly focused on economic aspect without incorporating impact on environment and society.

Considering the above mentioned identified gaps, this article proposes that a set of suitable indicators should be identified and considered for sustainability evaluation of biomass gasification systems in Indian context. Detailed description of these indicators is provided in following section.

7 Selection of Indicators for Proposed Study

There is no common consensus related to selection of sustainability criteria and indicators for assessing the sustainability of distributed power generation systems as these are dependent on variety of factors including location, size, etc. (Buchholz et al. 2009). Based on the literature review and scope of present study, a total of 18 indicators belonging to five dimension of sustainability are proposed as depicted in Fig. 4. The selected indicators including mix of qualitative and quantitative parameters represent economic, environmental, technical, resource, and social dimensions of sustainability.

Under economic criteria, capital cost of technology (Rs/kW) and cost of electricity generation (Rs/kWh) are considered since upfront requirement of higher capital cost for deployment of renewable energy technologies may limit developers to invest in distributed power generation systems for rural electrification. Further,

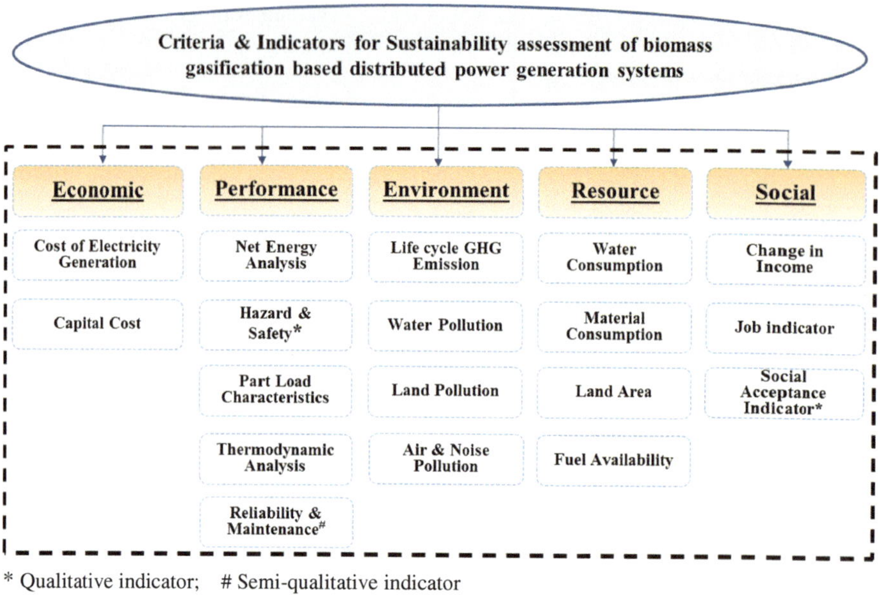

Fig. 4 Proposed criteria and indicators for sustainability assessment of biomass gasification based distributed power generation systems

cost of electricity generation from probable distributed power generation systems has to be at par with the paying capacity or cost of electricity generation from reference system to make it sustainable on longer term.

The technical criteria in the proposed study encompasses life cycle net energy analysis, hazard and safety associated with the technology, part load performance characteristics, thermodynamic analysis, reliability, and maintenance. Prior studies on sustainability assessment have been limited in addressing these parameters which are found to be the reasons for limited success of many small scale distributed power generation systems installed in the field. In case of biomass gasification based power generation systems, there are maintenance related issues necessitating requirement of trained man power for day-to-day operation. Further, part load characteristic of technology is considered as indicator because part load operation of technology affects overall efficiency and therefore higher cost of electricity generation. Hazard and safety and reliability have also been considered, as technologies facing these issues may not be sustainable on longer run.

On environmental front, life cycle GHG emission, water pollution, land pollution, air and noise pollution are considered as sustainability indicators because any technology has to fulfill regulatory pollution norms to be acceptable. The fourth criterion of sustainability proposed in the current study is local resource availability without which no technology can be sustainable on longer run. Water consumption, material consumption, land area, fuel requirement are the indicators proposed to be used for the current study. In case of biomass gasification systems, if the biomass fuel is not available locally, then transporting it from some other location would add to fuel cost resulting into higher cost of electricity generation. This may adversely affect sustainability on economic front.

Finally, for any technology to be sustainable and socially acceptable, it should have positive impacts on society. In the present case, change in the income level, number of jobs created, and social acceptance indicators are the indicators proposed to be used for sustainability assessment of biomass gasification based distributed power generation systems.

8 Summary

This paper reviews sustainability criteria and indicators along with various sustainability assessment methodologies discussed in the literature for assessing sustainability of power generation technologies. In case of sustainability assessment of renewable energy systems, technical and resource criteria are also considered in addition to economic, environment, and social criteria. Sustainability assessment methodologies using single-criterion analysis are mostly focused on assessing the cost of electricity generation without considering their impact on environment and society. Single-criterion analysis is although simple but heavily depends on the accuracy of data. Limitations of single-criterion analysis are overcome using multi-criteria analysis which evaluates different technology alternatives considering multiple parameters. Several methods of multi-criteria decision-making are suggested in the literature. However, weighted sum method and analytical hierarchy process (AHP) are most widely used for sustainability assessment of power generation technologies. Prevailing assessment methodologies mostly use weights to rank importance of indicators and appear to be limited in providing general and transparent framework for sustainability assessment of renewable energy systems. Total of 18 indicators representing economic, technical, environment, resource, and social aspect of sustainability are identified for assessing the sustainability of biomass gasification based distributed power generation systems. The set of indicator set identified may be used for comparing sustainability of different distributed power generation systems in Indian context.

References

Afgan NH, Carvalho MG (2002) Multi-criteria assessment of new and renewable energy power plants. Energy 27:739–755

Afgan NH, Carvalho MG, Hovanov NV (2000) Energy system assessment with sustainability indicators. Energy Policy 28:603–612

Banerjee R (2006) Comparison of options for distributed generation in India. Energy Policy 34:101–111

Bhattacharyya SC (2012) Review of alternative methodologies for analysing off-grid electricity supply. Renew Sustain Energy Rev 16(1):677–694

Brundtland GH (1987) Our common future, World Commission on Environment and Development (WCED)

Buchholz T, Rametsteiner E, Volk TA, Luzadis VA (2009) Multi Criteria Analysis for bioenergy systems assessments. Energy Policy 37:484–495

Burgi P (2003) Multi-criteria analysis of a biomass gasification power plant. Swiss Federal Institute of Technology (ETH) Zürich. Diploma Thesis

Buytaert V, Muys B, Devriendt N, Pelkmans L, Kretzschmar JG, Samson R (2011) Towards integrated sustainability assessment for energetic use of biomass: A state of the art evaluation of assessment tools. Renew Sustain Energy Rev 15:3918–3933

Castellanos JG, Walker M, Poggio D, Pourkashanian M, Nimmo W (2015) Modelling an off-grid integrated renewable energy system for rural electrification in India using photovoltaics and anaerobic digestion. Renewable Energy 74:390–398

Chatzimouratidis AI, Pilavachi PA (2009) Technological, economic and sustainability evaluation of power plants using the Analytic Hierarchy Process. Energy Policy 37:778–787

Dale VH, Efroymson RA, Kline KL, Langholtz MH, Leiby PN, Oladosu GA, Davis MR, Downing ME, Hilliard MR (2013) Indicators for assessing socioeconomic sustainability of bioenergy systems: a short list of practical measures. Ecol Ind 26:87–102

Demirtas O (2013) Evaluating the best renewable energy technology for sustainable energy planning. Int J Energy Econ Policy 3:23–33

Dimakis AA, Arampatzis G, Assimacopoulos D (2011) Monitoring the sustainability of the Greek energy system. Energy for Sustainable Development

Evans A, Strezov V, Evans TJ (2010) Sustainability considerations for electricity generation from biomass. Renew Sustain Energy Rev 14:1419–1427

Evans A, Strezov V, Evans TJ (2009) Assessment of sustainability indicators for renewable energy technologies. Renew Sustain Energy Rev 13:1082–1088

Finkbeiner M, Schau EM, Lehmann A, Traverso M (2010) Towards life cycle sustainability assessment. Sustainability 2:3309–3322

Frangopoulos CA, Keramioti DE (2010) Multi-criteria evaluation of energy systems with sustainability considerations. Entropy 2(5):1006–1020

Harish SM, Morgan GM, Subrahmanian E (2014) When does unreliable grid supply become unacceptable policy? Costs of power supply and outages in rural India. Energy Policy 68:158–169

Manish S, Pillai IR, Banerjee R (2006) Sustainability analysis of renewables for climate change mitigation. Energy Sustain Dev X(4)

Mcbride AC, Dale VH, Baskaran LM, Downing ME, Eaton LM, Efroymson RA, Garten CT, Kline KL, Jager HI, Mulholland PJ, Parish ES, Schweizer PE, Storey JM (2011) Indicators to support environmental sustainability of bioenergy systems. Ecol Ind 11:1277–1289

Mukunda HS (2011) Understanding clean energy and fuels from biomass. Wiley India Pvt Ltd

Musango JK, Brent AC, Amigun B, Pretorius L, Müller H (2011) Technology sustainability assessment of biodiesel development in South Africa: a system dynamics approach. Energy 36:6922–6940

Myllyviita T, Leskinen P, Lähtinen K, Pasanen K, Sironen S, Kähkönen T, Sikanen L (2013) Sustainability assessment of wood-based bioenergy—A methodological framework and a case-study. Biomass Bioenergy 59:293–299

NEEDS (2009) Final report on sustainability assessment of advanced electricity supply options. Project no: 502687

Nouni MR, Mullick SC, Kandpal TC (2007) Biomass gasifier projects for decentralized power supply in India: A financial evaluation. Energy Policy 35(2):1373–1385

Nouni MR, Mullick SC, Kandpal TC (2009) Providing electricity access to remote areas in India: Niche areas for decentralized electricity supply. Renew Energy 34(2):430–434

Onat N, Bayar H (2010) The sustainability indicators of power production systems. Renew Sustain Energy Rev 14:3108–3115

Palit D, Malhotra R, Kumar A (2011) Sustainable model for financial viability of decentralized biomass gasifier based power projects. Energy Policy 39:4893–4901
Pohekar SD, Ramachandran M (2004) Application of multi-criteria decision making to sustainable energy planning—A review. Renew Sustain Energy Rev 8:365–381
Sadamichi Y, Kudoh Y, Sagisaka M, Chen SS, Elauria JC, Gheewala SH, Hasanudin U, Romero J, Shi X, Sharma VK (2012) Sustainability assessment methodology of biomass utilization in East Area. J Jpn Inst Energy 91:960–968
Saunders AM, Pope J (2013) Conceptualising and managing trade-offs in sustainability assessment. Environ Impact Assess Rev 38:54–63
Stein EW (2013) A comprehensive multi-criteria model to rank electric energy production technologies. Renew Sustain Energy Rev 22:640–654
Streimikiene D (2010) Comparative assessment of future power generation technologies based on carbon price development. Renew Sustain Energy Rev 14:1283–1292
Varun PR, Bhatt IK (2009) Energy, economics and environmental impacts of renewable energy systems. Renew Sustain Energy Rev 13:2716–2721
Varun PR, Bhatt IK (2010) A figure of merit for evaluating sustainability of renewable energy systems. Renew Sustain Energy Rev 14:1640–1643
Wang J, Jing Y, Zhang C, Zhao J (2009) Review on multi-criteria decision analysis aid in sustainable energy. Renew Sustain Energy Rev 13:2263–2278

Overview of Biogas Reforming Technologies for Hydrogen Production: Advantages and Challenges

Priyanshu Verma and Sujoy Kumar Samanta

Abstract In the past two decades, production of biogas from biomass degradation has drawn the attention of several researchers. Biogas is produced during anaerobic degradation of plant and animal wastes, basically consisting of higher concentrations of methane (CH_4), carbon dioxide (CO_2), and trace amounts of hydrogen sulfide (H_2S). This biogas is an extremely potential and interesting source for the production of hydrogen gas (H_2). Hydrogen gas finds tremendous quantum of applications as an essential raw material to meet the several H_2 demands such as high temperature fuel cell, combustion engine, petrochemical and fertilizer industries, mostly ammonia production. Traditionally, large-scale production of H_2 gas involves a thermal reforming process that uses light hydrocarbons, mainly natural gas. Biogas which is regarded as a renewable source of methane, reduces the excessive burden on natural gas. It can also help to reduce the greenhouse gas emissions. However, the present methods used for biogas reforming have several technological limitations, which may depend on the quality of biogas produced, the conversion efficiency of the process, and specific requirements for the integration of H_2 production, purification, transportation, and application. This study reviews several biogas reforming methods, the types of catalyst used, the advantages and disadvantages offered by each route during the processing.

Keywords Biogas · Methane · Hydrogen · Reforming processes · Catalyst

1 Introduction

Energy crisis is one of the major problems for sustainable development. This is magnified especially due to the increase in global population, and consequently the increase in energy demands as well as the gradual exhaustion of natural resources.

P. Verma (✉) · S.K. Samanta
Department of Chemical and Biochemical Engineering,
Indian Institute of Technology, Patna 801103, India
e-mail: priyanshuvermamickey@gmail.com

S. Kumar et al. (eds.), *Proceedings of the First International Conference on Recent Advances in Bioenergy Research*, Springer Proceedings in Energy,
DOI 10.1007/978-81-322-2773-1_17

These challenging issues are enforcing us to go for several alternative energy sources that are sustainable in nature. Since the major energy sources used at present are derived from unsustainable fossil resources, it is a better alternative way to utilize renewable energy sources for sustainable development (Deublein and Steinhauser 2008).

Presently, biogas is one of the main categories of renewable energy sources. Biogas is a mixture of different gases produced by the anaerobic digestion or fermentation of organic matter in the presence of microorganisms. It is primarily composed of methane (CH_4) and carbon dioxide (CO_2), combined with traces of other gases such as ammonia (NH_3), hydrogen (H_2), nitrogen (N_2), oxygen (O_2), hydrogen sulfide (H_2S), and water vapors (H_2O). Chemical composition of biogas has been shown in Table 1. It is generally produced in biowaste digesters, landfills, and sewage sludge that include animal wastes, agricultural wastes, animal dung, energy crop, etc. H_2S is the most undesirable component of biogas. Due to its toxicity and corrosiveness, it used to damage the vessels, tubes, and pipe fittings involved in the overall processes (Chattanathan et al. 2014). CO_2 and water vapors also act as impurities since it reduces the energy value of biogas (Horikawa et al. 2004). An ideal energy source must be cheap, clean, renewable, and sustainable in nature and biogas is one of them. However, CH_4 and CO_2, the main components of biogas, are the major greenhouse gases, which upon release into the atmosphere results in unfavorable conditions causing global warming. To overcome this problem, it must be transformed or converted into some efficient and clean fuel such as H_2 (Deublein and Steinhauser 2008; Gupta 2009).

Hydrogen has the highest energy content per unit mass (140 MJ/kg) in comparison to any other fuel. In addition, it is a clean fuel that does not leave any residue like ash after combustion. The comparable properties of hydrogen and other fuels have been shown in Table 2.

Traditionally, H_2 has been widely used in chemical industries, hydrogenation process, food processing, pharmaceuticals, and fertilizer industries (mainly for ammonia production) (Armor 1999). Recently, hydrogen has been heavily used in fuel cells that directly convert chemical energy into electrical energy without any involvement of mechanical operations (Hotza and Da Costa 2008). In recent studies, these fuel cells have achieved exceptional progress in the form of

Table 1 Chemical composition of biogas (Deublein and Steinhauser 2008; Gupta 2009)

Compound	Formula	Percentage
Methane	CH_4	55–70 vol.%
Carbon dioxide	CO_2	30–45 vol.%
Hydrogen sulfide	H_2S	500–4000 ppm
Ammonia	NH_3	100–800 ppm
Hydrogen	H_2	<1 vol.%
Nitrogen	N_2	<1 vol.%
Oxygen	O_2	<1 vol.%
Water vapor	H_2O	<1 vol.%

Table 2 Comparative properties of hydrogen and other fuels (Gupta 2009)

Properties	Units	Hydrogen	Methane	Propane	Methanol	Ethanol	Gasoline[b]
Chemical formula		H_2	CH_4	C_3H_8	CH_3OH	C_2H_5OH	C_xH_y ($x = 4-12$)
Molecular weight		2.02	16.04	44.1	32.04	46.07	100–105
Density (NTP)[a]	kg/m^3	0.08375	0.6682	1.865	791	789	751
Viscosity (NTP)[a]	Pa s	8.81×10^{-4}	1.10×10^{-3}	8.01×10^{-4}	9.18×10^{-2}	0.119	0.037–0.044
Normal boiling point	°C	−252.8	−161.5	−42.1	64.5	78.5	27–225
Flash point	°C	<−253	−188	−104	11	13	−43
Flammability range in air	vol.%	4.0–75.0	5.0–15.0	2.1–10.1	6.7–36.0	4.3–19	1.4–7.6
Auto ignition temperature in air	°C	585	540	490	385	423	230–480
Higher heating value (at 25 °C and 1 atm)	MJ/kg	141.86	55.53	50.36	19.96	29.84	47.50
Lower heating value (at 25 °C and 1 atm)	MJ/kg	119.93	50.02	45.60	18.05	26.95	44.50

[a]NTP = 20 °C and 1 atm

[b]Properties of a range of commercial grades

transportation, power generation, and installations (Alves et al. 2013). This system will facilitate many applications such as massive power generation and distribution, domestic employment of generators, better portability of electronic devices due to small size, and an auxiliary power source for vehicles. The conversion efficiency of fuel cells is found to be around 60 % (Faghri and Guo 2005; Chen et al. 2007), which is about two to three times more than that of the device presently employed in combustion vehicles (i.e., only about 20–30 %) (Alves et al. 2013). These special features make hydrogen energy as a superior alternative over all the alternative fuel categories available to compliment/substitute fossil fuels.

Presently, steam reforming of fossil fuels has been the most frequently used method of H_2 production which generally requires natural gas, naphtha, and other lighter hydrocarbons as raw materials (Holladay et al. 2009). Biogas reforming for H_2 production is an appealing and promising technique to convert biogas into H_2, especially due to the reduction in emission of greenhouse gases and its reliability (Herle et al. 2004; Hotza and Da Costa 2008; Levin and Chahine 2010; Papadias et al. 2012). Another advantage of using biogas as a feedstock for H_2 production is its availability (mostly local) which reduces the transportation costs. Therefore, utilization of biogas in reforming process has better environmental and economical advantages (Rand and Dell 2008). Three major techniques used for conversion of CH_4 present in biogas to H_2 are steam reforming, dry reforming, and partial oxidation that can be used for the production of liquefied hydrogen (Galvagno et al. 2013).

Basically, biogas reforming involves CO_2 reforming as well as dry reforming of CH_4. Since this process requires the stripping of oxygen from carbon dioxide and stripping of hydrogen from methane, there are significant chances of coke deposition on the catalyst surface. That is why most of the reforming studies have been based on the broad and intense analysis of different catalysts. It used to minimize the costs and the total energy involved in the complete process (Muradov et al. 2008).

The development, modification, and utilization of several catalysts having a higher catalytic activity and stability in reforming processes can control the higher temperature needs and increase the reaction rate. In addition, it can also slower the carbon deposition rate and the poisoning of the catalyst, which are the major problems encountered during reforming processes (Alves et al. 2013).

The present study is focused on the limitations of several biogas reforming technologies used for the production of hydrogen gas. Furthermore, this review also highlights the advanced possibilities of H_2 production using biogas as a renewable resource.

2 Several Hydrogen Production Processes from Methane or Biogas

The most practiced methane/biogas reforming processes for hydrogen production are steam reforming (SR), dry reforming (DR), partial oxidation reforming (POR), auto-thermal reforming (ATR), and dry oxidation reforming (DOR). There are some other unconventional processes for the production of H_2 from methane such as thermal plasma reforming, catalytic decomposition, and solar reforming. Several industries uses SR, POR, and DR methods for the large-scale production of H_2 that uses natural gas (≈95 % of CH_4) or naphtha as major sources of hydrocarbons (Shiga 1998; Alves et al. 2013).

Most of the studies based on biogas reforming to produce H_2 reported several exhaustive analyses using mixtures of CH_4 and CO_2 to simulate laboratory biogas compositions. Even some studies also used the high purity CH_4 (>99 %) for lab scale reforming processes. Only a few studies reported usage of actual biogas, directly produced from the anaerobic digestion of residual biomass (real biogas), in their experiments (Araki et al. 2009, 2010). Thus, to settle on the reforming process using biogas, its composition must be measured and broadly classified into three different situations: (i) in actual form, with 55–70 % of CH_4, 30–45 % of CO_2 and 500–4000 ppm of H_2S (Table 1); (ii) partially treated for H_2S removal; and (iii) purified for "bio-methane" enrichment (93–96 % of CH_4, 4–7 % of CO_2 and <20 ppm of H_2S) (Alves et al. 2013). It is investigated that, after these three sequential steps, the probability of CH_4 content higher than 95 % could be very less using economically feasible treatment procedures (Effendi et al. 2005; Kolbitsch et al. 2008; Ohkubo et al. 2010; Lau et al. 2011).

Prior to the reforming process, pretreatment of biogas is required for the removal of corrosive or poisonous species present in biogas stream that can be conducted by two different methods: (i) physicochemical treatment (including absorption in aqueous solutions, chemical adsorption of H_2S on solid adsorbents such as zeolites, modified zeolites, activated carbon, and specific metal surfaces with the formation of metal sulfide and scrubbing with specific solvents); and (ii) biological treatment (removal of impurities by living microorganisms such as species of chemotrophic fungi and bacteria which used to act as oxidants in the bio-trickling filter, sulfur bio-filter, and bioaugmentation unit that can convert impurities into lighter and less harmful forms) (Holladay et al. 2009; Alves et al. 2013).

Generally, reforming processes for H_2 production are performed under low pressure (mostly below 1 atm) in tubular fixed-bed or fluidized-bed reactors, in a wide range of temperature 600–1000 °C (endothermic and reversible reactions) (Goransson et al. 2011; Alves et al. 2013). It also involves predominant catalytic reactions that are mostly combined (see Table 3).

These reforming processes mainly use natural gas as a raw material (methane source). However, the biogas also can be used as a methane source in various reforming processes (based on its composition and purity level). That is why most of the researches involving methane reforming can also be adapted or modified for biogas reforming (Galvagno et al. 2013). Name of chemical reactions are cited and their respective enthalpy of reaction at 298 K (25 °C) has also been shown in Table 3.

Table 3 Chemical reaction involved in reforming processes (Alves et al. 2013)

Reaction Eqs.	Name of reaction	Type of reaction	$\Delta H_{298\ K}$ (kJ/mol)
Equation 1	SR	$CH_4 + H_2O \rightarrow CO + 3H_2$	206.2
Equation 2	Gas-shift reaction	$CO + H_2O \rightarrow CO_2 + H_2$	−41.2
Equation 3	Combined reaction	$CH_4 + 2H_2O \rightarrow CO_2 + 4H_2$	165
Equation 4	Methane cracking	$CH_4 \rightarrow C + 2H_2$	74.9
Equation 5	Boudouard reaction	$2CO \rightarrow C + CO_2$	−172.4
Equation 6	Reduction of CO	$CO + H_2 \rightarrow C + H_2O$	−131.3
Equation 7	POR	$CH_4 + ½\ O_2 \rightarrow CO + 2H_2$	−35.6
Equation 8	Complete oxidation	$CH_4 + 2O_2 \rightarrow CO_2 + 2H_2O$	−801.7
Equation 9	ATR	$CH_4 + ½\ x\ O_2 + yCO_2 + (1 - x - y)\ H_2O \rightarrow (y + 1)\ CO + (3 - x - y)\ H_2$	≈ 0
Equation 10	DR	$CH_4 + CO_2 \rightarrow 2CO + 2H_2$	247.4
Equation 11	DOR	$CH_4 + \beta CO_2 + (1 - \beta)/2\ O_2 \rightarrow (1 + \beta)CO + 2H_2$	$(285\beta - 38)$ $0 \leq \beta \leq 1$

3 Overview of Different Conventional Reforming Processes

3.1 Steam Reforming (SR)

In SR, methane reacts with water vapors in the presence of a suitable catalyst that produces CO and H_2 (Eq. 1, Table 3). The reaction is highly endothermic in nature and requires temperatures in the range of 650–850 °C, to achieve maximum H_2 yield (between 60 and 70 %). SR requires substantial energy expenditure and despite that it is the most used industrial methods to produce H_2. In SR, the H_2/CO ratio is three that results in higher productivity of hydrogen. In order to remove CO, the water-gas shift reaction (also known as "Shift reaction") is employed (Eq. 2, Table 3) that requires temperature in the order of 300–450 °C and specific catalysts based on Cu, Fe, Mo, and Fe–Pd alloys that increases the production of additional hydrogen (Effendi et al. 2005; Kolbitsch et al. 2008). The reaction of methane SR associated with the shift reaction (Combined reaction) has been shown in Eq. 3 (Table 3).

These severe conditions required for SR used to promote the parallel carbon formation reactions such as methane decomposition reaction (Eqs. 4, Table 3), Boudouard reaction (Eq. 5, Table 3), and disproportionation reduction reaction of CO (Eq. 6, Table 3). It also promotes the carbon deposition on the catalyst surface causing deactivation of it. If the carbon deposited on catalyst surface is in the form of nanotubes then that is advantageous since that lowers the carbon dispersed over the catalyst surface and preserves the catalyst activity for a longer duration. The catalysts typically employed in the SR process could be made of either transition metals (such as Pt, Ni) or noble metals (such as Rh, Pd). Among these, the Ni has the lower cost advantage. On the contrary, Ni has the greater deactivation susceptibility by the coke formation (due to the high temperature processing) that makes the Pd and Pt catalysts more fascinating with regards to the stability of the catalyst. In case of limited mass transfer, Rh is employed as it has a better catalytic activity which is much better than that of Ni and offers lower coke formation (Iulianelli et al. 2010; Zhai et al. 2011). Basic supports containing promoter elements such as Mg, Ca, and K can also be utilized to minimize the carbon deposition on the catalyst surface, since they promote the gasification of carbon species by the carbon–water steam reaction (reverse of Eq. 6, Table 3), due to the increased adsorption of water. The referred metals (Ni, Pt, Rh, Ru, etc.) and matrices (such as SiO_2, ZrO_2, Al_2O_3, etc.) can be used for the preparation of active sites in support catalysts (Amin et al. 2012; Roh et al. 2012; Alves et al. 2013).

To obtain highly pure H_2 gas, the CO_2 and CO formed in SR process are required to be effectively separated from the syngas. Similarly, these species (CO_2 and/or CO) are required to be separated for other processes as well (POR, ATR, DR, DOR). Typically, the H_2 production unit of the SR process comprises of three integrated sections namely reformer (2 or more), conversion reactor (shift reaction) and a separator.

Some of the studies show significant improvement in the separation process with the application of selective membrane filters or reactors (Mahecha-Botero et al. 2009). The membrane reactor permits all reactions to occur in a single vessel. In the membrane system, SR methane reaction and shift reaction (Eqs. 1 and 2 respectively, Table 3) occur simultaneously within the reactor that contains a catalyst bed. In addition, membrane provides better ability to shift the chemical equilibrium of the reaction that increases the production of hydrogen. The shifting of chemical equilibrium enables the reforming reaction to be carried out at much lower temperatures (below 500 °C) without any specific conversion loss (Lin et al. 2003; Lu et al. 2007; Sato et al. 2010).

3.2 *Partial Oxidation Reforming (POR)*

Partial oxidation reforming (POR) is one of the best alternative processes to produce H_2 with reduced energy costs due to its moderately exothermic nature (Eq. 7, Table 3). This is contrary to the SR process which is highly endothermic in nature. In POR process, methane is partially oxidized with O_2 at the atmospheric pressure and CO and H_2 (collectively called "syngas or synthesis gas") are produced. It requires temperatures in the range of 700–900 °C that reduces the coke formation and ensures the complete conversion of methane (H_2/CO ratio is two). In case of slight decrease in CO selectivity, the methane reacts with oxygen and produces CO_2

(Eq. 8, Table 3) that promotes the complete combustion and results in a sudden increase in temperature. That causes the formation of hot spots in the reactor bed and deposit coke on the catalyst surface (Alves et al. 2013).

Some researchers have focused on the development of highly active and stable catalyst for biogas POR. Some catalysts such as solid solutions of Ca–Sr–Ti–Ni, NiO–MgO, Ni–Mg–Cr–La–O, and mixed metal oxides are reported to be highly selective as well as active catalysts at elevated temperature ranges (above 800 °C) with lower coke formation (Ruckenstein and Hu 1999; Corbo and Migliardini 2007; Rogatis et al. 2009).

3.3 Auto-Thermal Reforming (ATR)

ATR is an intensive endothermic process. That is why it requires internal heating of the reactor, as compared to the external heating for efficient and economical hydrogen production. Although, the POR of CH_4 has the advantage over ATR since it is exothermic (Eq. 7, Table 3) in nature, it generally offers lower H_2/CO ratio than that offered by ATR. ATR is the combination of already discussed reforming technologies (SR and POR) that results in the sum of reactions described by Eqs. 3, 7 and 8 in Table 3 (Souza and Schmal 2005; Cai et al. 2006; Simeone et al. 2008). It also takes place in the presence of carbon dioxide as shown in Eq. 9 in Table 3 (Araki et al. 2009, 2010). ATR offers several advantages such as it speeds up the processing of reactor, increases H_2/O_2 ratio in comparison to the POR. Moreover, the combination of SR and POR improves the control over temperature during the processing and consequently the formation of hot-spots is also reduced that inhibits the catalyst deactivation.

In ATR process, methane partial oxidation reaction occurs in parallel with the methane steam reforming that makes the overall process self-sustainable and less energy expensive. That is why most of the researchers favor adiabatic SR of biogas under auto-thermal condition and in combination with the partial oxidation reaction. It produces syngas with a H_2/CO ratio between 2.0 and 3.5, using O_2/CH_4 and H_2O/CH_4 ratios between 0.25–0.55 and 1.0–2.5, respectively (Cai et al. 2006; Mosayebi et al. 2012). It is also observed that the temperature of the different oxidation zones can affect the product composition selectivity. In this process, the total oxidation reaction (Eq. 8, Table 3) is favored by reducing the temperature, whereas partial oxidation reaction (Eq. 7, Table 3) is favored by increase in temperature (Galvagno et al. 2013).

3.4 Dry Reforming (DR)

DR is a process in which CH_4 present in biogas reacts with CO_2 and produce H_2 and CO (Eq. 10, Table 3). This reaction is very appealing as per environmental views, since it utilizes CH_4 and CO_2 as feed (which are potent greenhouse gases). However, the researchers observed that due to the endothermic nature of the reaction involved in the DR process, it requires a large amount of energy input in the form of heat, and production of that required heat increases the CO_2 emissions (due to the burning of other fossil fuels). For industrial application of biogas DR, it must have to satisfy the requirements of several manufacturing processes of liquid hydrocarbons and oxygenated compounds to efficiently produce syngas yielding a H_2/CO ratio ~ 1 (Eltejaei et al. 2012).

Based on earlier findings, the major reaction (Eq. 10, Table 3) can be completed by competing with the other parallel reactions that shift the equilibrium conversion of CO_2 in CH_4. These reactions are reverse gas–water shift (Eq. 2 reverse), Boudouard reaction (decomposition of carbon monoxide, Eq. 5 in Table 3), and decomposition of CH_4 (Eq. 4, Table 3). DR process uses a CH_4/CO_2 ratio between 1 and 1.5 and it mainly occurs at temperatures between 700 and 900 °C (Alves et al. 2013). This process gives around 50 % H_2 yield (Serrano-Lotina et al. 2012; Eltejaei et al. 2012; Lucredio et al. 2012).

In DR process, if the carbon removal is slower than the methane decomposition (Eq. 4, Table 3) rate then a serious problem of coke formation arises that results in the deactivation of the catalyst and blockage of the reactor and pipes due to the coke accumulation. Low temperature favors the Boudouard reaction (decomposition of CO, Eq. 5 in Table 3) and it produces carbon. Since, this reaction is exothermic in nature, it promotes the methane cracking reaction also (Eq. 4, Table 3), which produces an additional amount of carbons. Therefore, the coke formation step is highlighted as the major drawback of DR process and it also affects the active surface of the catalyst. Therefore, several studies have been carried out for the development of suitable catalysts that can prevent the carbon deposition, be highly selective in nature and thermally stable during the production of H_2 (Barrai et al. 2007; Eltejaei et al. 2012).

Several metals as the most active catalysts have been reported in the recent studies that belong to the groups 8, 9, and 10 of the Periodic Table, such as Rh, Ru, and Pt. However, these catalysts are not suitable for industrial applications as they are very expensive and limited in nature. Therefore, other catalysts have been developed such as Co and Ni catalysts with lower price and greater availability that have higher coke susceptibility as well (San-Jose-Alonso et al. 2009; Xu et al. 2009; Bereketidou and Goula 2012).

3.5 *Dry Oxidation Reforming (DOR)*

DOR is a specific method to control the carbon deposition on the surface of the catalysts. It is the combination of dry reforming and partial oxidation reforming, using a parallel feeding of oxygen with methane and CO_2. DOR offers some additional advantages such as improved methane conversion, reduced total energy involved in the process, increased deactivation resistance of the catalyst, enhanced catalyst stability, and increased hydrogen yield at lower temperature. The H_2/CO ratio could be controlled with the modulation of oxygen to meet the flow requirements (Lau et al. 2011). CH_4–DOR reaction has been shown in Eq. 11 (Table 3), where β is the stoichiometric fraction of CO_2 fed with traditional DR. It is clearly understood that the theoretical H_2/CO ratio, given by $2/(1 + \beta)$, is greater for DOR process for $\beta < 1$ than that for non-oxidative DR process ($\beta = 1$) (Alves et al. 2013).

In CH_4–DOR process, changing of the process variables such as reaction temperature and/or relative concentration of oxygen feed helps to control the H_2/CO ratio and the nature of the reaction (endothermic or exothermic). Basically, the exothermic nature is increased with the increase in feed O_2 concentration at a given temperature. Because the addition of oxygen determines the oxidation process (complete oxidation occurs when $\beta = 0$, Eqs. 7 and 8; or traditional DR reaction occurs when $\beta = 1$, Eq. 10, Table 3) and accordingly, the amount of heat consumed or released. It may be noted that, POR process (Eq. 7, Table 3) is exothermic, whereas traditional DR process is endothermic (Eq. 10, Table 3) in nature. Hence, DOR process could be utilized as a better alternative over traditional DR process using different degrees of oxygen input. Consequently, biogas can be converted to syngas at higher energy efficiency using less external energy (Effendi et al. 2002; Avraam et al. 2010).

Studies of different catalysts and their association with other metals such as bimetallic catalyst, have been used to enhance the stability of the catalysts (Xu et al. 2009; Lau et al. 2011). In this process, the addition of oxygen reduces the coke formation on the catalyst surface due to the better control over the partial oxidation of the novel carbon species, CH_4 and CO_2 for CO formation (Lucredio et al. 2012; Alves et al. 2013). Different H_2 production routes and reaction conditions using methane, biogas, or real biogas as a major substrate are summarized in Table 4.

The above mentioned reforming technologies have proven their effectiveness in the form of better conversion efficiency of biogas/methane (X_{CH_4}) into a useful hydrogen gas. The technical comparisons of these biogas reforming methods are summarized in Table 5.

4 Current Status of Biogas Reforming Technologies

Biogas, being a renewable and abundant resource, has attracted a large number of researchers to study the particular field and utilize this resource better for the production and development of clean fuel such as hydrogen. Their investigations are primarily focused on the efficient removal of H_2S which upgrades the biogas for further processing (Tippayawong and Thanompongchart 2010; Micoli et al. 2014; Sisani et al. 2014; Díaz et al. 2015; Liu et al. 2015).

Xu et al. (2004) developed a model for polymer electrolyte fuel cell (PEFC) usable H_2 gas production from desulfurized biogas produced through anaerobic

Table 4 A summary of studies based on H_2 production from methane or biogas

Process	Reactor	T (°C)	Catalyst	H_2/CO	X_{CH_4} (%)	References
SR[a]	Fixed-bed	700	Ni/Al_2O_3	–	85	Maluf and Assaf 2009
SR[a]	Fixed-bed	525	$Rh/Ce_\alpha Zr_{1-\alpha}O_2$	–	79	Halabi et al. 2010
SR[b]	Fluidized-bed	850	Ni/Al_2O_3	2.1	98	Effendi et al. 2005
SR[b]	Fixed-bed	715	Ru/Al_2O_3	2.7	90	Avraam et al. 2010
SR[b]	Fixed-bed	750	$Ni/CaO–Al_2O_3$	2.5	95	Kolbitsch et al. 2008
SR[b]	Fixed-bed	750	Ni/Al_2O_3	2.0	85	Effendi et al. 2002
SR[b]	Fluidized-bed	750	Ni/Al_2O_3	2.2	96	Effendi et al. 2002
POR[a]	Fixed-bed	800	Pt/CeO_2	2.0	85	Corbo and Migliardini 2007
POR[a]	Fixed-bed	850	NiO/MgO	2.0	87	Ruckenstein and Hu 1999
POR[a]	Fixed-bed	700	Ni/Al_2O_3	2.0	100	Rogatis et al. 2009
ATR[a]	Fixed-bed	700	$Ni/MgAl_2O_4$	3.2	92	Mosayebi et al. 2012
ATR[a]	Fixed-bed	800	$Pt/ZrO_2/Al_2O_3$	2.0	100	Souza and Schmal et al. 2005
ATR[a]	Fixed-bed	700	Rh/Al_2O_3	3.5	95	Simeone et al. 2008
ATR[a]	Fixed-bed	750	$Ni/Cu_5Zr_{10}Ce_{20}Al_{65}O_8$	3.9	100	Cai et al. 2006
ATR[c]	Fixed-bed	750	Ni/Cordierite	2.6	90	Araki et al. 2010
ATR[c]	Fixed-bed	850	Ni/Insulating (Si, Mg, Al)	2.8	95	Araki et al. 2009
ATR[b]	Fixed-bed	800	Ni/SBA-15 (molecular sieve)	1.4	92	Barrai et al. 2007
DR[b]	Fixed-bed	860	$Ni/CeO_2–Al_2O_3$	1.3	90	Bereketidou and Goula 2012
DR[b]	Fixed-bed	750	$Rh/NiLa/\gamma–Al_2O_3$	0.9	70	Lucredio et al. 2012
DR[b]	Fixed-bed	900	$Ni–Co/La_2O_3/Al_2O_3$	1.01	88	Xu et al. 2010
DR[b]	Fixed-bed	800	Reformax®-250	0.98	67	Chattanathan et al. 2014
DR[b]	Fixed-bed	850	Ni/Y (zeolite)	0.9	97	Sharifi et al. 2014
DOR[b]	Fixed-bed	750	$Rh-NiLa/\gamma–Al_2O_3$	1.0	86	Lucredio et al. 2012
DOR[b]	Fluidized-bed	900	$Pt–Rh/Ce–ZrO_2–Al_2O_3$	1.0	100	Lau et al. 2011

[a]Methane
[b]Synthetic biogas
[c]Real biogas
—Not reported

digestion. They used a gas producer, consisting of a steam reformer, two water-gas shift reactors at high and low reaction temperatures, respectively, and a selective CO oxidizer. The authors obtained that H_2-rich gas contained usually 70 % H_2, 30 % CO_2, and residual CH_4.

Purwanto and Akiyama (2006) studied H_2 production using hot slag in a packed bed reactor. They used continuous flow of CH_4 and CO_2 (biogas) into the packed bed of hot slag at a constant flow-rate and atmospheric pressure. The authors

Table 5 Comparison of biogas reforming methods (Yang et al. 2014)

Methods	Advantages	Challenges	Remarks
SR	Produces highly pure hydrogen; low carbon formation	Needs to remove H_2S and add oxidizing agents; high operating temperature and energy demand; catalyst can be expensive	It is widely used for hydrogen production
POR	High energy efficiency; relatively low operating temperature	May completely oxidize methane to CO_2 and H_2O; limited industrial application	It can be combined with other reforming methods
ATR	Produces high-purity hydrogen fuel; high energy efficiency	Complex process control; needs multiple catalysts; relatively unstable	It can use both CH_4 and CO_2 in biogas
DR	High conversion efficiency	Carbon formation; moderate selectivity; side reaction consumes hydrogen; high operating temperature and energy demand; catalyst can be expensive	It can also use CH_4 and CO_2 in biogas
DOR	Less carbon deposition; reduced overall energy requirement; better conversion efficiency	Complex process control; produce higher CO_2	It is the combination of DR and POR

observed that slag acted as not only a thermal media, but also a good catalyst, for better decomposition. The product gases were mainly H_2 and CO with/without solid C deposition on the surface of slag that depends on the reaction temperature. Increased temperature led to the large hydrogen generation with the higher methane conversion (about 96 %).

Chun et al. (2008) examined the reforming characteristics and optimum operating conditions for the hydrogen production through the plasmatron-assisted CH_4-reforming process. The authors also conducted some studies related to methane conversion rate and parametric screening to increase the productivity. They obtained the optimal methane conversion rate as 99.2 %, H_2/CO ratio as 6.6, thermal efficiency of the reformer as 63.6 %, and the H_2 yield as 78.8 %.

Bensaid et al. (2010) investigated the applicability of molten carbonate fuel cells (MCFCs) for power generation from landfill biogas, which was also coupled with the in situ production of hydrogen. They had performed the system modeling of the plant in steady state conditions, with the aim of assessing the overall power efficiency. Due to the exothermic nature of MCFCs (working temperature was around 650 °C), it showed compatibility to the steam reforming for the production of syngas. Moreover, they utilized the high temperature flue gases from the MCFC for power generation through the turbine. The authors reported that hydrogen was produced through a pressure swing adsorption system, whose feed could be from either the MCFC anode outlet or a split of the reformate before the anode inlet. The overall net power efficiency of the system was found almost similar for both mechanisms, 56 and 55 %, respectively.

Lin et al. (2012) proposed a distinct method to produce hydrogen from biogas reforming using mesoporous $Ni_{2x}Ce_{1-x}O_2$ catalysts. These catalysts were prepared through reverse precipitation method. The authors obtained the higher H_2/CO ratio and the lower H_2/CO_2 ratio than commercial catalyst in the steam reforming reaction with temperature range 500–900 °C. They also observed that the methane conversion was increased with increase in nickel content in all experiments.

Rueangjitt et al. (2012) assessed the possibility of upgradation of biogas through multi-staged AC (Alternating current) gliding arc system that could be further utilized for production of H_2 gas. They found that sudden increment in the stage numbers of plasma reactors would increase both CH_4 and CO_2 conversions. The combination of the plasma reforming and partial oxidation provided an exceptional improvement to the overall process performance achieved by the authors.

Meyer et al. (2014) proposed a novel sorption enhanced reforming (SER) technology which could also capture CO_2. This technology has the potential to produce renewable hydrogen and also provide bio-CO_2 that can be used for the production of biomethane through the Sabatier process. The authors designed and simulated a small SER prototype using a dual bubbling fluidized-bed reactor system with the help of advanced modeling and simulation tools.

Italiano et al. (2015) studied the synthesis of nanocrystalline Ni/CeO_2 catalysts for the biogas oxy-steam-reforming (OSR) used for the production of hydrogen. The study was mainly focused on the characterization of newly synthesized catalyst. In addition, they also observed that the OSR of biogas at 800 °C with the newly synthesized catalyst offered high H_2 yield (above 85 %).

There are also many biological processes that have been studied for the hydrogen production such as enzymatic hydrogen production and bio-hydrogen production through bio-abundant polysaccharides (Kovács et al. 2004; Redwood et al. 2009; Chang et al. 2012; Esquivel-Elizondo et al. 2014). Fermentation is the integral part of biohydrogen production that can be easily operated and the process is eco-friendly too (Wongtanet et al. 2007). However, low bio-H_2 production yield and slow processing rate have been reported as its main drawbacks. To overcome these limitations several strategies have been followed in the past such as genetic modifications and adjusting the reaction conditions. Zhang et al. (2007) proposed an unnatural enzymatic pathway which could produce 12 mol of hydrogen per mole of glucose unit of polysaccharides and water. In this process, the energy required for splitting of the water molecule was obtained from carbohydrates.

5 Future Research Directions

For the broader hydrogen production through biogas reforming technologies, progressive efforts are needed to a greater extent. Based on a substantial literature survey and recommendation or suggestion of several researchers, the following directions for potential research routes may be recommended:

- Proper design and development of process and reactor used for biogas reforming provides the efficient production of hydrogen with less overall energy expenditure. A scale-up study of the advanced processes is required to estimate the return on the investment and operating costs of an industrial-scale production plant.
- Operating parameters such as catalyst characteristics and throughput, optimum temperature, presence of impurities and its concentration have been found to determine the limit to which an effective hydrogen production is achieved.
- CFD modeling studies of advanced and combined biogas reforming technologies are still open areas of research that could be utilized to validate the existing experimental results.
- Some researchers favor biological methods of hydrogen production over chemical methods because of the possibility to use sunlight, CO_2, and organic wastes as substrates for environmentally benign conversions under moderate conditions that require more attention.

6 Conclusion

Biogas reforming is a promising process to produce green hydrogen as well as to reduce the overburden on natural gas. The major problems raised in the biogas reforming processes, reviewed by most of the researchers, are related to coke formation on the catalyst surface and the poisoning due to sulfur, which may cause the deactivation of the catalyst and reduce the H_2 production. Mostly, Ni-based catalysts are used in different reforming processes. A large amount of research works is focused on the evaluation of effectiveness of the catalyst support and the addition of promoter elements to encounter the coke formation issues. Several studies show the importance of biogas purification to eliminate or reduce the corrosive species such as H_2S. This step plays a vital role in the maximization of productivity and durability of the process and reduces the maintenance cost. The physicochemical methods including chemical adsorption and absorption processes have been used substantially, and several routes involving biological degradation with the help of living microorganisms (that consume contaminants) such as bacteria and fungi are also highlighted.

References

Alves HJ, Junior CB, Niklevicz RR, Frigo EP, Frigo MS, Coimbra-Araujo CH (2013) Overview of hydrogen production technologies from biogas and the application in fuel cells. Int J Hydrogen Energy 38:5215–5225

Amin AM, Croiset E, Constantinou C, Epling W (2012) Methane cracking using Ni supported on porous and non-porous alumina catalysts. Int J Hydrogen Energy 37:9038–9048

Araki S, Hino N, Mori T, Hikazudani S (2009) Durability of a Ni based monolithic catalyst in autothermal reforming of biogas. Int J Hydrogen Energy 34:4727–4734

Araki S, Hino N, Mori T, Hikazudani S (2010) Autothermal reforming of biogas over a monolithic catalyst. J Nat Gas Chem 19:477–481

Armor JN (1999) The multiple roles for catalysis in the production of H_2. Appl Catal A Gen 176:159–176

Avraam DG, Halkides TI, Liguras DK, Bereketidou AO, Goula MA (2010) An experimental and theoretical approach for the biogas steam reforming reaction. Int J Hydrogen Energy 35:9818–9827

Barrai F, Jackson T, Whitmore N, Castaldi MJ (2007) The role of carbon deposition on precious metal catalyst activity during dry reforming of biogas. Catal Today 129:391–396

Bensaid S, Russo N, Fino D (2010) Power and hydrogen co-generation from biogas. Energy Fuels 24(9):4743–4747

Bereketidou OA, Goula MA (2012) Biogas reforming for syngas production over nickel supported on ceria-alumina catalysts. Catal Today 195(1):93–100

Cai X, Dong X, Lin W (2006) Auto-thermal reforming of methane over Ni catalysts supported on $CuO–ZrO_2–CeO_2–Al_2O_3$. J Nat Gas Chem 15:122–126

Chang S, Li J, Liu F, Yu Z (2012) Effect of different gas releasing methods on anaerobic fermentative hydrogen production in batch cultures. Front Environ Sci Eng China 6(6):901–906

Chattanathan SA, Adhikari S, McVey M, Fasina O (2014) Hydrogen production from biogas reforming and the effect of H_2S on CH_4 conversion. Int J Hydrogen Energy 39:19905–19911

Chen Z, Grace JR, Lim CJ, Li A (2007) Experimental studies of pure hydrogen production in a commercialized fluidized-bed membrane reactor with SMR and ATR catalysts. Int J Hydrogen Energy 32:2359–2366

Chun YN, Song HW, Kim SC, Lim MS (2008) Hydrogen-rich gas production from biogas reforming using plasmatron. Energy Fuels 22(1):123–127

Corbo P, Migliardini F (2007) Hydrogen production by catalytic partial oxidation of methane and propane on Ni and Pt catalysts. Int J Hydrogen Energy 32:55–66

Deublein D, Steinhauser A (2008) Biogas from waste and renewable resources. Wiley-VCH Verlag GmbH & Co, KGaA, Weinheim

Díaz I, Ramos I, Fdz-Polanco M (2015) Economic analysis of microaerobic removal of H_2S from biogas in full-scale sludge digesters. Bioresour Technol 192:280–286

Effendi A, Hellgardt K, Zhang ZG, Yoshida T (2005) Optimising H_2 production from model biogas via combined steam reforming and CO shift reactions. Fuel 84:869–874

Effendi A, Zhang ZG, Hellgardt K, Honda K, Yoshida T (2002) Steam reforming of a clean model biogas over Ni/Al_2O_3 in fluidized and fixed-bed reactors. Catal Today 77:181–189

Eltejaei H, Bozorgzadeh HR, Towfighi J, Omidkhah MR, Rezaei M, Zanganeh R et al (2012) Methane dry reforming on $Ni/Ce_{0.75}Zr_{0.25}O_2–MgAl_2O_4$ and $Ni/Ce_{0.75}Zr_{0.25}O_2$-γ-alumina: effects of support composition and water addition. Int J Hydrogen Energy 37:4107–4118

Esquivel-Elizondo S, Chairez I, Salgado E, Aranda JS, Baquerizo G, Garcia-Peña EI (2014) Controlled continuous bio-hydrogen production using different biogas release strategies. Appl Biochem Biotechnol 173(7):1737–1751

Faghri A, Guo Z (2005) Challenges and opportunities of thermal management issues related to fuel cell technology and modeling: review. Int J Heat Mass Transf 48:3891–3920

Galvagno A, Chiodo V, Urbani F, Freni F (2013) Biogas as hydrogen source for fuel cell applications. Int J Hydrogen Energy 38:3913–3920

Goransson K, Soderlind U, He J, Zhang W (2011) Review of syngas production via biomass DFBGs. Renew Sust Energ Rev 15:482–492
Gupta RB (2009) Hydrogen Fuel: Production, transport and storage. CRC Press, Taylor and Francis Group, Boca Raton
Halabi MH, De Croon MHJM, Van Der Schaaf J, Cobden PD, Schouten JC (2010) Intrinsic kinetics of low temperature catalytic methane-steam reforming and water-gas shift over $Rh/Ce_{\alpha}Zr_{1-\alpha}O_2$ catalyst. Appl Catal A Gen 389(1–2):80–91
Herle JV, Membrez Y, Bucheli O (2004) Biogas as a fuel source for SOFC co-generators. J Power Sources 127:300–312
Holladay JD, Hu J, King DL, Wang Y (2009) An overview of hydrogen production technologies. Catal Today 139:244–260
Horikawa MS, Rossi F, Gimenes ML, Costa CMM, Silva MGC (2004) Chemical absorption of H_2S for biogas purification. Braz J Chem Eng 21(3):415–422
Hotza D, Da Costa JCD (2008) Fuel cells development and hydrogen production from renewable resources in Brazil. Int J Hydrogen Energy 33:4915–4935
Italiano C, Vita A, Fabiano C, Laganà M, Pino L (2015) Bio-hydrogen production by oxidative steam reforming of biogas over nanocrystalline Ni/CeO_2 catalysts. Int J Hydrogen Energy, 1–8. (Article in press)
Iulianelli A, Manzolini G, Falco M, Campanari S, Longo T, Liguori S et al (2010) H_2 production by low pressure methane steam reforming in a Pd-Ag membrane reactor over a Ni-based catalyst: experimental and modeling. Int J Hydrogen Energy 35:11514–11524
Kolbitsch P, Pfeifer C, Hofbauer H (2008) Catalytic steam reforming of model biogas. Fuel 87:701–706
Kovács KL, Kovács ÁT, Maróti G et al (2004) Improvement of biohydrogen production and intensification of biogas formation. Rev Environ Sci Biotechnol 3(4):321–330
Lau CS, Tsolankis A, Wyszynski ML (2011) Biogas upgrade to syngas (H_2–CO) via dry and oxidative reforming. Int J Hydrogen Energy 36:397–404
Levin DB, Chahine R (2010) Challenges for renewable hydrogen production from biomass. Int J Hydrogen Energy 35:4962–4969
Lin KH, Chang HF, Chang ACC (2012) Biogas reforming for hydrogen production over mesoporous $Ni_{2x}Ce_{1-x}O_2$ catalysts. Int J Hydrogen Energy 37(20):15696–15703
Lin Y, Liu S, Chuanga C, Chub Y (2003) Effect of incipient removal of hydrogen through palladium membrane on the conversion of methane steam reforming: experimental and modeling. Catal Today 82:127–139
Liu C, Zhang R, Wei S et al (2015) Selective removal of H_2S from biogas using a regenerable hybrid TiO_2/zeolite composite. Fuel 157:183–190
Lu GQ, Diniz-Costa JC, Dukec M, Giessler S, Socolowe R, Williams RH et al (2007) Inorganic membranes for hydrogen production and purification: a critical review and perspective. J Colloid Interface Sci 314:589–603
Lucredio AF, Assaf JM, Assaf EM (2012) Reforming of a model biogas on Ni and Rh–Ni catalysts: effect of adding La. Fuel Process Tech. 102:124–131
Mahecha-Botero A, Chen Z, Grace JR et al (2009) Comparison of fluidized bed flow regimes for steam methane reforming in membrane reactors: A simulation study. Chem Eng Sci 64 (16):3598–3613
Maluf SS, Assaf EM (2009) Ni catalysts with Mo promoter for methane steam reforming. Fuel 88:1547–1553
Meyer J, Mastin J, Pinilla CS (2014) Sustainable hydrogen production from biogas using sorption-enhanced reforming. Energy Procedia 63(1876):6800–6814
Micoli L, Bagnasco G, Turco M (2014) H_2S removal from biogas for fuelling MCFCs: New adsorbing materials. Int J Hydrogen Energy 39(4):1783–1787
Mosayebi Z, Rezaei M, Ravandi AB, Hadian N (2012) Auto-thermal reforming of methane over nickel catalysts supported on nanocrystalline $MgAl_2O_4$ with high surface area. Int J Hydrogen Energy 37:1236–1242

Muradov N, Smith F, T-Raissi A (2008) Hydrogen production by catalytic processing of renewable methane-rich gases. Int J Hydrogen Energy 33:2023–2035

Ohkubo T, Hideshima Y, Shudo Y (2010) Estimation of hydrogen output from a full-scale plant for production of hydrogen from biogas. Int J Hydrogen Energy 35:13021–13027

Papadias DD, Ahmed S, Kumar R (2012) Fuel quality issues with biogas energy—an economic analysis for a stationary fuel cell system. Energy 44:257–277

Purwanto H, Akiyama T (2006) Hydrogen production from biogas using hot slag. Int J Hydrogen Energy 31:491–495

Rand DAJ, Dell RM (2008) Hydrogen energy: challenges and prospects. RSC Press, Cambridge

Redwood MD, Paterson-Beedle M, Macaskie LE (2009) Integrating dark and light bio-hydrogen production strategies: towards the hydrogen economy. Rev Environ Sci Biotechnol 8(2):149–185

Rogatis L, Montini T, Cognigni A, Olivi L, Fornasiero P (2009) Methane partial oxidation on NiCu-based catalysts. Catal Today 145:176–185

Roh HS, Eum IH, Jeong DW (2012) Low temperature steam reforming of methane over Ni–$Ce_{(1-x)}Zr_{(x)}O_2$ catalysts under severe conditions. Renew Energy 42:212–216

Ruckenstein E, Hu YH (1999) Methane partial oxidation over NiO/MgO solid solution catalysts. Appl Catal A Gen 183:85–92

Rueangjitt N, Akarawitoo C, Chavadej S (2012) Production of hydrogen-rich syngas from biogas reforming with partial oxidation using a multi-stage AC gliding arc system. Plasma Chem Plasma Process 32(3):583–596

San-José-Alonso D, Juan-Juan J, Illan-Gomes MJ, Roman-Martinez MC (2009) Ni, Co and bimetallic Ni–Co catalysts for the dry reforming of methane. Appl Catal A Gen, 371:54–59

Sato T, Suzuki T, Aketa M, Ishiyama Y, Mimura K, Itoh N (2010) Steam reforming of biogas mixtures with a palladium membrane reactor system. Chem Eng Sci 65:451–457

Serrano-Lotina A, Martin AJ, Folgado MA, Daza L (2012) Dry reforming of methane to syngas over La-promoted hydrotalcite clay-derived catalysts. Int J Hydrogen Energy 37:12342–12350

Sharifi M, Haghighi M, Abdollahifar M (2014) Hydrogen production via reforming of biogas over nanostructured Ni/Y catalyst: Effect of ultrasound irradiation and Ni-content on catalyst properties and performance. Mater Res Bull 60:328–340

Shiga H (1998) Large-scale hydrogen production from biogas. Int J Hydrogen Energy 23(97):631–640

Simeone M, Salemme L, Allouis C (2008) Reactor temperature profile during auto-thermal methane reforming on Rh/Al_2O_3 catalyst by IR imaging. Int J Hydrogen Energy 33:4798–4808

Sisani E, Cinti G, Discepoli G, Penchini D, Desideri U, Marmottini F (2014) Adsorptive removal of H_2S in biogas conditions for high temperature fuel cell systems. Int J Hydrogen Energy 39 (36):21753–21766

Souza MMVM, Schmal M (2005) Autothermal reforming of methane over $Pt/ZrO_2/Al_2O_3$ catalysts. Appl Catal A Gen 281:19–24

Tippayawong N, Thanompongchart P (2010) Biogas quality upgrade by simultaneous removal of CO_2 and H_2S in a packed column reactor. Energy 35(12):4531–4535

Wongtanet J, Sang BI, Lee SM, Pak D (2007) Biohydrogen Production by Fermentative Process in Continuous Stirred-Tank Reactor. Int J Green Energy 4(4):385–395

Xu G, Chen X, Honda K, Zhang ZG (2004) Producing H_2-rich gas from simulated biogas and applying the gas to a 50 W PEFC stack. AIChE J 50(10):2467–2480

Xu J, Zhou W, Li Z, Wang J, Ma J (2009) Biogas reforming for hydrogen production over nickel and cobalt bimetallic catalysts. Int J Hydrogen Energy 34:6646–6654

Xu J, Zhou W, Li Z, Wang J, Ma J (2010) Biogas reforming for hydrogen production over a Ni–Co bimetallic catalyst: Effect of operating conditions. Int J Hydrogen Energy 35:13013–13020

Yang L, Ge X, Wan C, Yu F, Li Y (2014) Progress and perspectives in converting biogas to transportation fuels. Renew Sustain Energy Rev 40:1133–1152

Zhai X, Ding S, Liu Z, Jin Y, Cheng Y (2011) Catalytic performance of Ni catalysts for steam reforming of methane at high space velocity. Int J Hydrogen Energy 36:482–489

Zhang YHP, Evans BR, Mielenz JR, Hopkins RC, Adams MWW (2007) High-Yield hydrogen production from starch and water by a synthetic enzymatic pathway. PLoS ONE 2(5):e456. doi:10.1371/journal.pone.0000456

Modifications in Improved Cookstove for Efficient Design

Amit Ranjan Verma, Rajendra Prasad, Virendra Kumar Vijay and Ratnesh Tiwari

Abstract In India, majority of the population (around 85 %) living in rural areas still use traditional biomass fuels in form of wood, agriculture residues, animal dung, etc., for meeting their energy needs. These fuels are burnt in mud (traditional) cookstoves for cooking and other thermal applications. Inefficient burning of biomass in traditional cookstoves poses serious health hazards. Experiments have been carried out on low cost, improved forced draft cookstove, and modifications were made to increase the thermal efficiency and to lower smoke emissions by varying parameters such as height of grate from base, variations in air flow rate, height between top plate and vessel, etc., without making major changes in the structure of cookstove such as diameter and air inlet points. Modification in air flow rate leads to increased thermal efficiency with low smoke emissions. All measurements were made according to the Bureau of Indian Standards (IS 13152: Part 1-Portable Solid Bio-Mass Cookstove (Chulha).

Keywords Cookstove · Improved cookstove · Thermal efficiency · Forced draft cookstove · Bureau of Indian Standards

1 Introduction

Biomass has been a key resource for human being in meeting the fuel requirement for cooking and space heating since ancient time. Though the gaseous fuels, i.e., LPG and natural gas have replaced biomass in developed countries and most of urban homes in the developing world, half of the world's population and about 90 % of rural households in developing countries are primarily dependent on coal or biomass, such as wood, crop residues, cattle dung, and charcoal for their cooking and heating needs (WRI 1998–1999). According to National Census 2011, 67 % of

A.R. Verma (✉) · Rajendra Prasad · V.K. Vijay · Ratnesh Tiwari
Improved Biomass Cookstove Laboratory, Centre for Rural Development and Technology, Indian Institute of Technology Delhi, Hauz Khas, New Delhi 110016, India
e-mail: arviitd@gmail.com

S. Kumar et al. (eds.), *Proceedings of the First International Conference on Recent Advances in Bioenergy Research*, Springer Proceedings in Energy,
DOI 10.1007/978-81-322-2773-1_18

all households in India used solid fuels, mostly biomass. Firewood, crop residue, and cow dung cake (all biomass fuels) have a penetration of 62.5, 12.3, and 10.9 %, respectively. LPG penetration in rural households stands at 11.4 % followed by kerosene, coal/charcoal, and electricity at 0.7, 2.9, and 0.1 %, respectively (Indian Census 2011).

The majority of household energy requirement is in various cooking needs. Traditional cookstoves or *chulhas*, having overall thermal efficiencies less than 10 % are generally practiced in rural households in the developing countries, are main sources of large quantities of pollutants resulting various health hazards. Large quantity of biomass residues such as agricultural residues, animal dung, forest waste, firewood, etc., is available in India. The biomass is considered as sustainable and carbon-neutral source of energy (Kothari et al. 2008). Moreover, the use of biomass feedstocks in an inefficient traditional cookstoves leads to generation of increased indoor air pollution and wastage of stored chemical energy in biomass. These less efficient traditional cookstoves consume large amount of fuelwood and create pressure on the region's forest cover (Bhattacharya and Leon 2005). To increase the performance of cookstove several modifications in cookstove have been carried out by many researchers. Galitsky (2006) made several design modifications in existing design of Technology and Action for Rural Advancement (TARA) cookstove to improve stove performance in windy weather operating condition. These modifications included a windshield around the upper stove body and addition of a metal ring to prevent direct air flow through the stove body and to provide stability; stakes were also incorporated in the design. These modifications were incorporated in a prototype of cookstove called as Berkeley-Darfur Stove (BDS). BDS cookstove was found to consume less than 60 % of the fuelwood with thermal efficiency of 43 % and produced noticeably lower smoke compared to the three-stone fire. The time needed for cooking using the BDS cookstove was about the same as that needed over the three-stone fire. They had also modified locally available mud-and-dung "ITDG" cookstove, called as the modified stove "Avi". They also incorporated cast iron grate placed over the opening cut out of the bottom, which improved the combustion efficiency and reduced smoke level. The new design also included vertical ventilation channels carved into the inner walls of the stove and three mud knobs were added to the top to permit enhanced combustion air flow even when a tight-fitting large pot or a flat metal plate is being used for cooking. However, the experimental results were not found good enough, as the cookstove was consuming about 90 % of the fuel compared to the three-stone fire cookstove, and smoke emission was found significantly higher. Furthermore, the cookstove took more time to cook the food than the three-stone fire. Guillaume and LaRue (1995) experimentally verified that the swirl effect improved the combustion of volatiles by increasing air–fuel mixing and the residence time of the reactant gasses. Bhandari et al. (1988) incorporated four different types of swirling device in the throat area of cookstove combustion chamber and found that the strong swirling effect had increased the cookstove's efficiency to about 42 %.

This paper presents the studies on the parameters of existing force draft biomass cookstove and necessary modifications required in design variables which were improved the performance of biomass cookstove in terms of power output, thermal efficiency, and emission level.

2 Materials and Methods

2.1 Parametric Studies on Existing Improved Biomass Cookstove

Parametric studies were conducted in existing low cost improved biomass cookstove (IMPMUD model). Parameters of studies were those which affect the performance of cookstove, namely burning rate of fuel wood, air supply required for combustion, and power supplied to the fan using standard methods/protocols (Verma et al. 2014).

2.2 Selection of Suitable Material for Cookstoves

Through several past studies it was concluded that ceramics are superior for longer cooking periods. If the choice is between metal and clay, the latter is preferred for cooking times' longer than 15 min Krishna Prasad and Sangen (1983). On the basis of heat transfer calculations, preference was given to ceramics as a construction material. However, there are important drawbacks of ceramics, i.e., lack of suitable clays and the fragility of the final product. Ceramics are not strong enough for some cooking jobs. This has led to total rejection of ceramics in many cases and the choice is between metal and mud.

3 Dimensioning the IMPMUD Stove Elements

3.1 The Combustion Chamber

The combustion chamber can be thought of as being built-up from two separate parts with height Hcc_1 and Hcc_2. The first is occupied by the fuel bed where charcoal burns and the pyrolysis takes place; the second part is occupied by the flames. The fuel bed volume ($Hcc_1 \times Acc$) depends on the power output which is required, the wood mass charged, the packing density (χ), and the burning characteristics of the wood Bussmann PJT (1988). The maximum power density can be defined as the maximum power per m^2 fuel bed area can be obtained from a steadily burning fire. A power density of 180 kW/m^2 was obtained experimentally in the literature from free-burning fires of varying diameter, without grate, using oven-dry woodblocks of 15 × 15 × 50 mm^3. The power density could be increased by 50 % when using a grate. In the design work, the power density could be chosen as 200 kW/m^2. The cross-sectional area of the combustion chamber and the combustion chamber height are then given by

$$A_{CC} = P_{\text{Design}}/200$$
$$H_{cc1} = \Delta M_{\text{f}}/(\chi.\rho_f)(A_{cc})$$

where, A_{cc} is the cross-sectional area of combustion chamber in m^2; P_{Desin} is the design power of cookstove in kW; H_{cc1} is the height of the fuel bed combustion chamber in m; ΔM_{f} is the mass of wood charged in kg; and χ is packing density of fuel in combustion chamber.

3.2 The Fuel Loading Opening

A closed combustion chamber facilitates the control of incoming combustion air. However, the demands put forward by stove users often do not allow for such a configuration: the stove must accept long pieces of fuelwood. Consequently, the aeration of the fuel bed becomes a problem. The least that can be done is to keep the size of the fuel loading opening small. The criteria as shown in Table 1 could be used. Performance can be improved with a metal flap which closes the combustion chamber when fuel wood does not project out of the stove.

3.3 Primary and Secondary Air Holes

The aeration of the combustion chamber in a closed stove fire is very effective. The inlet air holes form the main flow resistance. The aeration of the fuel bed is a problem for cookstove having an open combustion chamber. The only measure to be taken into consideration is to keep the total area of both primary and secondary air holes to equal (including fuel loading opening).

Table 1 Criteria used for dimensioning the fuel loading opening

Height	Smaller than the combustion chamber height larger than two stick diameters: 2 × 30 mm
Base	Smaller than the combustion chamber diameter larger than three stick diameters: 3 × 30 mm

3.4 The Grate

The purpose of the grate is twofold. First, it is used to concentrate the fire during flaming combustion. Second, the grate is used to promote the combustion of charcoal formed during pyrolysis. For both reasons it is necessary that the flow resistances of the grate and other air entrance passages are of the same order of magnitude. The grate can easily regain its function if its porosity and the number of primary air holes underneath are increased. In optimizing the above stove it had been suggested, first, to increase the number and size of the primary air holes to equal the total cross-sectional area of the fuel loading opening; and second, to increase the porosity of the grate from 5 to 10 % to the required 50 %.

3.5 Flow Path Resistances

The experiments conducted by various researchers clearly showed the importance of reducing the air flow. Preferably this has to be done at the air inlets which make it possible to control primary and secondary air separately. For stoves having a large fuel loading opening this is not possible. The flow resistance must be created downstream of the combustion chamber. Each stove can concentrate flow resistance in the annular gap between pan and shield.

3.6 Design Description of IMPMUD

Dimensions of cookstove are as follows; (1) inner diameter of combustion chamber = 13.0 cm, (2) outer diameter of combustion chamber = 15.0 cm, (3) height of combustion chamber = 18.0 cm, and (4) pan support height = 3.0 cm. Figure 1 shows the top and front view of the cookstove. The computer aided design of domestic IMPMUD cookstove is shown in Fig. 2.

3.7 Parametric Study

Various factors which affect the performance of cookstove have been experimentally studied in this section.

(1) Determination of optimum burning rate: cookstove was experimentally tested with different burning rate and it was found that burning rate of 1.7 kg fuel/h is optimum for this IMPMUD cookstove.
(2) Determination of optimum air flow through primary and secondary holes for complete combustion of fuel: In all forced draft cookstove optimum air is

Fig. 1 Top and front view of IMPMUD stove (*L* to *R*, respectively)

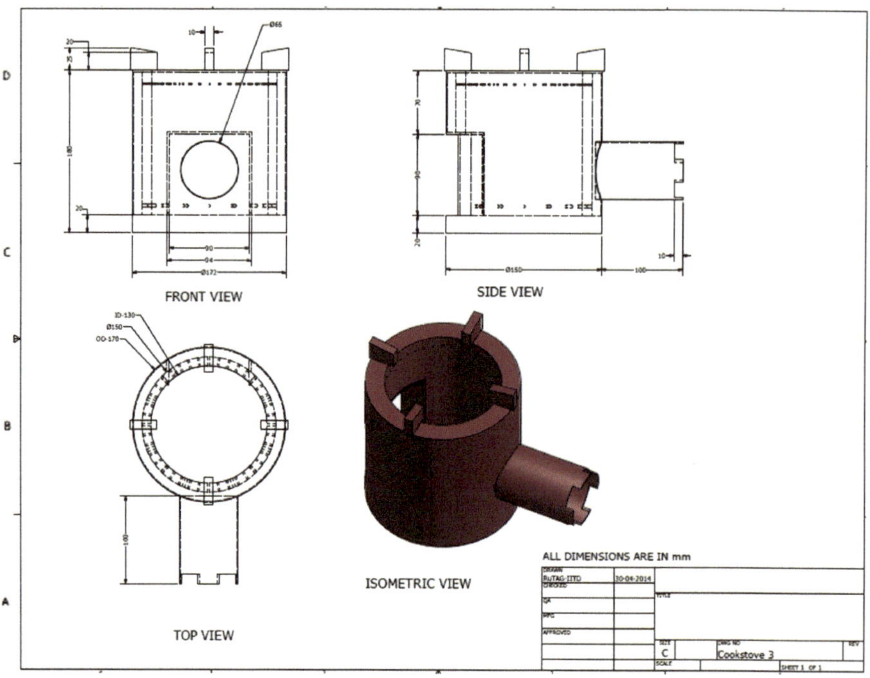

Fig. 2 Computer aided design of domestic IMPMUD cookstove

suppled through an external exhaust fan, but it is difficult to find a suitable fan which can supply optimum air in combustion chamber for clear combustion. Experimentally fans with different input power had been used for suitable air flow and fan of power 3.6 was suitable to provide air for complete combustion.

(3) Determination of optimum pan support: Pan support height or gap between top plate of cookstove and pot bottom. The height of the support has a considerable effect on the efficiency of the cookstove, and hence needs optimization. Stove with small support height although have high efficiency, but they also reduce the amount of gas that can flow through the gap and thus limit the firepower. In case of larger fire either smoke will pour out the stove door, or else the fire will be choked and suffer poor combustion or simply not build up to desire power. Experimentally it had been found pan support of height 2.5 cm is optimum.

4 Results

4.1 Performance and Emissions Results

The power output, thermal efficiency, and emissions, including both particulate matter (PM) and carbon monoxide (CO), performance of cookstove are given in Table 2. Furthermore, the thermal efficiency and emission performances for forced and natural draft cookstoves as specified by the Bureau of Indian Standards (BIS 2013) are given Table 3 for performance comparison analysis.

As per the results obtained from performance test, the cookstove performance could not meet the desirable limit of thermal efficiency and emissions. Thermal efficiency was low either because of low combustion efficiency or due to poor heat transfer between cookstove top plate and pot (pan) bottom. Emissions were too high because of poor combustion of fuel inside combustion chamber which is either because of insufficient amount of air was blown in side of combustion chamber through blower or excess air was supplied through blower than the required amount of air for complete combustion. Hence, it has been decided to perform parametric study to find out parameters which are affecting the cookstove performance and

Table 2 Performance of IMPMUD cookstove (without modification)

S. No.	Power output (kW)	Thermal efficiency (%)	PM (mg/MJ_d)	CO (g/MJ_d)
1	1.04	21.3	600.5	7.1
2	1.16	23.3	578.9	7.0
Average	1.1	22.3	589.7	7.05

Table 3 BIS limits for forced and natural draft cookstoves

Parameters	Forced draft	Natural draft
Thermal efficiency	≥35 %	≥25 %
PM in mg/MJ_d	≤150	≤350
CO in g/MJ_d	≤5	≤5

Table 4 Performance of IMPMUD forced draft cookstove (with incorporated modifications)

S. No.	Power output (kW)	Thermal efficiency (%)	PM (mg/MJ_d)	CO (g/MJ_d)
1	2.96	35.67	101.11	1.48
2	3.01	36.29	97.46	1.36
3	2.94	35.47	101.13	1.07
4	3.02	36.49	106.42	1.69
5	3.01	36.36	96.40	1.45
6	3.04	37.11	81.55	1.22
Average	2.99	36.23	97.34	1.38

further made necessary modify in existing parametric design so that it could meet the BIS requirement.

Modifications in IMPMUD forced draft cookstove have been done according to the results of parametric study and modified IMPMUD was again tested with BIS protocol. The results of modified IMPMUD forced draft biomass cookstove are presented in Table 4. It is clearly evident that the average thermal efficiency of the cookstove was found as 36.23 % which is significantly higher than the BIS limit (≥35 %). The emission of particulate matter (PM) and carbon monoxide (CO) was found to be 97.34 mg/MJ_d and 1.38 g/MJ_d, respectively. The emissions were lower than the BIS limits as given in Table 3.

5 Conclusions

The experimental study had revealed that the modifications can be done in existing cookstoves for enhanced thermal efficiency and lower emissions by varying parameters such as fuel burning rate, air supply and power supply of fan, and height between top plate and vessel without making major changes in the structure of existing cookstoves (such as diameter and secondary air inlet points, etc.). With above modifications cookstove efficiency was increased up to 63 % and there was decrease in CO emission up to 80 % and PM emission up to 84 %.

References

Bhandari S, Gopi S, Date AW (1988) Investigation of CTARA wood-burning stove. Part 1. Experimental investigation. Sadhana 13(4):271–293

Bhattacharya SC, Leon MA (2005) Prospects for biomass gasifiers for cooking applications in Asia'. http://www.retsasia.ait.ac.th/Publications/WRERC2005/AIT-gasifierstoveforcooking-final.pdf

Bureau of Indian Standard (BIS) on Solid Biomass Chulha (2013) Specification CIS 1315Z (Part 1): 1991, First revision (2013)

Bussmann PJT (1988) Woodstoves Theory and applications in developing countries', Ph. D. thesis, Eindhoven, University of Technology, Eindhoven, pp 1–174

Census of India (2011) "Fuel Used for Cooking" http://censusindia.gov.in/Census_Data_2011/Census_data_finder/HH_Series/Fuel_used_for_cooking.htm

Galitsky C, Gadgil A, Jacobs M, Lee Y (2006) Fuel efficient stoves for Darfur IDP camps: report of field trip to North and South Darfur. Lawrence Berkeley National Laboratory Report LBNL-59540

Guillaume DW, LaRue JC (1995) Combustion enhancement using induced swirl. Exp Fluids 20:59–60

Kothari DP, Singal KC, Ranjan R (2008) Biomass energy, renewable energy sources and emerging technologies, pp 284–307. P.H.I. Learning India

Krishna Prasad K, Sangen E (1983) Technical aspects of wood burning cookstoves, a report from the wood burning Stove Group, Eindhoven University of Technology Eindhoven, Netherlands

Verma AR, Prasad R, Vijay VK, Tiwari R, Mal R (2014) Modifications in traditional Indian cookstove and improved cookstove for efficient design. 2014-ETHOS conference, Kirkland, Washinton, DC

WRI (World Resources Institute) (1998–1999) UNEP, UNDP, World Bank. World resources: a guide to global environment. Oxford University Press, Oxford, UK

Part IV
Biomass Management

Thermochemical Characterization of Pine Needles as a Potential Source of Energy

Anil Kumar Varma and Prasenjit Mondal

Abstract Present paper deals with the thermochemical characterization of pine needles to assess its suitability as an energy feedstock. Proximate and ultimate analyses revealed that pine needles contain low moisture, ash, nitrogen, and sulfur content, as well as high fixed carbon and volatile matter content. The higher heating value (HHV) is determined as 19.22 MJ kg^{-1} using bomb calorimeter as well as the lower heating value (LHV) is 17.66 by using HHV value. The empirical formula of pine needle is found to be $CH_{1.83}O_{0.74}N_{0.008}$ on the basis of ultimate analysis. Fourier transform infrared (FTIR) spectra is developed to identify the basic functional groups in pine needles. Thermal degradation analysis has been done by using a thermogravimetric analyzer from 32 to 1000 °C in a nitrogen atmosphere at a heating rate 10 °C min^{-1}. In addition, the thermal decomposition behavior of pine needles has been determined through TGA/DTG/DTA analysis. The chemical characteristics and thermal behavior show that pine needles have good energy potential for exploitation through pyrolysis.

Keywords Biomass · Pine needles · Pyrolysis · FTIR · Thermogravimetric analysis

1 Introduction

The world's energy market is mainly dependent on the fossil fuels such as coal, petroleum, and natural gas as a prime source of energy. Since millions of years are required to form fossil fuels in the earth, their reserves are finite and it is predicted that these sources of energy will deplete within the next 40–50 years (Saidur et al. 2011). Further, due to continuous degradation in qualities these fossil fuels are

A.K. Varma · Prasenjit Mondal (✉)
Department of Chemical Engineering, Indian Institute of Technology Roorkee, Roorkee 247667, Uttarakhand, India
e-mail: mondal2001@gmail.com

S. Kumar et al. (eds.), *Proceedings of the First International Conference on Recent Advances in Bioenergy Research*, Springer Proceedings in Energy,
DOI 10.1007/978-81-322-2773-1_19

creating environmental problems such as global warming and acid rain. Thus, applications of renewable energy sources are being investigated around the world to meet the energy security and environmental sustainability. Biomass is a clean fuel because it has a very low content of sulfur, nitrogen, and ash, which gives lower emissions of SO_x, NO_x, and soot than conventional fossil fuels. Moreover, biomass itself is carbon neutral and its application in an engineered way may add carbon credit.

In India, pine needles are available in large quantities in the Himalayan region of Uttarakhand. Himachal Pradesh, etc. Around 340,000 hectares of pine forests cover is available in Uttarakhand (Bisht et al. 2014), which produces around 2.06 million tons of pine needle annually (Web 1). Pine needles are needle-shaped leaves falling from tree mainly in the summer season in the month of March to July. Due to their highly inflammable nature, it is a serious threat to forest fires. These fires create environmental pollution, damage the fertile top layer of soil, and destroy grazing grounds for cattle. So, their collection and disposal are very important in environmental point of view. Normally, collection of pine needles is done by local people and it creates opportunity for earning to the common mass of the hilly region. Due to creation of awareness among the people, collection of pine needle is increasing gradually for use as an energy source. In 2012 nearly 600 tons of pine needle was collected in Hamirpur district of Himachal Pradesh, which is around 12 times more than the corresponding collection in 2011 (Web 2). Further, in comparison to other biomass such as sugar cane bagasse, semi-dried banana leaves, rice husk, elephant grass (Fernandes et al. 2013), etc., this biomass has better potential as a feedstock for bioenergy through suitable thermochemical conversion process such as pyrolysis to fulfill the energy requirement of the hilly region. Pyrolysis is a thermochemical conversion process to convert biomass into volatile fraction composed of condensable organic vapors (bio-oil), carbon-rich solid waste (char), and noncondensable gasses (Fernandes et al. 2013). This process is very attractive option considering the current initiatives of seeking new renewable sources for clean energy production and may be applied for this feedstock. However, to understand the requirement of operating conditions, reactor configuration/design, etc., the thermochemical properties of the biomass are essential (Ceylan and Topçu 2014). In the previous literatures, there is not much evidence available regarding thermal characterization of pine needles.

The aim of this study is to characterize the pine needle through proximate and ultimate analyses, find the empirical formula for pine needles from ultimate analysis data and determine its heating values. It also includes FTIR analysis to find out the functional groups in the sample as well as thermal analysis such as (TG/DTG/DTA) to evaluate the thermochemical conversion of pine needles. Suitablity of this biomass as a feedstock for pyrolysis has also been assured.

2 Materials and Methods

2.1 Materials

Pine needles were collected from the campus of Indian Institute of Technology Roorkee, India. The collected pine needles were sun dried for 2 days, the dried biomass were stored in polyethylene bags. 1 kg of this biomass was further oven dried at 105 °C for 2 h, ground in Wiley grinder and powdery sample having particle size 250–500 μm was prepared by using IS sieves. Figure 1 shows the photograph of pine needles before grinding and after grinding.

2.2 Methods

Proximate analysis was done as per procedure laid in the Bureau of Indian Standards code IS: 1350 (PART 1)-1984 to determine the moisture (M), volatile matter (VM), ash (As), and fixed carbon (FC) of the sample. The fixed carbon was obtained by subtracting the percentage of volatile matter, moisture, and ash from 100.

The ultimate analysis was used to find out the elemental composition such as carbon, hydrogen, nitrogen, sulfur, and oxygen in the biomass sample and analysis was done in Elementar Vario EL III model. The higher heating value (HHV) of the pine needles was determined in a Parr 6300 Bomb calorimeter and this experimental value of HHV was compared with the theoretical value determined using the following Eq. (1) (Channiwala and Parikh 2001).

$$\mathrm{HHV}\left(\mathrm{MJ\,kg^{-1}}\right) = 0.3941\mathrm{C} + 1.1783\mathrm{H} + 0.1005\mathrm{S} - 0.1034\mathrm{O} - 0.0151\mathrm{N} - 0.0211\mathrm{As} \quad (1)$$

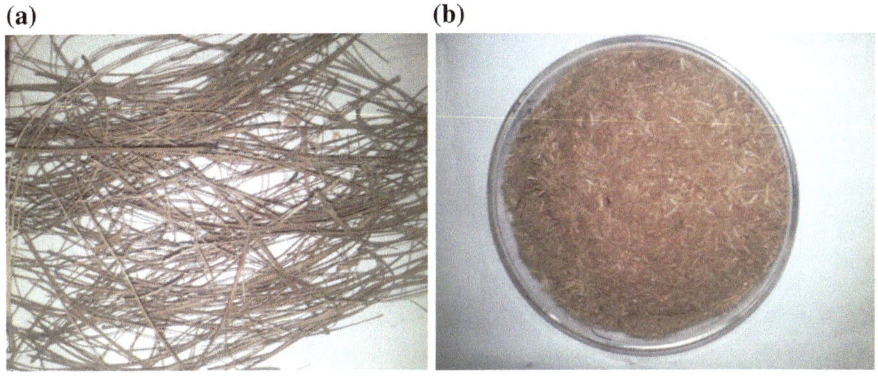

Fig. 1 Photograph of pine needles. **a** Before grinding. **b** After grinding

where C, H, S, O, N, and As represent the carbon, hydrogen, sulfur, oxygen, nitrogen, and ash content of the sample in wt. (%).

The lower heating value (LHV) of pine needles was calculated by Eq. (2) (Basu 2010).

$$LHV\left(MJ\,kg^{-1}\right) = HHV\left(MJ\,kg^{-1}\right) - hg\,(9H/100 + M/100) \tag{2}$$

where LHV, HHV, H, M, and hg are the lower heating value, higher heating value, hydrogen percentage, moisture percent, and latent heat of steam (2.26 MJ kg^{-1}), respectively.

FTIR analysis of the pine needles sample was done in FTIR spectrometer (Thermo Nicollet, Model Magna 760). Pellet technique was used for this analysis, where dried samples were embedded in KBr pellets with a ratio of 1:10. The spectra were recorded in the wavenumber range of 4000–500 cm^{-1}.

The thermal decomposition behavior of pine needles was determined by using a thermal analyzer (Model–SII 6300 EXSTAR). The thermal analysis, such as TG (thermogravimetric), DTG (derivative thermo-gravimetric) and DTA (differential thermal analysis) were done simultaneously. The sample was analyzed in an Al_2O_3 crucible at a heating rate of 10 °C min^{-1} from 32 to 1000 °C in nitrogen (N_2) atmosphere with a flow rate of 200 ml min^{-1} as sweeping gas. The mass of sample for the experiment was taken about 10 mg.

3 Results and Discussion

3.1 Characteristics of Pine Needles

The result from proximate analysis, ultimate analysis, and heating values of the present pine needles sample are shown in Table 1. As evident in Table 1 the moisture content of pine needle is 5.03 wt%, which is less than 10 wt%, the upper limit of moisture content for pyrolysis preferably (Asadullah et al. 2007). High moisture content in biomass requires prior drying to remove the moisture and reduces the energy efficiency of the system. Further, high moisture content also increases water content in the bio-oil and reduces its heating value. From Table 1 it is also evident that the volatile matter content in pine needle sample is high (77.55 wt%). High amount of volatile matter facilitates the thermal decomposition and such biomass is better for the pyrolysis. The result shows that the fixed carbon content is 13.77 wt%, which is comparable to the other biomass such as semi-dried banana leaves (12.5 wt%), rice husk (14.6 wt%), elephant grass 15 wt%, and briquette 14.2 wt% (Fernandes et al. 2013; Wilson et al. 2011; García et al. 2012; Mesa-Perez et al. 2005). The ash content of above biomass samples falls in the range of 5.9–11.7 %, which is higher than that of the present biomass, i.e., 3.38 wt%. High ash content in the biomass has a negative effect on the thermochemical conversion processes. Ash content in the biomass affects the handling and processing costs in thermochemical conversion

Table 1 Proximate and ultimate analyses of pine needles

Proximate analysis (wt%)	
Moisture	5.03
Volatile matter	77.55
Fixed carbon[a]	13.77
Ash content	3.38
Ultimate analysis (wt%)	
Carbon	46.38
Hydrogen	7.08
Oxygen[a]	46.02
Nitrogen	0.45
Sulfur	0.07
Empirical formula	$CH_{1.83}O_{0.74}N_{0.008}$
H/C molar ratio	1.83
O/C molar ratio	0.74
HHV (MJ kg^{-1})	19.22
HHV by correlation (MJ kg^{-1})	19.70
LHV (MJ kg^{-1})	17.66

[a]By difference

process. Other problems associated with the high ash content are reactor fouling, sintering, corrosion, and slagging in pyrolysis process (Fernandes et al. 2013). Low ash and high volatile content in the present biomass make it good feedstock for pyrolysis. Ultimate analysis of pine needle shows carbon, hydrogen, oxygen, nitrogen, and sulfur content are 46.38, 7.08, 46.02, 0.45, and 0.07 wt%, respectively. This result indicates that the sulfur and nitrogen content in pine needle is very low, so the emission of toxic gasses like So_x and No_x is expected to be very less during thermochemical conversion process. The empirical formula for the present pine needle is found to be $CH_{1.83}O_{0.74}\ N_{0.008}$. The H/C and O/C molar ratios as calculated from the ultimate analysis are 1.83 and 0.74, respectively. The HHV of the pine needles is found to be 19.22 MJ kg^{-1} and calculated HHV from Eq. (1) is 19.70 MJ kg^{-1}. The difference between experimental and calculated values of HHV is ~0.48 MJ kg^{-1}. The HHV of pine needles is close to other biomass such as switchgrass 19.6 MJ kg^{-1} (Masnadi et al. 2014), semi-dried banana leaves 19.8 MJ kg^{-1} (Fernandes et al. 2013), sugar cane bagasse 19.2 MJ kg^{-1} (Asadullah et al. 2007), cedar wood 19.26 MJ kg^{-1} (Asadullah et al. 2004), etc. The lower heating value (LHV) of the pine needle calculated from Eq. (2) is 17.66 MJ kg^{-1}. Low moisture and ash content, as well as high volatile content and HHV (19.22 MJ kg^{-1}) suggest pine needles as a good potential for production of bio-oil through pyrolysis.

The FTIR analysis is used to identify the characteristic functional groups and chemical constituents present in the sample, which are important for determining the composition and distribution of pyrolysis products (Nyakuma et al. 2014). Presence of hemicelluloses, cellulose, and lignin in some biomass such as nonedible

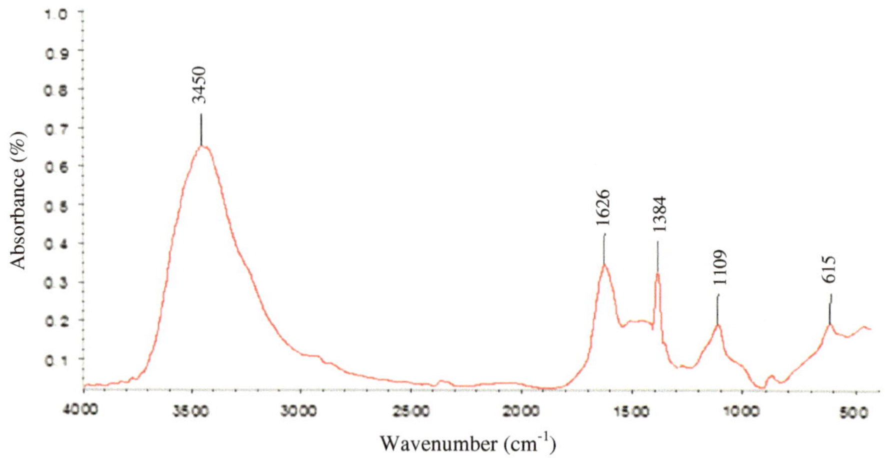

Fig. 2 FTIR spectra of pine needle

oil seeds as mahua, karanja, niger, and linseed (Shadangi and Mohanty 2014), empty fruit bunch briquettes (Nyakuma et al. 2014), orange waste (Lopez-Velazquez et al. 2013), mesua ferrea L.deoiled cake (Chutia et al. 2013), date palm (Sait et al. 2012), have recently been confirmed by FTIR analysis. For the present case Fig. 2 shows the FTIR spectra of pine needle in the wavenumber range of 4000–500 cm^{-1}. The first strong peak at 3450 cm^{-1} as shown in Fig. 2 denotes the O–H stretching vibration due to the presence of carbohydrates and lignin (Chutia et al. 2013). The peak around 1626 cm^{-1} is related to the C=C stretching vibration of aromatic ring, which corresponds to lignin (Asadieraghi and Wan Daud 2014). The peak at 1384 cm^{-1} is due to the presence of C–H stretching (Chutia et al. 2013). The peaks between the regions from 1200 to 900 cm^{-1} are due to the presence of C–O, C–C, C–O–C, and C–O–P stretching vibrations of polysaccharides, which are related to hemicelluloses and cellulose components (Asadieraghi and Wan Daud 2014). The FTIR analysis of pine needles shows the presence of functional groups associated with the cellulose, hemicelluloses, and lignin.

3.2 *Thermogravimetric Analysis of Pine Needles*

Figure 3 shows the TG (thermogravimetric), DTG (derivative thermogravimetric), and DTA (differential thermal analysis) plots of pine needles, where TG curve shows the mass loss profile as function of temperature, DTG curve denotes the rate of mass loss as function of temperature, and DTA curve denotes the differential thermal analysis of sample, it shows the energy change with temperature during thermal analysis of sample (Doshi et al. 2014). The entire pyrolysis process of present pine needles can be divided into four stages. The first stage starts at 32 °C

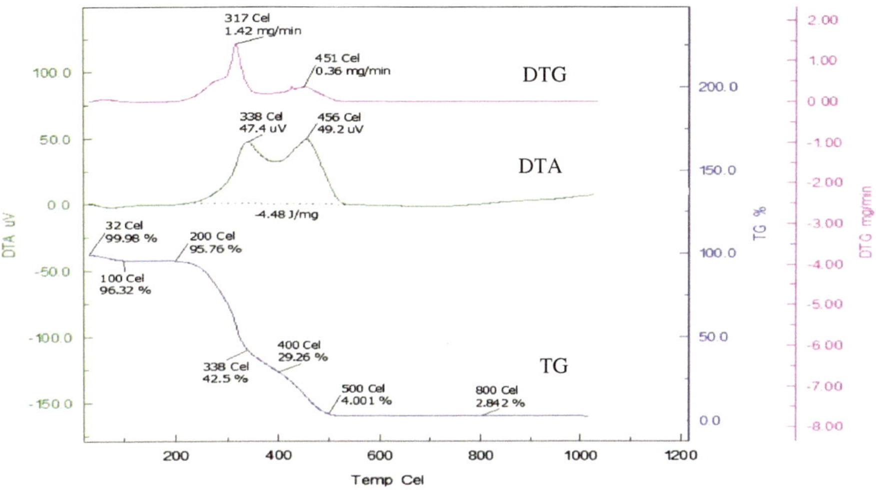

Fig. 3 TG/DTG/DTA curves of pine needle

and ends at 200 °C. As evident from TG curve the mass loss in this stage is ~5 % due to the removal of moisture and very light volatile components. (Shadangi and Mohanty 2014), which is also represented in the DTG curve (weak peak). Very small deviation in the DTA peak (absorption of heat) and proximate analysis is also in agreement with this fact. Second stage starts at 200 °C and ends at 340 °C with mass loss of ~53 % as evident from TG curve. At this stage DTG curve shows the maximum mass loss rate of ~1.42 mg min^{-1} at 317 °C, it is due to the degradation of hemicellulose and cellulose (Jeguirim and Trouvé 2009). This stage is stated as active pyrolysis stage. The third stage starts at 340 °C and finishes at 500 °C with mass loss of ~38 % from TG curve. This stage is also denoted in DTG curve with a mass loss rate of 0.36 mg min^{-1} at 451 °C. The DTA curve gives the energy change of ~47.4 µv, ~49.2 µv at 338 and 456 °C, respectively, and energy release during this stage is 4.48 J mg^{-1} from DTA curve. This stage, also denoted as an active pyrolysis stage, is due to the degradation of cellulose. The fourth stage starts at 500 °C and continues up to 1000 °C, the mass loss occured in this stage is only ~2 % and there is very less variation in DTG and DTA curve. This stage is referred as passive pyrolysis. The mass loss observed in the second, third, and fourth stage can also be explained by hemicellulose, cellulose, and lignin decomposition. From the above discussion it is evident that around 91 % of mass loss takes place within the temperature range of 500 °C, which indicates the suitability of the sample for application as pyrolysis feedstcok.

4 Conclusions

Low moisture and ash content, as well as high volatile content and HHV (19.22 MJ kg^{-1}) suggest pine needles as a good potential for production of bio-oil through pyrolysis. Low content of nitrogen and sulfur reduces threat of toxic gasses (NO_x and SO_x) emission. An empirical formula originated from ultimate analysis for pine needle is given as $CH_{1.83}O_{0.74}N_{0.008}$. The FTIR result shows the presence of hemicellulose, cellulose, and lignin in the biomass sample. The results of thermogravimetric analysis show the active pyrolysis in the temperature range of 200–500 °C, so the results obtained in thermochemical characterization of the pine needles indicate that it has good potential for energy generation through pyrolysis process.

References

Asadieraghi M, Wan Daud WMA (2014) Characterization of lignocellulosic biomass thermal degradation and physiochemical structure: Effects of demineralization by diverse acid solutions. Energy Convers Manag 82:71–82

Asadullah M, Miyazawa T, Ito S, Kunimori K, Koyama S, Tomishige K (2004) A comparison of $Rh/CeO_2/SiO_2$ catalysts with steam reforming catalysts, dolomite and inert materials as bed materials in low throughput fluidized bed gasification systems. Biomass Bioenergy 26:269–279

Asadullah M, Rahman MA, Ali MM, Rahman MS, Motin MA, Sultan MB, Alam MR (2007) Production of bio-oil from fixed bed pyrolysis of bagasse. Fuel 86:2514–2520

Basu P (2010) Biomass gasification and pyrolysis: practical design and theory. Associated Press for Elsevier Inc., U.K

Bisht AS, Singh S, Kumar SR (2014) Use of pine needle in energy generation application. Int J Res App Sci Eng Technol 2(11):59–63

Ceylan S, Topçu Y (2014) Pyrolysis kinetics of hazelnut husk using thermogravimetric analysis. Biores Tech 156:182–188

Channiwala SA, Parikh PP (2001) A unified correlation for estimating HHV of solid, liquid and gaseous fuels. Fuel 81:1051–1063

Chutia RS, Kataki R, Bhaskar T (2013) Thermogravimetric and decomposition kinetic studies of Mesua ferrea L. deoiled cake. Bioresour Technol 139:66–72

Doshi P, Srivastava G, Pathak G, Dikshit M (2014) Physicochemical and thermal characterization of nonedible oilseed residual waste as sustainable solid biofuel. Waste Manag 34 (10):1836–1846

Fernandes ERK, Marangoni C, Souza O, Sellin N (2013) Thermochemical characterization of banana leaves as a potential energy source. Energy Convers Manag 75:603–608

García R, Pizarro C, Lavín AV, Bueno JL (2012) Characterization of Spanish biomass wastes for energy use. Bioresour Technol 103:249–258

Jeguirim M, Trouvé G (2009) Pyrolysis characteristics and kinetics of Arundo donax using thermogravimetric analysis. Biores Tech 100(17):4026–4031

Lopez-Velazquez MA, Santes V, Balmaseda J, Torres-Garcia E (2013) Pyrolysis of orange waste: a thermo-kinetic study. J Anal Appl Pyrolysis 99:170–177

Masnadi MS, Habibi R, Kopyscinski J, Hill JM, Bi X, Lim CJ, Ellis N, Grace JR (2014) Fuel characterization and co-pyrolysis kinetics of biomass and fossil fuels. Fuel 117:1204–1214

Mesa-Perez JM, Cortez LAB, Rocha JD, Brossard-Perez LE, Olivares-Gómez E (2005) Unidimensional heat transfer analysis of elephant grass and sugar cane bagasse slow pyrolysis in a fixed bed reactor. Fuel Process Technol 86:565–575

Nyakuma BB, Johari A, Ahmad A, Amran T, Abdullah T (2014) Thermogravimetric Analysis of the fuel properties of empty fruit bunch briquettes. Jurnal Teknologi 3:79–82

Saidur R, Abdelaziz EA, Demirbas A, Hossain MS, Mekhilef S (2011) A review on biomass as a fuel for boilers. Renew Sustain Energy Rev 15(5):2262–2289

Sait HH, Hussain A, Salema AA, Ani FN (2012) Pyrolysis and combustion kinetics date palm biomass using thermogravimetric analysis. Biores Tech 118:382–389

Shadangi KP, Mohanty K (2014) Kinetic study and thermal analysis of the pyrolysis of non edible oilseed powders by thermogravimetric and differential scanning calorimetric analysis. Renew Energy 63:337–344

Web1: http://www.ureda.uk.gov.in/pages/display/142-pine-needle-based-project

Web2:http://archive.indianexpress.com/news/best-of-waste-pine-needles-boost-rural-economy/989247

Wilson L, Yang W, Blasiak W, John GR, Mhilu CF (2011) Thermal characterization of tropical biomass feedstocks. Energy Convers Manag 52:191–198

Biomass Supply Chain Management: Perspectives and Challenges

Yogender Singh Yadav and Y.K. Yadav

Abstract In India, biomass is abundant renewable energy source available in terms of about 686 MT gross agricultural residues annually with surplus of 234 MT for bioenergy production. Biomass-based bioenergy is having major emphasis for potential solution in renewable energy programmes across India. The bioenergy and biofuel produced from biomass is reliable which provide eco-friendly energy source under clean development mechanism but it also depends on many reasons which also includes primarily as biomass assessment and supply chain management. Biomass supply chain is the association of material between the initial source and the end-user. Characteristic of biomass supply chain is typically comprised of some distinct processes as harvesting, collection, transportation, pretreatment, storage and end use, which is crucial for success of bioenergy production. There are several issues in this aspect especially for transportation and storage such as seasonal availability, production cost and efficient conversion technologies. Research and development is essential to optimize the biomass supply chain management for maximise the utilization capacity with minimizing the inputs. The information presented in this paper would be useful for biomass supply chain management-based energy programmes in India. In this facet, multi-biomass approach is sustainable in present scenario for more advantageous with relatively available renewable energy alternative.

Keywords Biomass · Supply chain · Bioenergy · Biofuels · Renewable energy

---

Y.S. Yadav (✉) · Y.K. Yadav
Sardar Swaran Singh National Institute of Bio-Energy,
Kapurthala 144601, Punjab, India
e-mail: yogender784@yahoo.co.in

S. Kumar et al. (eds.), *Proceedings of the First International Conference on Recent Advances in Bioenergy Research*, Springer Proceedings in Energy,
DOI 10.1007/978-81-322-2773-1_20

1 Introduction

The increased economic and industrial development results are having severe context for available energy segment in India. Although India is known as rapid growing economy in the world but basic energy requirement for the people is yet to fulfil (Hiloidhari et al. 2014). The demand of energy in commercial sector is increasing with the same speed as it is produced. India is an agriculture dominant nation having more strength to frame biomass-based bioenergy programmes to meet the present energy requirement. Although, agriculture is regarded as the backbone of Indian economy and it contributes only 17 % to country GDP and it is also a source of subsistence for nearly 60 % of its population. About 60 % of land area of the country is under various agricultural practices (http://data.worldbank.org). Biomass is extensively used in various traditional and rural enterprises (Kishore et al. 2004) and the advanced utilization of biomass takes the advantages of modern biomass conversion technologies (combustion, pyrolysis, gasification, fermentation and anaerobic digestion) for production of heat and electricity, liquid and gaseous transportation fuel, biogas, etc.

Many researchers studied the aspects of biomass supply chain management and estimated that the biomass used for production of energy in domestic and industrial is yet not fully exploited in biomass harvesting, transport and processing (Nielsen et al. 2007). In this context, the logistics and transport sector accounts about one-third of the total energy consumption which is almost fully dependent on fossil fuels and this consumption of energy from biomass would significantly lead to energy independency (Nielsen et al. 2007). Biomass supply chain having certain important barriers for biomass utilization for generation of bioenergy/biofuel such as energy production cost and the technology used. Hence it is an essential need to pretend and optimize biomass supply chain for better management applications with significant reduced cost and efficient logistics operations.

2 Biomass to Bioenergy

The various biomasses available can be used for potential bioenergy production with several bioenergy applications in India. In this context, agricultural residue obtained during many agricultural operations is being considered as potential biomass feed stock for bioenergy. Agricultural residue comprises of organic materials as biomass produced from tillage to post-harvest operations (Singh and Gu 2010). This biomass potential can be demarcated as gross residue potential as the total amount of residue produced while surplus residue potential is the residue left after any competing uses (such as feed, animal bedding, heating and cooking fuel and organic fertilizer). The surplus fraction can be used for bioenergy generation. Gross residue potential of a particular crop depends upon three parameters such as area covered by the crop, yield of the crop and residue to product ratio

(RPR) of crop (Hiloidhari and Baruah 2011). There is some significant availability of biomass which is observed apart from agricultural residues such process industries, vegetable market places, road side sweepings and plantations which can be considered as potential bioenergy generation source.

In this perspective, biomass utilization (including, energy crops, biosolids, crop and agricultural residues, municipal waste and industrial by-products and wastes) has an emerging scope among other available resources and is reflected as a viable alternative for energy production, including various potential conversion processes (Parikka 2004; Yamamoto et al. 2001). The most relevant processes available and commonly used are thermal, biochemical and chemical conversions. The thermal conversion processes mainly comprise of direct combustion, pyrolysis and gasification, the biochemical conversion is primarily based on fermentation and anaerobic digestion with the end product ethanol and biogas, chemical conversion consists of transesterification and other significant processes which convert plant and vegetable oils to biodiesel and biofuel.

3 Biomass Supply Chain

Typically biomass supply chain can be defined as movement of material or goods between the source and end-user. It mainly comprises of four commercial entities such as raw material supplier, distribution centre, manufacturer and the customer (Beamon 1998). Biomass supply chain management mainly emphasis on integration of all such entities so that end product is to be produced and distributed in the right quantity at right time to right location with providing preferred quality and service at minimum cost (Simchi-Levi et al. 2003). The biomass supply chain of crop residue essentially consists of the biomass supplier, storage, pretreatment, transportation and conversion of biomass to bioenergy (Becher and Kaltschmitt 1994; Vlachos et al. 2008). The complex biomass supply chain involves two interdependent and interconnected processes as production and control processes and distribution and logistical processes. The production planning and control processes consist of planting, harvesting, baling and pretreatment of biomass, whereas distribution and logistical processes mainly contains storage and transportation (Fiedler et al. 2007). The mutual integration of both the processes of biomass supply chain plays a crucial role for competing bioenergy industry.

The biomass supply chain also deals with biomass supply and availability uncertainty to determine commercially economic viability for the industry. Because, any bioenergy sector industry needs cost efficient year-around uniform and reliable supply of desired biomass feedstock. Also, there are numerous sources of variability in any biomass supply chain, i.e. weather conditions, seasonal availability, biomass characteristics, suppliers, transportation and distribution and government policies (Iakovou et al. 2010; Cundiff et al. 1997; Gold and Seuring 2011).

Generally, biomass supply chain models concentrate on the objective to minimize the production cost along with providing a better supply chain structure

(De-Mol et al. 1997). This can help to insight potential future production need of bioenergy from biomass for planning purpose (Kim et al. 2011; Akgul et al. 2010). Conventionally, biomass was used for energy production through direct combustion in areas which are close to its production sites. The increased attentiveness and complexity of supply chain prescript the need to develop planned and optimized biomass supply chain management system (Meixell and Gargeya 2005; Min and Zhou 2002; Sarmiento and Nagi 1999; Vidal and Goetschalckx 1997).

4 Indian Scenario of Biomass Supply Chain Management

In India, agriculture is being considered as main source of livelihood among major population, hence biomass of agro residues and other sources is readily available to overcome the regional limitations for bioenergy production as in case of other renewable resources. In present scenario of climate change, India confronts to severe energy shortages with energy insecurity and also limits its ability to adapt to these impacts. The present renewable energy installed capacity of the country accounts to nearly 32,000 MW with large contribution is of wind power and solar. Further, biomass considered as third resource in present pattern in which conversion of biomass for heat and power generation is the most common produced form of bioenergy.

According to MNRE's Renewable Energy Plan 2022, the target allocated is of 1,75,000 MW out of which 99,533 MW ($\sim$57 %) for solar, 60,000 MW ($\sim$34 %) for wind energy, 5000 MW ($\sim$3 %) for small hydro and 10,000 MW ($\sim$6 %) for biomass energy. Efficient biomass supply chain management is crucial step for success of any biofuels generation system, as the biofuel price is considered as an important factor for its economic viability. The performance of the any supply chain also depends on integration and coordination among all entities along with positive flow of information and its products.

5 Typical Layout of Biomass Supply Chain

A typical biomass supply chain includes several distinct processes, i.e. harvesting, handling, storage, transportation and utilization for bioenergy production (Fig. 1). The selection of these discrete processes for bioenergy production are critical and strategic decisions that many researchers have studied for specific biomass as raw materials, i.e. switchgrass (Cundiff and Marsh 1996), cotton plant residues (Gemtos and Tsiricoglou 1992) corn stover (Shinners et al. 2007). The relevant research reviewed for different biomass concludes that there is a need to further optimization of each step of the process in the biomass supply chain for efficient planning to make it sustainable.

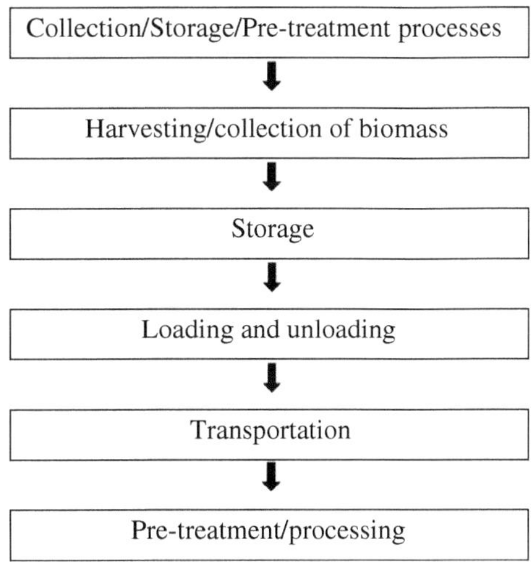

Fig. 1 Layout of biomass supply chain

The availability of biomass is varying seasonally due to specific harvesting time but the requirement at the bioenergy generation is continuous and year-round basis which results the necessity to store the biomass at peak season. The biomass storage is crucial linkage in respective supply chain to choose low-cost storage solutions in the present situation. Several researchers stated that on-field biomass storage has the advantage of low cost (Huisman et al. 1997; Sokhansanj et al. 2006) but on the other way there will be significant material loss due to uncontrolled climatic condition, thus leading to potential problems in the biomass power plant processes. Additionally, health and safety issues also exist with danger of spores and fungus formation (Nilsson 1999). Hence, on-farm storage of the biomass may not be allowed to the farmers exceeding the significant storage time period (Sokhansanj et al. 2006).

Logistics and transportation of biomass mainly depends upon type of biomass, available form, quantity, proposed customer and destination distance. It can be further characterized in two steps as from field to farm or temporary storage and from farm to power generation plant in which main mode of transport system is road as it offers great flexibility between distances traveled (Mitchell et al. 1999). Further processing of biomass is essential to enhance its efficiency that can be take place at any stage of biomass supply chain but often preferred before transportation.

6 Factors Influencing the Biomass Supply Chain

The biomass supply chain represents several distinctive characteristics and also is dependent on many variables, hence it is important to consider the significant factors that affect the optimal biomass chain. These factors can be described as follows.

6.1 Seasonal Availability

The time period available for different biomass types is limited which can be determined through harvesting period and replanting time of the fields (Skoulou and Zabaniotou 2007). Hence for year-round operation of the bioenergy generation plant there is a need of storing significant amount of biomass, this problem may be avoided if year-round biomass to be used or by using multi-biomass approach to overcome these problems.

6.2 Weather

Weather is somewhat impossible to predict in long term but possible to planning carefully for year-round process to have less effect from it. There is a great influence on bioenergy generation indirectly because it reduces the yield of the biomass and quality significantly at the time of harvesting due to bad working conditions of field.

6.3 Collection and Handling Equipment

Almost all types of biomass materials require personalized harvesting, collection and handling equipment which lead to a typical structure of the supply chain. The procurement of biomass from field in available form generally determines the production and operational costs of the respective bioenergy generation system in any biomass supply chain.

6.4 Low-Density Material

Biomass primarily deals with low density due to occupancy of larger volume which results increased transportation and handling cost and more storage space requirement. This can be enhanced through low heating value which is partly due to the increased moisture of types biomass obtained in the respective biomass supply chain (Allen et al. 1998).

6.5 Communication Issues Between Industry and Suppliers

The lack in communication among industry and supplier can lead to failure of any biomass generation system with less efficient biomass supply chain. This is also due to deficiency in real-time knowledge processes. The use of a common database and information technologies of suppliers and producers help to make it up to date. The process of information sharing and communication with state-of-art keep can provide today's difference between success and failure of any system.

6.6 *Sustainability*

Sustainability of any operation is critical and important must be considered during designing, planning and executing the biomass supply chain for bioenergy production. Researchers have studied and represented the biomass generation system and investigate the resulting profiles of bioenergy supply chains (Frombo et al. 2009; Linton et al. 2007). Elghali et al. (2007) also propose a sustainable framework to assess the bioenergy systems to provide better practical advice for policy-makers, planners commercially using multi-criteria decision-based analysis.

7 Supply Chain Design and Planning

Design and planning of any supply chain decides the optimal use of available resources and infrastructure for better bioenergy production practices in order to demand and supply forecasting of any system. This has been recognized earlier also in management and science to coordinate the planning and management with economic viability of supply chain (Fig. 2). The need of different approaches to be explored for focusing the supply chain management has been identified as under.

7.1 *Modelling Approach*

In modelling approach of any biomass supply chain mathematical modelling is common to identify the prescribed problem. The mathematical models are set of equations which describe real-world phenomena in more efficient manner to optimize the given set of variables (Keramati and Eldabi 2011). The available models can be categorized mainly as stochastic, deterministic, hybrid and information-driven models. The stochastic models, also called as probabilistic models, having uncertain and random parameters, whereas deterministic models are fixed with certain parameters. The hybrid models are comprised of both deterministic and stochastic model elements but IT-driven models integrate and coordinate various phases of supply chain (Christensen and Koppenjan 2010).

In present situation of technology advancement special simulation and algorithm approach should be developed to overcome the real challenges. Optimization through simulation makes the process fast and less expensive to assess large

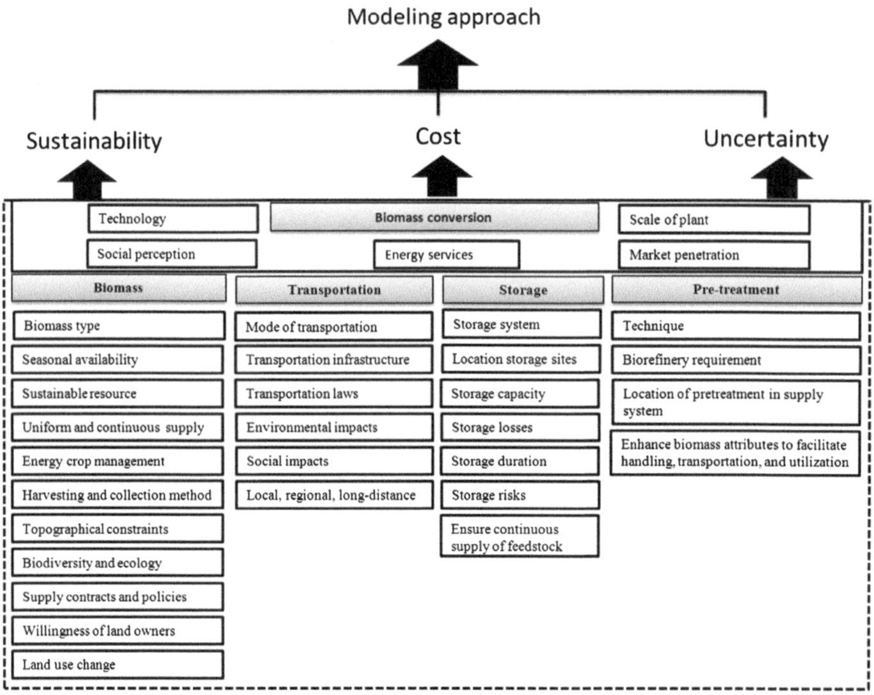

Fig. 2 Factor considered for developing modelling approach for biomass supply chain (*Source* Sharma et al. 2013)

number of dependent variables to provide the more optimal decision. This technique also consists of unique 'What-if' analysis which is very popular among the industry and corporate sector to identify the uncertainty of the process (Sharma et al. 2013).

Since, biomass supply chain is complex and difficult to optimize through the mathematical modelling due to incapable to identify the uncertain parameters and dynamics of the supply chain through the pure mathematical optimization models. While combining the simulation may provide the appropriate solution of the issues related to the biomass supply chain modelling. Consequently, there is a need to develop techniques that result both in handing of complexity of problems along with providing solution for economic viable, environmental and social aspects to the biomass supply chain system (Fig. 2).

Many researchers have studied about optimisation-based decision algorithm of supply chain to manage the complex problems starting facility selection to product distribution (Brown et al. 1987; Timpe and Kallrath 2000; Neiro and Pinto 2004; Camm et al. 1997; Voudouris 1996; Pinto et al. 2000; Karimi et al. 2005). They also focused the process operation issues through combining integer programming, network optimisation and geographical information systems (GIS) to improve efficiency and responsiveness of supply chain system. The novel feature of any mathematical model can accommodate different timescales and distribution of variable length for production with better resolution before planning of the supply chain system for efficient logistics and production operations to enhance its performance.

7.2 *Multi-biomass Approach*

The concept of multi-biomass approach is not much familiar to the researcher till now and research carried out is also limited even with many direct and indirect advantages. This method covers simultaneous use of different types of available biomass (Nilsson and Hansson 2001). Researchers also reached to the conclusion that adopting this technique leads to decrease in system operation cost up to 20 % as compared to use of single biomass (Papadopoulos and Katsigiannis 2002). There will be significant saving in the product cost with the smooth running of biomass power generation system round the year and additional reduction in requirement of equipment and labour facilities. The major technical challenge involved in this approach is efficient biomass to energy conversion technology to use multi-biomass with distinct fuel characteristics according to the seasonal availability. The storage and logistics may become complex as dissimilarities in availability, handling, storage and transportation and processing of biomass sources (Faaij et al. 1997). Now available optimisation and simulation algorithms are capable to resolve the issues of farm and industrial relevance such as process representation, process integration, modelling tools, model nonlinearities and performance measures.

8 Biomass Supply Chain Management

Supply chain management includes the processes for efficient utilization of resources through coordinating and integrating suppliers and manufacturers to produce and distribute the products to satisfy the service level requirements (Simchi-Levi et al. 1999). The coordination and integration of multiple functions at a time is complex in supply chain context. But technological advancement and applicability of information technology especially software applications are greatly reducing the strategic costs and planning of biomass supply chains. It has been recognized through information sharing, process automation and relationship management to facilitate the supply chain management system. The increasing use of information technology in present business context further improved the available supply chain systems through real-time association, accessibility and 24/7 availability global market (Lancioni et al. 2003) which helps the business entities to manage with uncontrollable factors affecting the demand in terms of prices, lack of resources, and production.

9 Issues Related to Biomass Supply Chain Management

There are several issues related to biomass supply chain management to keep it competitive and continuous operation, which is basically categorised in terms of feedstock costs and supply, harvesting and collection, storage and transportation, pretreatment, conversion process, infrastructure, economic feasibility, regulations and policies, design and planning, etc. (Fig. 3).

9.1 Harvesting and Postharvest Operations

The literature suggested that most relevant harvesting and postharvest operations of biomass carried out manual or mechanical drying and bailing (Rauch 2007). Bailing is a specific technology in harvesting process for agricultural and crop residues to improve density of biomass and makes it convenient for handling, transport and storage through bailer equipment. Researchers suggest that advancement in harvest technologies should be primarily focused by considering the need and the sustainability of bioenergy production system, further limited harvesting period and seasonal availability causes unutilised machinery and resources with additional

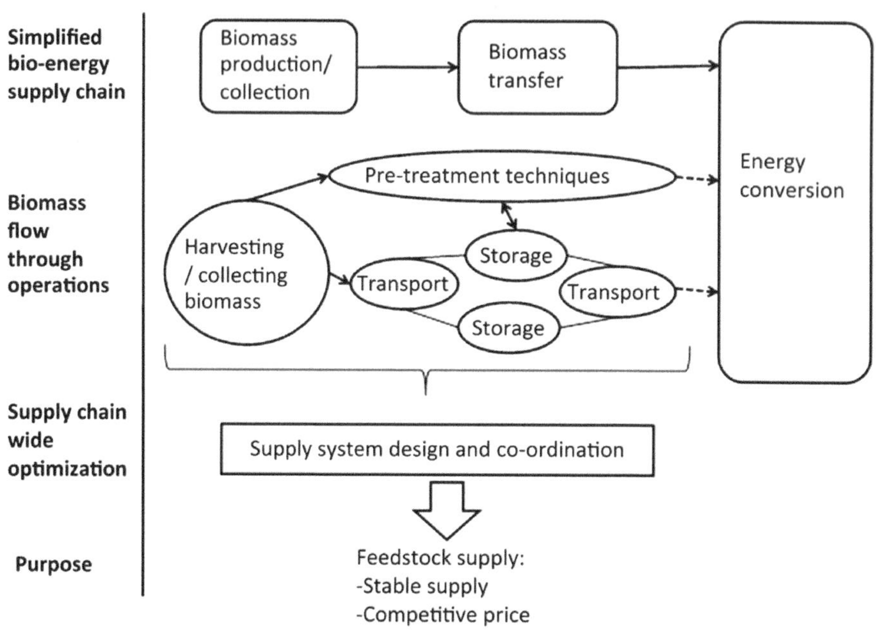

Fig. 3 Facilitating biomass flow along the bioenergy chain (*Source* Gold and Seuring 2011)

increase in production cost (Forsberg 2000; Hess et al. 2007; Madlener and Bachhiesl 2007; Hamelinck et al. 2005; Dunnett et al. 2007).

9.2 Storage Throughout the Bioenergy Chain

The storage of biomass feedstock is essential concern due to round the year operation of the biomass power generation plant. In which, less time available for harvesting with large geographical distributions resulting the need of convenient storage facilities can ensure the constant supply of feedstock bioenergy power plants (Sims and Venturi 2004; Hess et al. 2007; Uslu et al. 2008; Gronalt and Rauch 2007). The risk involved for storage of biomass feedstock is significant weight loss and quality of stored biomass which mainly depends on the climate conditions and biomass procession stage.

9.3 Transport in the Biomass Supply Chain

Transport issues related to the biomass supply chain can be categorised as legal and infrastructural framework but up to some extent social and environmental factor are also significant impact of transport process such as low density of biomass fuels as compared to the fossil fuels (Mayfield et al. 2007). Usually, the conversion of biomass to bioenergy is also not identical therefore, there is a need of environmental and social controls required in the transport determinations (Kumar et al. 2006).

9.4 Pretreatment Analysis

Pretreatment analysis for the available standard techniques may be considered according to purpose and need of the process such as drying, pelletization, torrefaction and pyrolysis etc. (Dunnett et al. 2007; Uslu et al. 2008).

9.5 Design of the Bioenergy Production System

The supply chain is the core issue for any bioenergy generation system which can be described through available tools and strategies for enhancing the performance of the system. The bioenergy power generation systems exist in various sizes and production capacity to consider as decisive parameter for bioenergy production system designing which are generally small as compared to fossil fuel power plants due to requirement of massive feed stock of relatively low calorific value (Elghali et al. 2007; Caputo et al. 2005). These systems deal with high complexity of various biomass resources, different market segments, conversion methods and different end use applications (Heaton et al. 2004; Gronalt and Rauch 2007).

Management of biomass supply chain may be available and can be further extended with use of more advance management approaches. The efficiency of the supply chain network can be better assessed by launching the suitable and optimal performance measures (Beamon 1998). This can be used to design development of a system to enhance performance of supply chain in a highly competitive environment with improved decisions such as number, size and location of bioenergy production system, production planning and scheduling, network connectivity, management and replenishment policies.

10 Conclusion

The overview of the biomass supply chain management suggested that knowledge gap exists to get an optimal biomass supply chain considering technical, societal, business, economic and environmental aspects. Supply chain management in any industrial organization implicates combination of the activities from raw material to end use, whereas coordination of all aspects of the supply chain is paramount to success. As evident from the above considerations, India has huge potential to explore the possibilities to convert the surplus biomass to bioenergy and in this aspect multi-agricultural biomass approach can be an attractive option for biomass based power plants for bioenergy generation. The factors considered for sustainable management and through combining optimal processes and technologies can lead to an organized and integrative biomass supply chain.

References

Akgul O, Zamboni A, Bezzo F, Shah N, Papageorgiou LG (2010) Optimization-based approaches for bioethanol supply chains. Ind Eng Chem Res 50:4927–4938

Allen J, Browne M, Hunter A, Boyd J, Palmer H (1998) Logistics management and costs of biomass fuel supply. Int J Phys Distrib Logistics Manage 28:463–477

Beamon BM (1998) Supply chain design and analysis: models and methods. Int J Prod Econ 55:281–294

Becher S, Kaltschmitt M (1994) Logistic chains of solid biomass-classification and chain analysis. Biomass for energy, environment, agriculture and industry. In: Proceedings of the 8th European biomass conference, Vienna, Austria, pp 401–408

Brown GG, Graves GW, Honczarenko MD (1987) Design and operation of a multicommodity production–distribution system using primal goal decomposition. Manage Sci 33:1469–1480

Camm JD, Chorman TE, Dill FA, Evans JR, Sweeney DJ, Wegryn GW (1997) Blending OR/MS, judgment and GIS: restructuring P&G's supply chain. Interfaces 27:128–142

Caputo AC, Palumbo M, Pelagagge PM, Scacchia F (2005) Economics of biomass energy utilization in combustion and gasification plants: effects of logistic variables. Biomass Bioenergy 28:35–51

Christensen CA, Koppenjan G (2010) Planting and managing switchgrass as a dedicated energy crops. Blade Energy Crops

Cundiff JS, Marsh LS (1996) Harvest and storage costs for bales of switchgrass in the southeastern united states. Bioresour Technol 56:95–101

Cundiff JS, Dias N, Sherali HD (1997) A linear programming approach for designing a herbaceous biomass delivery system. Bioresour Technol 59:47–55

De-Mol RM, Jogems MAH, Beek PV, Gigler JK (1997) Simulation and optimization of the logistics of biomass fuel collection. Neth J Agric Sci 45

Dunnett A, Adjiman C, Shah N (2007) Biomass to heat supply chains: applications of process optimization. Process Saf Environ Prot 85:419–429

Elghali L, Clift R, Sinclair P, Panoutsou C, Bauen A (2007) Developing a sustainability framework for the assessment of bioenergy systems. Energy Policy 35:6075–6083

Faaij A, Van Ree R, Waldheim L, Olsson E, Oudhuis A, Van Wijk A, Daey-Ouwens C, Turkenburg W (1997) Gasification of biomass wastes and residues for electricity production. Biomass Bioenergy 12:387–407

Fiedler P, Lange M, Schultze M (2007) Supply logistics for the industrialized use of biomass—principles and planning approach. In: International symposium on logistics and industrial informatics. Wildau, Germany, pp 41–46

Forsberg G (2000) Biomass energy transport: analysis of bioenergy transport chains using life cycle inventory method. Biomass Bioenergy 19:17–30

Frombo F, Minciardi R, Robba M, Rosso F, Sacile R (2009) Planning woody biomass logistics for energy production: a strategic decision model. Biomass Bioenergy 33:372–383

Gemtos TA, Tsiricoglou T (1992) Harvesting of cotton residue for energy production. Biomass Bioenergy 16:51–59

Gold S, Seuring S (2011) Supply chain and logistics issues of bio-energy production. J Clean Prod 19:32–42

Gronalt M, Rauch P (2007) Designing a regional forest fuel supply network. Biomass Bioenergy 31:393–402

Hamelinck CN, Suurs RAA, Faaij APC (2005) International bioenergy transport cost and energy balance. Biomass Bioenergy 29:114–134

Heaton EA, Clifton-Brown J, Voigt TB, Jones MB, Long SP (2004) Miscanthus for renewable energy generation: European Union experience and projections for Illinois. Mitig Adapt Strat Glob Change 9:433–451

Hess JR, Wright CT, Kenney KL (2007) Cellulosic biomass feedstocks and logistics for ethanol production. Biofuels, Bioprod Biorefin 1:181–190

Hiloidhari M, Baruah DC (2011) Crop residue biomass for decentralized electrical power generation in rural areas (part 1): investigation of spatial availability. Renew Sustain Energy Rev 15:1885–1892

Hiloidhari M, Das D, Baruah DC (2014) Bioenergy potential from crop residue biomass in India. Renew Sustain Energy Rev 32:504–512. http://data.worldbank.org

Huisman W, Venturi P, Molenaar J (1997) Costs of supply chains of Miscanthus giganteus. Ind Crops Production 6:353–366

Iakovou E, Karagiannidis A, Vlachos D, Toka A, Malamakis A (2010) Waste biomass to energy supply chain management: a critical synthesis. Waste Manag 30:1860–1870

Karimi IA, Sharafali M, Mahalingam H (2005) Scheduling tank container movements for chemical logistics. AIChE J 51:178–197

Keramati A, Eldabi T (2011) Supply chain integration: modelling approach. In: Proceeding of European, Mediterranean and Middle Eastern conference on information systems, Athens, Greece

Kim J, Realff MJ, Lee JH, Whittaker C, Furtner L (2011) Design of biomass processing network for biofuel production using an MILP model. Biomass Bioenergy 35:853–871

Kishore VVN, Bhandari PM, Gupta P (2004) Biomass energy technologies for rural infrastructure and village power—opportunities and challenges in the context of global climate change concerns. Energy Policy 32:801–810

Kumar A, Sokhansanj S, Flynn PC (2006) Development of a multi criteria assessment model for ranking biomass feedstock collection and transportation systems. Appl Biochem Biotechnol 129:71–87

Lancioni R, Schau HJ, Smith MF (2003) 23-Internet impacts on supply chain management. Inst Mark Manage 173–175

Linton JD, Klassen R, Jayaraman V (2007) Sustainable supply chains: an introduction. J Oper Manage 25:1075–1082

Madlener R, Bachhiesl M (2007) Socio-economic drivers of large urban biomass cogeneration: sustainable energy supply for Austria's capital Vienna. Energy Policy 35:1075–1087

Mayfield CA, Foster CD, Smith CT, Gan J, Fox S (2007) Opportunities, barriers and strategies for forest bioenergy and bio-based product development in the Southern United States. Biomass Bioenergy 31:631–637

Meixell MJ, Gargeya VB (2005) Global supply chain design: a literature review and critique. Transp Res Part E: Logistics Transp Rev 41:531–550

Min H, Zhou G (2002) Supply chain modeling: past, present and future. Comput Ind Eng 43:231–249

Mitchell CP, Stevens EA, Watters MP (1999) Short-rotation forestry—operations, productivity and costs based on experience gained in the UK. For Ecol Manage 121:123–136

Neiro SMS, Pinto JM (2004) A general modeling framework for the operational planning of petroleum supply chains. Comput Chem Eng 28:871–896

Nielsen JBH, Oleskowitz-Popiel P, Al Seadi T (2007) In: Proceedings of the 15th European biomass conference and exhibition—energy crop potentials for biorefinery in EU-27

Nilsson D (1999) SHAM—a simulation model for designing straw fuel delivery systems. Part 1: model description. Biomass Bioenergy 16:25–38

Nilsson D, Hansson PA (2001) Influence of various machinery combinations, fuel proportions and storage capacities on costs for co-handling of straw and reed canary grass to district heating plants. Biomass Bioenergy 20:247–260

Papadopoulos DP, Katsigiannis PA (2002) Biomass energy surveying and techno-economic assessment of suitable CHP system installations. Biomass Bioenergy 22:105–124

Parikka M (2004) Global biomass fuel resources. Biomass Bioenergy 27:613–620

Pinto JM, Joly M, Moro LFL (2000) Planning and scheduling models for refinery operations. Comput Chem Eng 24:2259–2276

Rauch P (2007) SWOT analyses and SWOT strategy formulation for forest owner cooperations in Austria. Eur J Forest Res 126:413–420

Sarmiento AM, Nagi R (1999) A review of integrated analysis of production–distribution system. IE Transactions 31:1061–1074

Sharma B, Ingalls RG, Jones CL, Khanchi A (2013) Biomass supply chain design and analysis: basis, overview, modeling, challenges and future. Renew Sustain Energy Rev 24:608–627

Shinners KJ, Binversie BN, Muck RE, Weimer PJ (2007) Comparison of wet and dry corn stover harvest and storage. Biomass Bioenergy 31:211–221

Simchi-Levi D, Kaminsky P, Simchi-Levi E (1999) Designing and managing the supply chain: concepts, strategies and cases. McGraw-Hill Publishers

Simchi-Levi D, Kaminsky P, Simchi-Levi E (2003) Designing and managing the supply chain: concepts, strategies and case studies. The McGraw-Hill Companies, Inc., New York

Sims REH, Venturi P (2004) All-year-round harvesting of short rotation coppice eucalyptus compared with the delivered costs of biomass from more conventional short season, harvesting systems. Biomass Bioenergy 26:27–37

Singh J, Gu S (2010) Biomass conversion to energy in India—a critique. Renew Sustain Energy Rev 14:1367–1378

Skoulou V, Zabaniotou A (2007) Investigation of agricultural and animal wastes in Greece and their allocation to potential application for energy production. Renew Sustain Energy Rev 11:1698–1719

Sokhansanj S, Kumar A, Turhollow AF (2006) Development and implementation of integrated biomass supply analysis and logistics model (IBSAL). Biomass Bioenergy 30:838–847

Timpe CH, Kallrath J (2000) Optimal planning in large multi-site production networks. Eur J Oper Res 126:422–435

Uslu A, Faaij APC, Bergman PCA (2008) Pre-treatment technologies and their effect on international bioenergy supply chain logistics: techno-economic evaluation of torrefaction, fast pyrolysis and pelletisation. Energy 33:1206–1223

Vidal CJ, Goetschalckx M (1997) Strategic production-distribution models: a critical review with emphasis on global supply chain models. Eur J Oper Res 98:1–18

Vlachos D, Karagiannidis EIA, Toka A (2008) A strategic supply chain management model for waste biomass networks. In: Proceedings of the 3rd international conference on manufacturing engineering. Chalkidiki, Greece, pp 797–804

Voudouris VT (1996) Mathematical programming techniques to debottleneck the supply chain of fine chemical industries. Comput Chem Eng 20:1269–1274

Yamamoto H, Fujino J, Yamaji K (2001) Evaluation of bioenergy potential with a multi-regional global-land-use-and-energy model. Biomass Bioenergy 21:185–203

Part V
Electrochemical Processes

Enhanced Power Generation in Microbial Fuel Cell Using MnO_2-Catalyzed Cathode Treating Fish Market Wastewater

Md.T. Noori, M.M. Ghangrekar, A. Mitra and C.K. Mukherjee

Abstract Development of cost effective metal-based catalyzed cathode for enhancing oxygen reduction reaction (ORR) is necessary to produce higher power in microbial fuel cell (MFC). In this study, γ-MnO_2 was synthesized using chemical co-precipitation method and tested in MFC as cathode catalyst. Two single chamber MFCs, one with γ-MnO_2 catalyzed cathode (MFC-M) coated with carbon supported MnO_2 (0.5 mg cm^{-2}) on cathode and other control MFC (MFC-C) coated without catalyst in cathode, were fabricated using clayware cylinders. Maximum power density of 1.66 and 0.5 W m^{-3} with a coulombic efficiency (CE) of 9 ± 0.65 % and 4.6 ± 0.26 % was obtained in MFC-M and MFC-C, respectively. The average organic matter removal efficiencies in terms of chemical oxygen demand (COD) were found to be 84.4 ± 3.4 % and 75 ± 1.4 % and total nitrogen removal efficiencies of 66.5 ± 2.56 % and 52.8 ± 3.17 % were achieved in MFC-M and MFC-C, respectively. MFC-M demonstrated enhanced ORR and subsequent increase in electrochemical activity on cathode, thereby improving power generation and substrate degradation. The present study shows the applicability of cost-effective metal-based (γ-MnO_2) catalyst for scaling-up of air-cathode MFCs to treat fish market wastewater, which can also be used for treatment of other organic matter containing wastewater.

Keywords Air-cathode · Catalysis · Fish market wastewater · Microbial fuel cell · γ-MnO_2 catalyst

Md.T. Noori · A. Mitra · C.K. Mukherjee
Department of Agricultural and Food Engineering,
Indian Institute of Technology Kharagpur, Kharagpur 721302, India

M.M. Ghangrekar (✉)
Department of Civil Engineering, Indian Institute of Technology Kharagpur,
Kharagpur 721302, India
e-mail: ghangrekar@civil.iitkgp.ernet.in

S. Kumar et al. (eds.), *Proceedings of the First International Conference on Recent Advances in Bioenergy Research*, Springer Proceedings in Energy,
DOI 10.1007/978-81-322-2773-1_21

1 Introduction

India is the second largest inland fish producer in the world after China (FAO 2012). The major fish consuming states in India are West Bengal, Kerala, Goa, Tripura, and Orissa (Anon 2001). Wastewater generated from fish market contains organic matter measured in terms of chemical oxygen demand (COD), protein, and fats of similar concentration as in fish processing industry wastewater (González 1996). Consequently it is necessary to treat this wastewater before discharging into the water body. On the other hand, global energy shortage and environmental pollution are affecting human population in wide extent. Hence, the problem needs to be addressed urgently. Many researchers reported the suitability of microbial fuel cell (MFC) to solve these two problems simultaneously (Logan 2008; Behera et al. 2010a). MFC is a type of fuel cell in which bacteria acts as biocatalyst, oxidize organic and inorganic matter, and produces electricity (Gabr et al. 2009; Rahimnejad et al. 2011). A typical MFC consists of an anode and a cathode chamber separated by proton exchange membrane (Ieropoulos et al. 2008; Behera et al. 2010b).

Performance of MFC depends on several factors such as microorganism, the cathode catalyst, and the architecture (Liu and Logan 2004; Aelterman et al. 2006b; Oh and Logan 2006). Most commonly used catalyst in fuel cells is Platinum (Pt). However, the high cost of Pt makes it unfeasible for large-scale commercial application of MFC. Therefore, finding or development of alternative low cost catalyst to Pt is highly necessary (Rosen and Dincer 1999). Extensive efforts have been made to find/develop low cost transition metal-based electro-catalyst, metal oxides, and organic polymers composite transition metal oxides (Bashyam and Zelenay 2006; Meng et al. 2013; Ghoreishi et al. 2014).

Manganese metal-based oxide (MnO_x) cathode catalyst is getting greater concern among researchers because of low cost, abandoned availability, and environmental friendly (Mao et al. 2003). Zhang et al. (2009) reported that Pt could be replaced by MnO_2 catalyst in MFC. MnO_2-treated graphite cathode showed higher power generation as compared with non-treated graphite in MFC (Clauwaert et al. 2007). There are five distinct crystallographic phases as α, δ, γ, λ, β of MnO_2 as reported in the literature; whereas the catalytic activity in presence of air follows β- < λ- < γ- < δ- = α-MnO_2 (Cao et al. 2003). Synthesis of different phases of MnO_2 using hydrothermal, solo-gel, and chemical co-precipitation method is widely reported (Chu and Zhang 2009; Xiao et al. 2009). The chemical co-precipitation method is simple, high yield, and the reaction condition is easy to control among other methods.

In view of high catalytic activity and simple synthesis method, γ-MnO_2 was synthesized in this study using chemical co-precipitation method, and tested in single chamber air-cathode MFC made with ceramic cylinder. Fish market wastewater was used as substrate. Performance of MFC having γ-MnO_2 as cathode catalyst was evaluated on the basis of operating voltage (OV), power generation, organic matter removal, total nitrogen removal, and coulombic efficiency (CE). In addition, results were also compared with control MFC (zero catalyst).

2 Materials and Methods

2.1 *Synthesis of MnO_2*

Potassium permanganate ($KMnO_4$) and manganese sulfate ($MnSO_4$) were used without further purification. MnO_2 was synthesized using chemical co-precipitation route reported elsewhere (Chu and Zhang 2009). In brief, 0.2 M potassium permanganate solution was added slowly in 0.3 M manganese sulfate solution in stirring condition at room temperature. Resultant precipitates were filtered and then air dried at room temperature for 24 h. Collected solids were washed several times with deionized water to remove soluble materials and then dried at 80 °C in air for 12 h.

2.2 *Characterization of as-synthesized MnO_2*

X-ray diffraction (XRD) was used to characterize the crystal structure of MnO_2. Diffractograms were obtained in 2θ range of 20°–70° using Diffraktometer D8 (Bruker Axs, Germany) diffractometer with Cu Kα radiation ($\lambda = 0.1541$ nm). FTIR analysis was performed on spectrophotometer (Spectrum Rx, PerkinElmer, Inc., USA). Morphological structure was observed using surface electron microscopy (SEM, ZEISS EVO 60, Carl ZEISS SMT, Germany) equipped with tungsten filament and Oxford EDS detector. Images were taken by applying electron beam having an acceleration voltage of 20.00 kV.

2.3 *Test MFC*

Two single chamber MFCs, viz., catalyzed cathode MFC (MFC-M) and control MFC (MFC-C) were constructed using ceramic cylinder (Behera et al. 2010a). Both MFCs had a similar working volume and cathode surface area of 900 ml and 370 cm^2, respectively. Cathodes were coated on the outer surface of earthen cylinder using ink-based carbon solution containing MnO_2 and carbon loading of 0.5 mg cm^{-2}, polyvinyl alcohol (5 % w/v) and 20 ml acetone. For performance comparison, cathode of MFC-C was made similarly without having MnO_2 in the solution. Stainless steel mesh having total projected surface area of 350 cm^2 was used as anode in both MFCs. MFCs were inoculated with anaerobic septic tank sludge and fish market wastewater was used as substrate. MFCs were operated under batch mode of 4 days batch cycle and fresh feed having COD of 2512 ± 69 mg l^{-1} was added at the beginning of each batch cycle.

2.4 Measurement and Calculation

Electrical parameters such as operating voltage (OV) and open-circuit voltage (OCV) were measured using data logger (Agilent Technology, Malaysia). Polarization studies were performed using a variable resistance box by varying resistance from 10,000 to 5 Ω and corresponding voltage (E) was recorded using a data logger connected with personal computer. Power density was calculated according to P (W m^{-3}) = E^2/RV, where E is voltage, R is resistance, and V is total volume of anode chamber. Chemical parameters such as COD and total nitrogen were determined using standard protocols (APHA 1998). CE was calculated according to CE, % = (C_h/C_{th}) * 100, where C_h = total harvested coulombs and C_{th} = theoretical coulombs available in the amount of COD consumed by microorganisms during each batch cycle (Logan et al. 2006).

3 Result and Discussions

3.1 Characterization of as Synthesizes MnO_2

Structural characterization of as synthesized MnO_2 was done using XRD. Peaks of diffractogram (Fig. 1a) appeared at 2θ = 37.14°, 42.48°, 56.30°, and 67.8°. The indexable broad peaks of 131, 300, 160, 003 as shown in Fig. 1a, confirm the plane of γ-MnO_2 phase of synthesized MnO_2 (JCPDF file no. 14-0644) having high crystalline structure.

In FTIR spectra (Fig. 1b), major peaks appeared at wave number 3400 cm^{-1} –654 cm^{-1}, which are associated to the absorbed water molecule and carbon dioxide because of high surface-to-volume ratio of MnO_2 (Zhang et al. 2005). Other peaks at 535 and 1125 cm^{-1} are associated to the vibration of Mn–O and Mn–OH

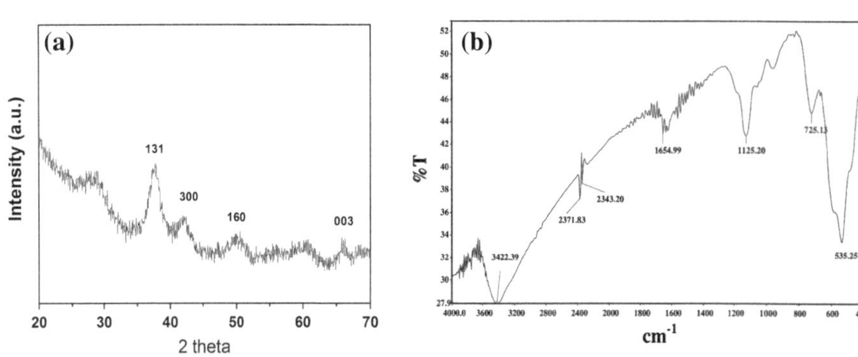

Fig. 1 Characteristics plot of MnO_2 synthesized using co-precipitation method. **a** XRD diffraction pattern. **b** FTIR spectra

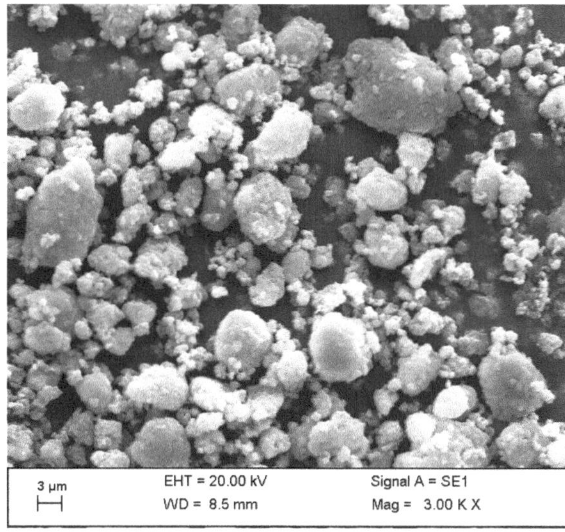

Fig. 2 SEM image of as synthesized γ-MnO_2

functional group, respectively. Figure 2 shows surface morphology of as synthesized MnO_2, which confirms growth of crystalline round shape MnO_2.

3.2 *Electricity Production*

MFCs were started giving voltage just after inoculation and OV across 100 Ω external resistance was found to be 10 and 54 mV in MFC-C and MFC-M, respectively. The OV increased with days of operation and became stable after two sequential fed-batch cycles (Fig. 3). The average OV after getting stable performance was found to be 215 ± 8.3 mV and 410 ± 5.49 mV, respectively, in MFC-C and MFC-M. The enhanced voltage generation in MnO_2 loaded MFC (MFC-M) as compared to control MFC (MFC-C) is because of higher oxygen reduction reaction (ORR) and enhanced electrochemical kinetics due to catalytic action of MnO_2 (Mao et al. 2003).

Polarization studies were performed after getting stable performance of MFCs. The voltages were recorded at external resistances varying from 10,000 to 5 Ω, and the corresponding polarization curves of MFCs are shown in Fig. 4. Higher voltage (604 mV) and power density (1.64 W m^{-3} at current density of 113 mA m^{-2}) was observed in MFC having MnO_2 as catalyst in cathode (MFC-M) than that of control MFC (without catalyst) with voltage and power density of 455 mV and 0.491 W m^{-3} (73 mA m^{-2}), respectively. Internal resistances of MFCs were calculated by the slope of voltage versus current graph (Logan et al. 2006). Lower internal resistance of 71 Ω in MFC-M as compared to MFC-C (93 Ω) showed less

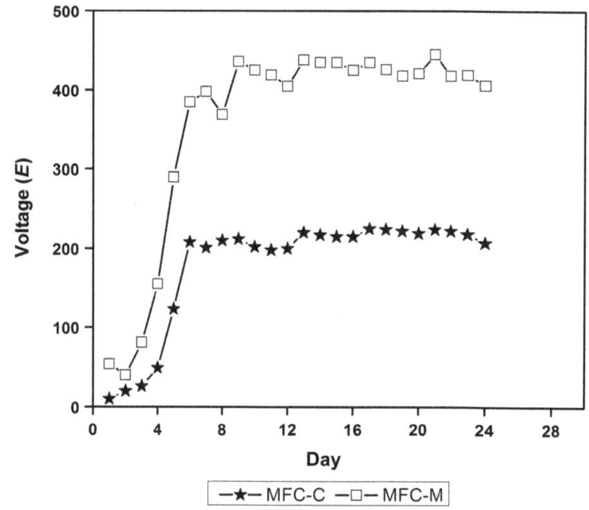

Fig. 3 Operating voltage generated from MFC-M and MFC-C during entire experiment

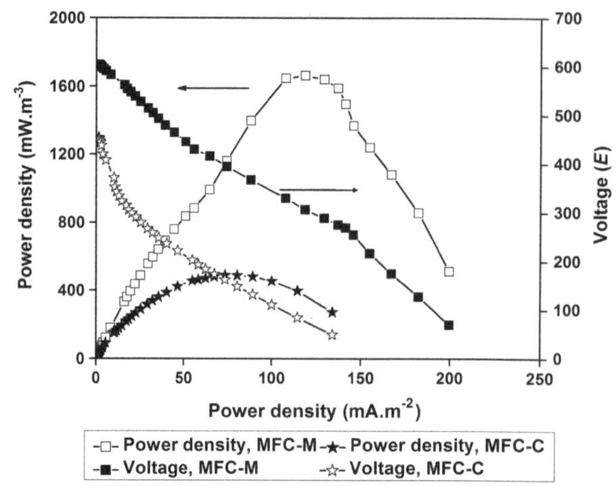

Fig. 4 Polarization and power density plot of MFC-M and MFC-C with respect to current density at higher to lower external resistance. *Arrow* shows the corresponding Y-axes of power density and cell potential

ORR overpotential loss and superior electrochemical activity in MnO_2 catalyzed cathode to produce higher power.

Moreover, power production in MFC-M was found higher than reported (0.77 mW m^{-3}) by Liu et al. (2010), which is due to improved crystal structure of MnO_2 (γ-phase). However, Khilari et al. (2013) reported 2.85 times higher power (4.68 W m^{-3}) than present work using graphene supported α-MnO_2 nanotube in cathode, which is possibly due to the superior electrical conductivity of graphene and higher ORR kinetics of α-MnO_2.

3.3 Wastewater Treatment and Coulombic Efficiency

MFC could be used as an effective wastewater treatment system with simultaneous generation of electrical energy (Aelterman et al. 2006a). The organic matter removal efficiency in terms of COD and total nitrogen (Fig. 5a) were calculated to observe wastewater treatment efficiency of MFCs, whereas CE (Fig. 5b) was estimated to evaluate electricity harvesting efficiency. The average influent COD concentration was 2512 ± 69 mg l^{-1} during entire experiment. COD and total nitrogen removal efficiencies were always higher in MFC-M as compared to MFC-C in all batch cycles. Average COD and total nitrogen removal efficiencies were found to be 84.4 ± 3.4 % and 66.5 ± 2.56 %, respectively, in MFC-M, whereas in MFC-C, the respective efficiencies were 75 ± 1.4 % and 52.88 ± 3.17 %. Enhanced organic matter removal efficiency in MFC-M is due to the higher ORR and electrochemical activity which tended the microorganism to utilize more substrate by maintaining pH balance in anodic chamber of MFC (Ghasemi et al. 2011).

The average CE of 9 ± 0.65 % and 4.6 ± 0.26 % was estimated for MFC-M and MFC-C, respectively. The enhanced CE of MFC-M than that of MFC-C is in agreement with the power generation (Fig. 4). The result shows the effectiveness of MnO_2 as a catalyst for cathode in MFC for enhancing power and wastewater treatment and its potential ability to replace costly catalyst for large commercial applications. Figure 5b shows the CE of MFCs during different batch cycles, where CE of MFC-M was found to be better in each batch cycle in terms of Coulombs recovery than MFC-C.

Figure 6 shows the overall biochemical degradation pathway of macromolecules to micro-molecules and finally to electrons and protons in anode using bacteria as biocatalyst (Noori et al. 2016); and cathodic oxygen reduction using MnO_2 catalyst. During substrate degradation, electrochemically active microorganisms convert chemical energy associated with the organic matter into electrons and protons.

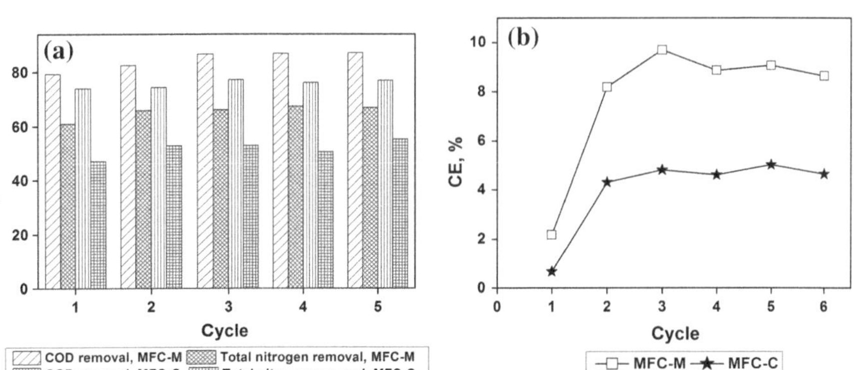

Fig. 5 Performance of MFCs in terms of organic matter removal and coulombic efficiency during each fed batch cycle: **a** COD removal and total nitrogen removal, and **b** coulombic efficiency

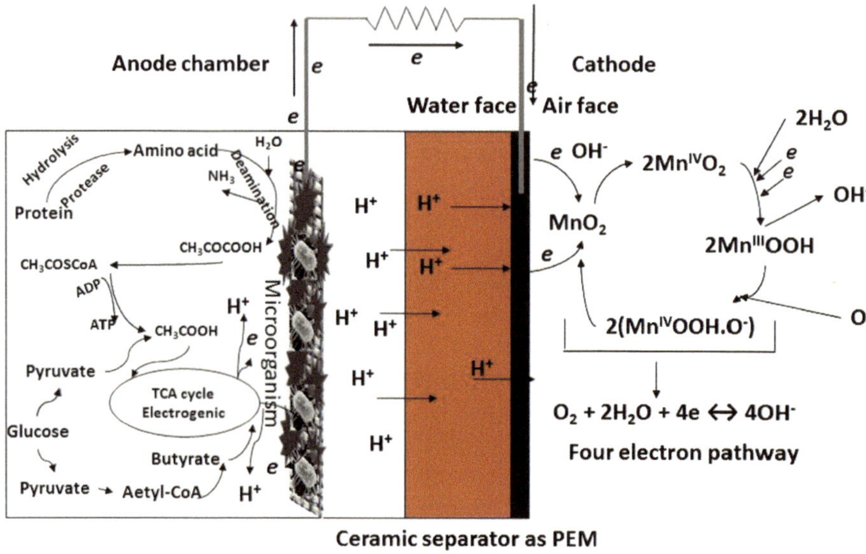

Fig. 6 A schematic diagram of overall working mechanism of MFC

Electrons pass through the external circuit and protons migrate from the proton exchange membrane to the cathode. In cathode, reduction reaction occurs in presence of terminal electron acceptor. The ORR in presence of MnO_2 catalysts favors four electron oxygen reduction pathways using the following reaction sequence:

$$2Mn^{(IV)}O_2 + 2H_2O + 2e \rightarrow 2Mn^{(III)}OOH + 2OH^- \quad (1)$$

$$2Mn^{(III)}OOH + O_2 + 2e \rightarrow 2\left(Mn^{(IV)}OOHO^-\right) \quad (2)$$

$$2\left(Mn^{(IV)}OOHO-\right) \rightarrow 2Mn^{(IV)}O_2 + 2OH^- \quad (3)$$

Overall reaction is as follows:

$$O_2 + 2H_2O + 4e \rightarrow 4OH^- \quad (4)$$

4 Conclusions

Effectiveness of low cost MnO_2, synthesized using chemical co-precipitation method was evaluated as cathode catalyst in single chamber MFC to treat fish market wastewater and simultaneous generation of electricity. The results showed significant increase in power generation up to 330 % in MFC-M having MnO_2 as catalyst in cathode as compared to control MFC-C. Moreover, COD removal, total nitrogen removal and CE was higher in MFC-M than that of MFC-C. The low cost and easy to synthesis γ-MnO_2 demonstrated greater potential to be used as cathode catalyst in MFC to treat fish market wastewater and valuable energy recovery.

Acknowledgments The grant received from Department of Science and Technology, Govt. of India (File No. DST/TSG/NTS/2010/61) to undertake this work is duly acknowledged.

References

Aelterman P, Rabaey K, Clauwaert P, Verstraete W (2006a) Microbial fuel cells for wastewater treatment. Water Sci Technol 54:9–15

Aelterman P, Rabaey K, Pham HT, Boon N, Verstraete W (2006b) Continuous electricity generation at high voltages and currents using stacked microbial fuel cells. Environ Sci Technol 40:3388–3394

Anon (2001) Handbook of fishery statistics 2000. In: Ministry of agriculture, Govt of India

APHA (1998) Standard methods for the examination of water and wastewater, 20

Bashyam R, Zelenay P (2006) A class of non-precious metal composite catalysts for fuel cells. Nature 443:63–66

Behera M, Jana PS, Ghangrekar MM (2010a) Performance evaluation of low cost microbial fuel cell fabricated using earthen pot with biotic and abiotic cathode. Bioresour Technol 101: 1183–1189

Behera M, Jana PS, More TT, Ghangrekar MM (2010b) Rice mill wastewater treatment in microbial fuel cells fabricated using proton exchange membrane and earthen pot at different pH. Bioelectrochemistry 79:228–233

Cao YL, Yang HX, Ai XP, Xiao LF (2003) The mechanism of oxygen reduction on MnO_2-catalyzed air cathode in alkaline solution. J Electroanal Chem 557:127–134

Chu X, Zhang H (2009) Catalytic decomposition of formaldehyde on nanometer manganese dioxide. Modern Appl Sci 3:P177

Clauwaert P, Van der Ha D, Boon N, Verbeken K, Verhaege M, Rabaey K, Verstraete W (2007) Open air biocathode enables effective electricity generation with microbial fuel cells. Environ Sci Technol 41:7564–7569

FAO (2012) The state of world fisheries and aquaculture. In: Nations Faaootu (ed) Food and agriculture organization of the United Nations, Rome

Gabr RM, Gad-Elrab SM, Abskharon RN, Hassan SH, Shoreit AA (2009) Biosorption of hexavalent chromium using biofilm of E. coli supported on granulated activated carbon. World J Microb Biotechnol 25:1695–1703

Ghasemi M, Shahgaldi S, Ismail M, Kim BH, Yaakob Z, Wan Daud WR (2011) Activated carbon nanofibers as an alternative cathode catalyst to platinum in a two-chamber microbial fuel cell. Int J Hydrogen Energ 36:13746–13752

Ghoreishi KB, Ghasemi M, Rahimnejad M, Yarmo MA, Daud WRW, Asim N, Ismail M (2014) Development and application of vanadium oxide/polyaniline composite as a novel cathode catalyst in microbial fuel cell. Int J Energy Res 38:70–77

González JF (1996) Water treatment in fishery industry. In: FAO fisheries technical paper

Ieropoulos I, Greenman J, Melhuish C (2008) Microbial fuel cells based on carbon veil electrodes: stack configuration and scalability. Int J Energy Res 32:1228–1240

Khilari S, Pandit S, Ghangrekar MM, Das D, Pradhan D (2013) Graphene supported α-MnO_2 nanotubes as a cathode catalyst for improved power generation and wastewater treatment in single-chambered microbial fuel cells. RSC Adv 3:7902–7911

Liu H, Logan BE (2004) Electricity generation using an air-cathode single chamber microbial fuel cell in the presence and absence of a proton exchange membrane. Environ Sci Technol 38:4040–4046

Liu X-W, Sun X-F, Huang Y-X, Sheng G-P, Zhou K, Zeng RJ, Dong F, Wang S-G, Xu A-W, Tong Z-H (2010) Nano-structured manganese oxide as a cathodic catalyst for enhanced oxygen reduction in a microbial fuel cell fed with a synthetic wastewater. Water Res 44:5298–5305

Logan BE (2008) Microbial fuel cells. Wiley

Logan BE, Hamelers B, Rozendal R, Schröder U, Keller J, Freguia S, Aelterman P, Verstraete W, Rabaey K (2006) Microbial fuel cells: methodology and technology. Environ Sci Technol 40:5181–5192

Mao L, Zhang D, Sotomura T, Nakatsu K, Koshiba N, Ohsaka T (2003) Mechanistic study of the reduction of oxygen in air electrode with manganese oxides as electrocatalysts. Electrochim Acta 48:1015–1021

Meng F, Yan X, Zhu Y, Si P (2013) Controllable synthesis of MnO_2/polyaniline nanocomposite and its electrochemical capacitive property. Nanoscale Res Lett 8:1–8

Noori MT, Ghangrekar MM, Mukherjee CK (2016) V_2O_5 microflower decorated cathode for enhancing power generation in air-cathode microbial fuel cell treating fish market wastewater. Int J Hydrogen Energ 41:3638–3645

Oh SE, Logan BE (2006) Proton exchange membrane and electrode surface areas as factors that affect power generation in microbial fuel cells. Appl Microbiol Biotechnol 70:162–169

Rahimnejad M, Ghoreyshi AA, Najafpour G, Jafary T (2011) Power generation from organic substrate in batch and continuous flow microbial fuel cell operations. Appl Energy 88: 3999–4004

Rosen MA, Dincer I (1999) Exergy analysis of waste emissions. Int J Energy Res 23:1153–1163

Xiao W, Wang D, Lou XW (2009) Shape-controlled synthesis of MnO_2 nanostructures with enhanced electrocatalytic activity for oxygen reduction. J Phys Chem C 114:1694–1700

Zhang YC, Qiao T, Hu XY, Zhou WD (2005) Simple hydrothermal preparation of γ-MnOOH nanowires and their low-temperature thermal conversion to β-MnO_2 nanowires. J Cryst Growth 280:652–657

Zhang L, Liu C, Zhuang L, Li W, Zhou S, Zhang J (2009) Manganese dioxide as an alternative cathodic catalyst to platinum in microbial fuel cells. Biosens Bioelectron 24:2825–2829

Bioelectricity Generation from Marine Algae *Chaetoceros* Using Microbial Fuel Cell

P.P. Rajesh and M.M. Ghangrekar

Abstract Marine algae *Chaetoceros* can be effectively used both as an electron donor and methanogen inhibitor in MFCs due to its high organic matter content and methanogen inhibition properties. Hexadecatrienoic acid present in *Chaetoceros* inhibits the growth of gram-positive methanogenic archaea via adsorption and disruption of cell membrane. Performance evaluation of two dual-chambered MFCs without any pretreatment of anaerobic sludge used as anodic inoculum was compared using *Chaetoceros* and acetate as substrate. MFC fed with *Chaetoceros* having a COD concentration of 2000 mg l^{-1}, could produce a maximum operating voltage of 495 mV and sustainable power density (PD) of 12.25 W m^{-3} after a period of 12 days from start-up. A maximum PD of 15.13 W m^{-3} and Coulombic efficiency (CE) of 47.25 % could be obtained in MFC fed with *Chaetoceros*; whereas MFC fed with acetate (2000 mg COD l^{-1}) produced a maximum PD of 3.64 W m^{-3} and a CE of 19.65 %. An average total chemical oxygen demand, carbohydrate and protein removal efficiencies of 73.57 ± 2.1 %, 75.77 ± 1.1 %, and 43.11 ± 1.4 % were obtained in MFC using *Chaetoceros* as substrate. Cyclic voltammetry study reveals a higher electron discharge on the anode surface of the MFC fed with *Chaetoceros*.

Keywords *Chaetoceros* · Coulombic efficiency · Marine algae · Methanogenesis · Microbial fuel cell

---

P.P. Rajesh
PK Sinha Centre for Bioenergy, Indian Institute of Technology Kharagpur, Kharagpur 721302, India

M.M. Ghangrekar (✉)
Department of Civil Engineering, Indian Institute of Technology Kharagpur, Kharagpur 721302, India
e-mail: ghangrekar@civil.iitkgp.ernet.in

S. Kumar et al. (eds.), *Proceedings of the First International Conference on Recent Advances in Bioenergy Research*, Springer Proceedings in Energy,
DOI 10.1007/978-81-322-2773-1_22

1 Introduction

Increasing energy demand and the escalating fuel cost have encouraged the search for alternative source of energy. Bioelectrical systems such as microbial fuel cell (MFC) offers a sustainable means of energy generation from organic matter. The use of algae biomass as a substrate in MFCs can provide a cost-effective means of electricity generation due to its low energy requirement for cultivation. In the anode of MFC, microbes oxidize the substrate and electrons are donated to the solid anode. These electrons move through an external circuit toward the cathode and combine with an electron acceptor to generate electricity. Various genera of bacteria present in the activated sludge such as *Alcaligenes faecalis*, *Enterococcus gallinarum*, *Pseudomonas aeruginosa,* and *Shewanella* sp. can degrade algae biomass and transfer electrons to the anode surface (Naim et al. 2013). It is possible that MFCs can be used as an energy efficient biological treatment process for algal organic matter degradation (Velasquez-Orta et al. 2009).

Methanogenesis is often considered as the greatest challenge in achieving on-field applications of MFCs due to the reduction in Coulombic efficiency (CE). The similar growth conditions as that of anode-respiring electrogens and ability to grow in planktonic form favors the growth of methanogenic archaea in MFCs. Inhibition of growth of methanogens in MFCs can lead to a reduction in energy loss to methanogenesis and the consequent improvement in CE of the system. Various pretreatment methods such as 2-bromoethanesulfonate dosing (Zhuang et al. 2012), heat and ultra-sonication treatment (Kim et al. 2005; More and Ghangrekar 2010); oxygen stress condition (Chae et al. 2010) and lauric acid dosing (Rajesh et al. 2014) have been tried in MFCs previously.

The marine algae *Chaetoceros* can be effectively used both as an electron donor and inhibitor for methanogens in MFCs due to its high organic matter content and methanogen inhibition properties. Long chain saturated fatty acid (SFA), hexadecatrienoic acid present in *Chaetoceros* inhibits the growth of methanogenic archaea via adsorption and disruption of cell membrane. Gram-negative bacteria are inhibited to a limited extent at higher fatty acid concentrations (Galbraith and Miller 1973). Utilization of *Chaetoceros* as a methanogen inhibitor enhances the performance of MFC (Rajesh et al. 2015).

The main goal of present study was to assess the potential of *Chaetoceros* to be used as substrate in MFC for producing bioelectricity. The effect of different concentration of *Chaetoceros* on the performance was evaluated. Performance of MFC fed with *Chaetoceros* was compared with control MFC fed with acetate as substrate. Cyclic voltammetry was performed to find the change in the redox activity at the anode surface, while using acetate and *Chaetoceros* as substrates.

2 Materials and Methods

2.1 MFC Construction and Operation

Experiments were performed in two dual chambered aqueous cathode MFCs with an anodic liquid volume of 250 ml. Baked clayware cylinders served as the anodic chambers of these MFCs and the 8 mm thick wall material of the cylinder acted as a separator between anodic and cathodic chambers as well as the cation exchange membrane (Behera et al. 2010). The anode and cathode electrodes of both MFCs were made up of carbon felt with a projected surface area of 192 and 260 cm^2, respectively. Concealed copper wire was used to connect both the electrodes through an external resistance of 100 Ω.

Mixed anaerobic sludge collected from the bottom of a septic tank without any pretreatment was used as the inoculum in the anodic chamber. Pure culture of *Chaetoceros* maintained in a commercial shrimp hatchery was collected for the present study by sieving 100 l of algae grown in a batch culture system using F/2 medium (Guillard and Ryther 1962). The marine algae were then sundried and provided as ground preparation. Dry algae biomass was pretreated using ultra-sonication method to release the content of the cell and provided in the anodic chamber with a concentrated solution of 200 ml so as to result in anolyte total COD concentration of 512, 942, 2000, and 2870 mg l^{-1} after mixing. As a control another identical MFC system was maintained with acetate as the substrate with other chemicals as reported by Behera et al. (2010). These MFCs were operated under controlled temperature varying from 33 to 37 °C in batch mode with a feeding interval of 4 days. At the end of each batch cycle, 200 ml of anodic fluid was replaced with 200 ml of fresh concentrated feed having pre-decided COD concentration so as to maintain the same concentration of substrate in MFC fed with algae and acetate in every batch operation.

2.2 Analysis and Calculations

Performance of MFC was evaluated in terms of voltage (V) and current (I) measured using a data acquisition unit (Agilent Technologies, Malaysia) and converted to power according to $P = V * I$, where P = power (W), I = current (A), and V = voltage (V). The electrode potentials of both anode and cathode were measured with an Ag/AgCl (+197 mV vs. SHE, Bioanalytical Systems Inc., USA) reference electrode. Open-circuit voltage (OCV) was measured under no current flow condition of the circuit. Power density (PD) and power per unit volume were calculated by normalizing power to the anode surface area and net liquid volume of anodic chamber, respectively. Polarization studies were carried out after attaining a stable cell potential by changing the external resistances from 20,000 to 5 Ω using the resistance box (GEC 05 R Decade Resistance Box). Cyclic voltammetry

(CV) was performed using Autolab PGSTAT 302 N potentiostat (Metrohm, The Netherland) and NOVA 1.9 software.

Total and soluble COD concentrations of the anolyte in MFCs were measured for total and soluble fractions of the bulk liquid by closed reflux colorimetric method as mentioned in standard methods (APHA 1998). Carbohydrate estimation was carried out by colorimetric method based on anthrone reagent (Gerhardt et al. 1994). Total and soluble protein was estimated using Bradford method (Kruger 1994). For the measurement of soluble fraction, the anolyte was pre-filtered using GF/C filter paper. The CE was calculated as the fraction of total coulombs actually transferred to the anode against that theoretically present in substrate for current generation over the time period (Logan et al. 2006).

3 Results and Discussion

3.1 *Organic Matter Removal in MFC Fed with* Chaetoceros

Algal organic matter (AOM) is composed primarily of carbohydrate, protein, and lipid. The fermentable macromolecules in algal organic matter such as protein and carbohydrate can be removed by syntrophic interaction between acidogenic bacteria and electrogenic species. A total COD removal of 73.57 ± 2.1 % and soluble COD removal of 85.25 ± 1.3 % could be achieved while using *Chaetoceros* (as 2000 mg l^{-1} COD concentration) as substrate in MFC (Fig. 1). Total COD removal was found to be less since it requires the degradation of particulate organic matter into soluble substrates by fermentative bacteria. Carbohydrates were removed to a greater extent as compared to protein. A total carbohydrate removal of 75.77 ± 1.1 % was observed in MFC, whereas total protein removal was limited to 43.11 ± 1.4 %. Both the particulate and soluble carbohydrate removal was found to be higher than that of particulate and soluble protein removal which suggests that

Fig. 1 Organic matter removal in MFC fed with algae as substrate at COD concentration of 2000 mg l^{-1}

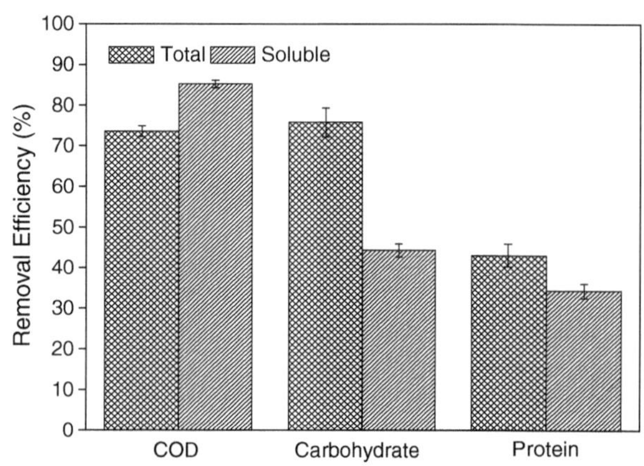

carbohydrate was properly used by microorganisms as compared to protein which has a more complex molecular structure. This observation indicates that microbial selection occurred in MFCs for carbohydrate and protein metabolism. The reason for the higher removal of carbohydrate might be due to the acclimatization of carbohydrate degrading microbial strains in MFCs.

3.2 *Electricity Generation in MFCs Fed with Algae and Acetate as Substrate*

A repeatable cycle of power generation was obtained by *Chaetoceros* as a substrate after three batch cycles. The long time incurred to stabilize the system suggests the longer period required for the fermentative and electrogenic bacteria to get acclimatize in MFC fed with algae. MFC fed with *Chaetoceros* (at 2000 mg l^{-1} COD concentration) generated a maximum operating voltage of 495 mV, whereas the voltage generation in MFC fed with acetate was limited to 280 mV. A sustainable PD of 128 mW m^{-2} could be obtained in MFC fed with algae, whereas the MFC fed with acetate could only produce a PD of 44.26 mW m^{-2}. Anode potential of the MFC fed with algae (−413 ± 8 mV; versus. Ag/AgCl) was found to be lower as compared to MFC fed with acetate (−336 ± 12 mV; vs. Ag/AgCl). This might be due to the enhanced utilization of substrate by electrogenic species in MFC fed with algae; whereas, the methanogens present in MFC fed with acetate accounted for substrate loss and consequent decrease in performance. Rajesh et al. (2015) reported that *Chaetoceros* dosing can enhance the electrogenesis in MFCs, while limiting the substrate loss to methanogens.

MFC fed with *Chaetoceros* produced an average current of 4.9 ± 0.8 mA after reaching a stabilized state of operation. A maximum CE of 47.25 % could be obtained in MFC using algae as substrate (Table 1). This result indicates that the algal organic matter present in *Chaetoceros* can be effectively used for electricity generation while reducing the substrate loss to methanogenesis. A maximum CE of

Table 1 Overall performance of the MFCs

Parameters	$MFC_{acetate}$	MFC_{algae}			
Feed COD concentration (mg l^{-1})	2000	512	942	2000	2870
Sustainable power density (W m^{-3})	3.64	3.20	7.22	12.25	11.52
Cell voltage (mV)	270 ± 10	370 ± 14	420 ± 11	480 ± 14	465 ± 12
Total COD removal efficiency (%)	88.31 ± 1.9	78.21 ± 2.7	75.21 ± 1.9	73.57 ± 2.1	68.91 ± 3.1
Soluble COD removal efficiency (%)		89.32 ± 1.7	86.14 ± 2.1	85.25 ± 1.3	82.13 ± 2.9
Maximum CE (%)	19.65	66.39	65.97	47.25	35.24

19.65 % could only be obtained in MFC fed with acetate. The CE obtained in this study using *Chaetoceros* as substrate was higher than that obtained with *Chlorella vulgaris* (28 %) and *Ulva lactuca* powder (23 %) as MFC feed stock (Velasquez-Orta et al. 2009).

3.3 Polarization of MFCs

Polarization was conducted by changing the external circuit load from 20,000 to 5 Ω once stabilized performance of MFC was observed. The MFC fed with *Chaetoceros* with COD concentration of 2000 mg l^{-1} showed a maximum power density of 159 mW m^{-2} and volumetric power density of 15.13 W m^{-3} (Fig. 2). The maximum power density in acetate fed MFC was lesser (36.14 mW m^{-2}). This might be due to the fact that part of the substrate was utilized for methanogenesis in MFC fed with acetate; whereas in MFC fed with algae, enhanced utilization of substrate by electrogens occurred. The maximum volumetric power density obtained in this study was considerably higher than the reported value using *Scenedesmus* (8.67 W m^{-3}) and *Chlorella vulgaris* (3.7 W m^{-3}) (Cui et al. 2014; Wang et al. 2012).

3.4 Cyclic Voltammetry

Bioelectrochemical behavior of the MFCs fed with acetate and algae was studied in situ by cyclic voltammetry. Voltammograms were recorded by applying a potential ramp of +1 to −1 V at a slow scan rate of 1 mV/s. The CV analysis was

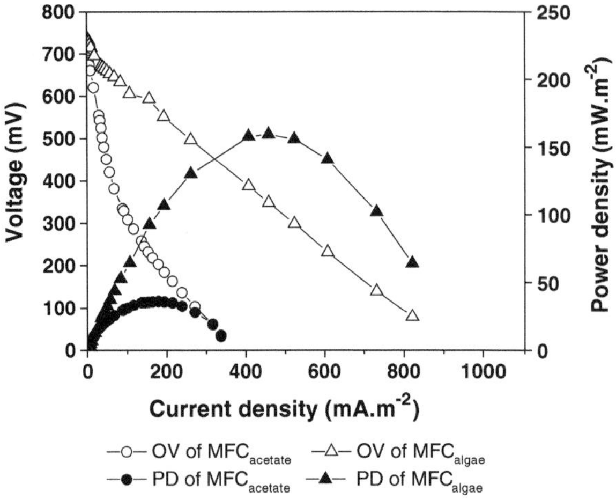

Fig. 2 Polarization curves of the MFCs fed with algae and acetate at a COD concentration of 2000 mg l^{-1}

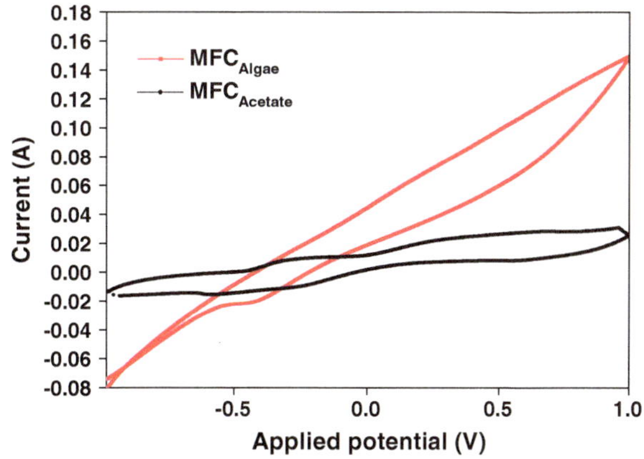

Fig. 3 CV of MFCs fed with algae and acetate at a COD concentration of 2000 mg l^{-1}

performed for anodic half cell with carbon felt as a working electrode, Ag/AgCl as a reference electrode (+197 vs. SHE) and graphite rod as counter electrode. MFC fed with algae (2000 mg COD l^{-1}) has noted higher oxidative current (150 mA) as compared to the MFC using acetate (2000 mg COD l^{-1}) as substrate (oxidative current of 30 mA) (Fig. 3). Higher current observed on the voltammograms can be correlated to higher electron discharge by the biofilm developed on the anode surface of MFC fed with algae. Higher electrogenic activity observed in MFC fed with algae than MFC using acetate can be correlated to the enhanced utilization of substrate by electrogens while limiting the energy loss to methanogenic species.

3.5 *Effect of* Chaetoceros *Concentration on MFC Performance*

The performance of MFC in terms of maximum power generation at different substrate (algae) concentration (512, 942, 2000 and 2870 mg COD l^{-1}) was evaluated. The cell voltage increased with the increase in COD concentration of marine algae from 512 mg l^{-1} (370 ± 14 mV) to 2000 mg l^{-1} (480 ± 14 mV) and remained more or less stable (465 ± 12 mV) at a COD concentration of 2870 mg l^{-1} (Fig. 4). The bioelectrochemical reaction and the subsequent electron transfer to the anode surface vary with the concentration of organic matter available for the microbes to degrade. A maximum sustainable PD of 12.25 W m^{-3} was obtained at a COD concentration of 2000 mg l^{-1} of *Chaetoceros*. The performance of MFC did not improve further and almost remained stable at COD concentration of 2000–2870 mg l^{-1}. Similar trend of effect of varying concentration of substrate on electricity generation was reported by Cui et al. (2014) with the microalgae *Scenedesmus* as feed. A maximum COD removal efficiency of 78.21 ± 2.7 % could be obtained with algal feed at COD concentration of 512 mg l^{-1}. With increase in concentration of the algae the COD removal efficiency was found to be decreasing.

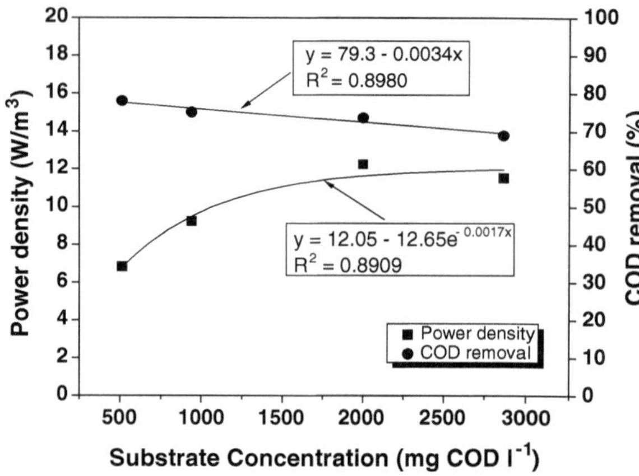

Fig. 4 Effect of *Chaetoceros* concentration in maximum power density and COD removal efficiency

The COD removal efficiency at the concentration of 2870 mg l^{-1} was found to be lesser (68.91 ± 3.1 %). These results indicate that maximum utilization of substrate for electrogenesis occurred at a COD concentration of 2000 mg l^{-1} and reached to an optimum performance of MFC.

4 Conclusion

Utilization of marine algae *Chaetoceros* as an electron donor and methanogenesis inhibitor in MFC system was explored. The use of microalgae biomass as a substrate showed higher coulombic efficiency than acetate as substrate. Polarization study and CV analysis showed a higher electrogenic activity on the anode surface using marine algae as substrate. Carbohydrate contained in *Chaetoceros* was degraded more than the protein present. Maximum power production was found to be at COD concentration of 2000 mg l^{-1} of *Chaetoceros*. A maximum coulombic efficiency of 47.25 % could be obtained using *Chaetoceros* as a substrate. This study reveals that the marine algae *Chaetoceros* can be used as an efficient and sustainable biofuel feed stock for electricity generation in MFCs.

Acknowledgments The grant received from Department of Science and Technology, Govt. of India (File No. DST/TSG/NTS/2010/61) to undertake this work is duly acknowledged.

References

APHA, AWWA, WPCF (1998) Standard methods for examination of water and Wastewater. American Public Health Association, Washington, DC

Behera M, Jana PS, Ghangrekar MM (2010) Performance evaluation of low cost microbial fuel cell fabricated using earthen pot with biotic and abiotic cathode. Bioresour Technol 101 (4):1183–1189

Chae KJ, Choi MJ, Kim KY, Ajayi FF, Park W, Kim CW, Kim IS (2010) Methanogenesis control by employing various environmental stress conditions in two-chambered microbial fuel cells. Bioresour Technol 101(14):5350–5357

Cui Y, Rashid N, Hu N, Rehman MSU, Han JI (2014) Electricity generation and microalgae cultivation in microbial fuel cell using microalgae-enriched anode and bio-cathode. Energy Convers Manag 79:674–680

Galbraith H, Miller TB (1973) Physicochemical effects of long chain fatty acids on bacterial cells and their protoplasts. J Appl Bacteriol 36(4):647–658

Gerhardt P (1994) Methods for general and molecular bacteriology. In: Wood WA, Krieg NR (eds) vol 1325. American Society for Microbiology, Washington, DC

Guillard RR, Ryther JH (1962) Studies of marine planktonic diatoms: i. cyclotella nana hustedt, and detonula confervacea (cleve) gran. Can J Microbiol 8(2):229–239

Kim JR, Min B, Logan BE (2005) Evaluation of procedures to acclimate a microbial fuel cell for electricity production. Appl Microbiol Biotechnol 68(1):23–30

Kruger NJ (1994) The Bradford method for protein quantitation. In: Basic protein and peptide protocols. Humana Press, pp 9–15

Logan BE, Hamelers B, Rozendal R, Schröder U, Keller J, Freguia S, Aelterman P, Verstraete W, Rabaey K (2006) Microbial fuel cells: methodology and technology. Environ Sci Technol 40 (17):5181–5192

More TT, Ghangrekar MM (2010) Improving performance of microbial fuel cell with ultrasonication pre-treatment of mixed anaerobic inoculum sludge. Bioresour Technol 101 (2):562–567

Naim R, Cui Yu-Feng, Muhammad SR, Jong-In H (2013) Enhanced electricity generation by using algae biomass and activated sludge in microbial fuel cell. Sci Total Environ 456–457:91–94

Rajesh PP, Noori MT, Ghangrekar MM (2014) Controlling methanogenesis and improving power production of microbial fuel cell by lauric acid dosing. Water Sci Technol 70(8):1363–1369

Rajesh PP, Jadhav DA, Ghangrekar MM (2015) Improving performance of microbial fuel cell while controlling methanogenesis by *Chaetoceros* pretreatment of anodic inoculum. Bioresour Technol 180:66–71

Velasquez-Orta SB, Curtis TP, Logan BE (2009) Energy from algae using microbial fuel cells. Biotechnol and Bioeng 103(6):1068–1076

Wang H, Liu D, Lu L, Zhao Z, Xu Y, Cui F (2012) Degradation of algal organic matter using microbial fuel cells and its association with trihalomethane precursor removal. Bioresour Technol 116:80–85

Zhuang L, Chen Q, Zhou S, Yuan Y, Yuan H (2012) Methanogenesis control using 2-bromoethanesulfonate for enhanced power recovery from sewage sludge in air-cathode microbial fuel cells. Int J Electrochem Sci 7:6512–6523

Selective Enrichment of Electrochemically Active Bacteria in Microbial Fuel Cell By Pre-treatment of Mixed Anaerobic Sludge to Be Used as Inoculum

B.R. Tiwari and M.M. Ghangrekar

Abstract Pre-treatment methods for seed sludge which are capable of selectively enriching the electrogenic bacteria in a microbial fuel cell (MFC) need to be employed in order to enhance the electricity production. Effect of combined pre-treatments such as acid-ultrasonication pre-treatment (AU), acid-heat pre-treatment (AH), and ultrasonication-heat pre-treatment (UH) on anaerobic mixed sludge to be used as inoculum in a dual chambered MFC was evaluated. All pre-treatments were found to have an incremental effect on the power generation and Coulombic efficiency. MFC inoculated with acid-ultrasonication pre-treated sludge achieved an increment of 520 % in power density (2.91 W m^{-3}) and 276 % in Coulombic efficiency (6.2 %) than that produced by control MFC operated with untreated sludge as inoculum. Acid-ultrasonication pre-treatment was found to be the most effective inoculum pre-treatment, followed by acid-heat pre-treatment and ultrasonication-heat pre-treatment.

Keywords Inoculum pre-treatment · Methanogenesis inhibition · Microbial fuel cell · Power density · Wastewater treatment

1 Introduction

The huge reserves of fossil fuels on which the world energy needs depend will be consumed in about a century. Again, the alternate sources of energy which will cater to the energy needs must not add to the already elevated levels of carbon dioxide in the atmosphere. Hence, we need to look out for an option which not only provides sufficient energy but also should be carbon neutral. In the present world, organic load in wastewater is no longer seen as waste, but instead as a valuable energy resource. The simultaneous production of energy and the degradation of contaminants in

B.R. Tiwari · M.M. Ghangrekar (✉)
Department of Civil Engineering, Indian Institute of Technology Kharagpur, Kharagpur 721302, India
e-mail: ghangrekar@civil.iitkgp.ernet.in

S. Kumar et al. (eds.), *Proceedings of the First International Conference on Recent Advances in Bioenergy Research*, Springer Proceedings in Energy,
DOI 10.1007/978-81-322-2773-1_23

wastewater by microbial fuel cells (MFCs) can provide both economic and environmental benefits (Logan and Regan 2006; Liu and Logan 2004).

MFC is a bioreactor in which biodegradable substrates are oxidized through catalytic reaction of a group of microorganism capable of harvesting electricity from degraded organic matter without external energy input (Logan et al. 2006). It consists of an anaerobic anode compartment and an aerobic cathode compartment separated by a proton exchange membrane. The cathode chamber is free from microorganisms and is filled with a catholyte, usually water, which is continuously aerated. The anode compartment is inoculated with anaerobic sludge which harbors bacteria which are electrogenic in origin. These bacteria get attached to the anode and oxidize the organic matter present in the wastewater into carbon dioxide, protons, and electrons. The anode acts as the electron acceptor and the electrons are transported to the cathode across an external circuit. The protons migrate across the proton exchange membrane into the cathode chamber (Behera et al. 2010). The protons and electrons combine with oxygen to produce water.

$$\text{Anode: } CH_3COOH + 2H_2O \rightarrow 2CO_2 + 8H^+ + 8e^- \quad \Delta E = 0.300\,V$$
$$\text{Cathode: } 4H^+ + 4e^- + O_2 \rightarrow 2H_2O \quad \Delta E = 0.805\,V$$

MFCs are an environment friendly technology because they reduce the load on the environment by consuming the organic matter present in the waste and in turn produce a green and clean energy. No intermediate steps are involved in the conversion of the substrates to electricity, so the conversion efficiency is also high (Ren et al. 2015). MFCs are anaerobic systems; thereby the estimated sludge production is less. However, power outputs from MFCs are insufficient to meet the day-to-day energy needs. So, there is a need to modify the present fuel cell architecture, electrode materials, and inoculum in order to maximize the current production (Malvankar et al. 2012).

The inoculum is the most essential component of a MFC as it is the only biotic factor influencing the power generation. Types of communities residing in the anode biofilm are responsible for maximum current output from MFCs. Mixed consortia from anaerobic sludge which are commonly used for inoculating MFCs are metabolically diverse (Du et al. 2007; Logan 2009) and consist of fermentative bacteria, methanogens, and sulfate reducers along with electrogenic bacteria (Angenent et al. 2002; Dollhopf et al. 2001; Wheatley 1990). Microscopic and PCR analysis have confirmed the presence of methanogens along with electrogens in the anode compartment of MFCs. As the growth conditions and feed dependency of electrogens and methanogens are similar, both groups of bacteria have been found to inhabit the anode compartment of MFCs (Chae et al. 2010; Zhuang et al. 2012). Utilization of a part of the substrate by methanogens is a direct loss in Coulombs to the MFC (Rabaey et al. 2010). The methanogens may also compete for the space on the anode, thus decreasing the density of electrogenic population on the anode. This will lead to decrease in the power generation from the MFCs. So, it would be worth finding ways to selectively enrich the anodic communities with bacteria which are electrogenic in origin and restrict the growth of methanogens (Kim et al. 2005).

Different methods have been adopted to enhance the performances of electrogens and limit the growth of methanogens in bioelectrochemical systems such as ultrasonication pre-treatment to inoculum (More and Ghangrekar 2010), 2-bromoethanesulfonic acid application (Chae et al. 2010), methanogenesis control by electrolytic oxygen production (Tice and Kim 2014), use of a ferric oxide-coated electrode (Kim et al. 2005), temperature stress (Liu et al. 2005), and oxygen stress (Chae et al. 2010). Although the above-mentioned methods were found to be effective to some extent, still each of them had some drawbacks. 2-bromoethanesulfonic is expensive and large quantities need to be administered throughout the operation of MFC when implemented for large scale (Chae et al. 2010).

In the present study, physiological differences between the methanogens and electrogens have been exploited to device a combination of three different physical pre-treatments, i.e., acid pre-treatment (pH 5.3), ultrasonication pre-treatment (40 kHz, 5 min), and heat pre-treatment (100 °C, 15 min). Methanogens are easily affected by variations in the pH even when slightly away from neutral pH (van Haandel and Lettinga 1994). Methane production rate decreases at pH below 6.3 or above 7.8 (van Haandel and Lettinga 1994). Thus, exposure of anaerobic sludge to pH below 6.0 would be effective in inhibiting methanogens. Ultrasonication was used as pre-treatment for inoculum in MFC, and this MFC performed better as compared to MFC inoculated with untreated sludge in terms of the power as well as organic matter removal (More and Ghangrekar 2010). Exposure to heat can lead to suppression of methanogens because of their inability to form protective spores (Zhu and Beland 2006). Heating inoculum at 100 °C for 15 min results in suppressing methanogens (Ghangrekar and Shinde 2007). This study was conducted to improve the performance of MFCs by inhibition of methanogenesis in MFCs under the application of combination of pre-treatments to the inoculum.

2 Materials and Methods

2.1 MFC Construction and Operation

Four identical earthen pots with wall thickness of about 6 mm were used as anodic chambers of four different MFCs inoculated with differently pre-treated sludge. The anodic chambers had an effective volume of 1.3 l and it was placed in the plastic jar of volume 10 l, where tap water was used as catholyte. The catholyte was continuously aerated by an aquarium air pump. Non-transparent polyacrylic sheet was used to seal the top of anode chamber. Electrode spacing of 20 mm was maintained

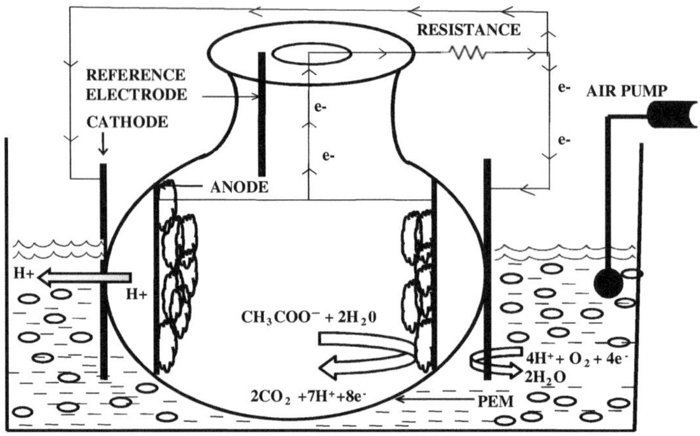

Fig. 1 Schematic diagram of the experimental setup

(Fig. 1). Stainless steel mesh of total surface area of 354.75 cm^2 and graphite plates having a surface area of 305 cm^2 were used as anode and cathode materials, respectively. Electrodes were connected externally with concealed copper wires across a resistance of 100 Ω. Acetate as a source of carbon having chemical oxygen demand (COD) of about 3000 mg l^{-1} was fed at an interval of 4 days and MFCs were operated in batch mode. Synthetic wastewater was prepared as per the composition given by Jadhav and Ghangrekar (2009).

2.2 *Inoculation and Pre-treatment*

Mixed anaerobic sludge having VSS of 16 g l^{-1}, collected from septic tank bottom, was used for inoculating the MFCs. The inoculum sludge was sieved through 1 mm opening sieve. Performance of three MFCs inoculated with different pre-treatments given to the mixed anaerobic sludge was evaluated along with a control MFC-WT, which was inoculated with mixed anaerobic sludge without any pre-treatment. Pre-treatments such as acid-ultrasonication pre-treatment (AU), acid-heat pre-treatment (AH), and ultrasonication-heat pre-treatment (UH) were applied.

Acid pre-treatment was administered by addition of 0.01N H_2SO_4 drop by drop to the inoculum sludge till it attained a final pH of 5.3. Heat pre-treatment was provided by heating the inoculum sludge at 100 °C for 15 min. Sonication pre-treatment was provided for a duration of 5 min using ultrasonic processor (Piezo-U-Sonic) having constant supplied power of 120 W and frequency of 40 kHz. In all the MFCs, 200 ml of sludge was added along with synthetic wastewater. Cathodic chamber did not receive any microbial addition.

2.3 Analysis and Calculation

Operating voltage (*V*) and current (*I*) were measured using a data acquisition unit (Agilent Technologies, Malaysia). Power density was calculated by normalizing power to net liquid volume of anode chamber. Polarization studies were carried out during stable power generation phase by changing the external resistances in steps from 20,000 to 1 Ω by using the resistance box (GEC 05 R Decade Resistance Box, Renown Systems, Kolkata, India). A time interval of 30 min was provided after each change of resistance. Cathode and anode potentials were measured during polarization using Ag/AgCl reference electrode (Bioanalytical Systems Inc., West Lafayette, Indiana, USA).

Exchange current density and charge transfer resistance values were calculated from Tafel curves plotted using the equation shown below:

$$\ln j = \ln j_0 + \beta F \eta (RT)^{-1} \tag{1}$$

where j_0 is the exchange current density (mA m^{-2}), j is the electrode current density (mA m^{-2}), β is the electron transfer coefficient, R is the ideal gas constant (8.31 J mol^{-1} K^{-1}), F is the Faraday constant (96,485 C mol^{-1} e^{-1}), T is the absolute temperature (K), and finally η is the activation over potential.

Coulombic efficiency (CE) of MFCs operated under batch mode, evaluated over a period of time *t*, was calculated as given by Logan et al. (2008).

$$\mathrm{CE} = \frac{M \int_0^t I \mathrm{d}t}{F b V \Delta \mathrm{COD}} \tag{2}$$

Where M is the molecular weight (32) of oxygen, F is Faraday's constant, b is the number of electrons exchanged (4) per mole of oxygen, V is the volume of liquid in the anodic chamber, and ΔCOD is the change in COD over time t.

Volatile suspended solids of the inoculum and wastewater COD were measured according to APHA standard methods (APHA 1998). The procedure described by Bhunia and Ghangrekar (2007) was adopted to calculate specific methanogenic activities of the sludge after receiving different pre-treatments. Methane gas production was measured by a liquid displacement system which was filled with 5 % NaOH solution. The pH was monitored using water quality bench meter (Cyber Scan, PCD 6500, Eutech Instruments, Ayer Rajah Crescent, Singapore).

3 Results and Discussion

3.1 *Effect of Pre-treatment of Inoculum Sludge on Electricity Generation*

During the stable stage, the operating voltage outputs from these MFCs were 535 ± 18 mV (MFC-AU), 492 ± 20 mV (MFC-AH), and 301 ± 11 mV (MFC-UH). The MFC inoculated without any treatment to the sludge (MFC-WT) recorded working voltage of 236 ± 15 mV. The operating voltages denote that all the MFCs inoculated with pre-treated sludge were capable of generating current higher than the MFC-WT. Highest operating voltage was recorded in the MFC-AU followed by MFC-AH and MFC-UH. More and Ghangrekar (2010) reported that ultrasonication pre-treatment limits the activity of gram positive methanogens and it was ineffective against gram negative bacteria. Most of the electrochemically active bacteria, i.e., *Geobactor metallireducens* (Lovley et al. 1993), *Shewanella oneidensis*, *Shewanella putrefaciens*, *Desulfuromonas acetoxidans* (Debabov 2008), and *Geobactor sulfurreducens* (Caccavo et al. 1994) are gram negative bacteria. Researchers have found that acidophilic conditions (pH 5.5) help to suppress the activity of methanogens in a MFC (Venkata Mohan et al. 2008). So when acid and ultrasonication pre-treatment were used in a combination, the voltage generated was the highest among all pre-treatments.

3.2 *Polarization of MFCs*

When current is passed across the electrodes in a MFC, then the respective electrode potential registers a shift from their equilibrium potentials, which is denoted by the term polarization (Raman and Lan 2012). The polarization study was conducted to visualize the difference in electron discharge process with respect to different inoculum pre-treatments. MFCs recorded a maximum power density of 2.915 W m^{-3} (V = 389 mV, R = 40 Ω) for MFC-AU, 2.339 W m^{-3} (V = 427 mV, R = 60 Ω) for MFC-AH, 811 mW m^{-3} (V = 251 mV, R = 60 Ω) for MFC-UH, and 559 mWm^{-3} (V = 225 mV, R = 70 Ω) for MFC-WT, respectively (Fig. 2). MFC-AU recorded the highest power density, which is 520 % higher than that demonstrated by the MFC-WT. Power density obtained follows the following trend—MFC-AU > MFC-AH > MFC-UH > MFC-WT. The trend establishes the fact that all pre-treatments work in synergistic manner to improve the power generation from the MFCs.

Since, methanogens are sensitive to changes in factors like temperature and pH, exposure of the inoculum to acidic pH, high temperature, and ultrasonication can be applied to suppress their activity and population. The different combinations of pre-treatments were found favorable to the electrogenic species and may have lead

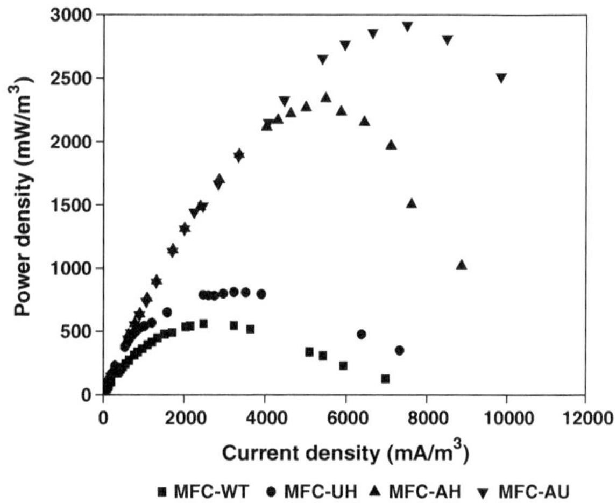

Fig. 2 Power density curves for MFCs inoculated with different pre-treated sludge

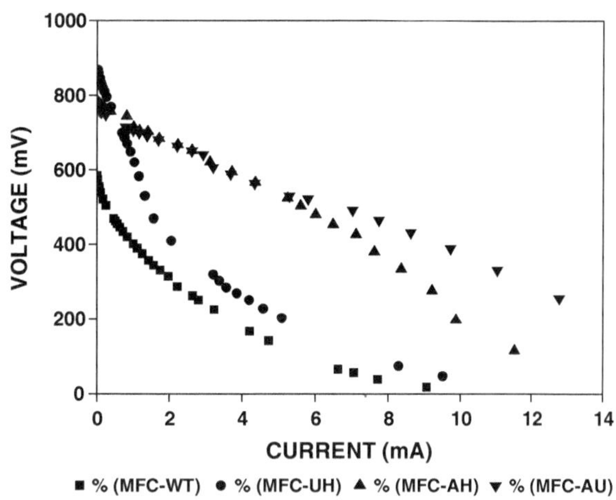

Fig. 3 Polarization curves for MFCs inoculated with different pre-treated sludge

to the inhibition of nonelectrogenic species, thereby recording higher power generation in MFCs with pre-treated sludge as inoculum.

Internal resistance was calculated from the polarization curves between the recorded voltage and current values for MFCs. Internal resistance values were 36, 49, 72, and 89 Ω for MFC-AU, MFC-AH, MFC-UH, and MFC-WT, respectively (Fig. 3). It is evident that the administration of pre-treatments has lead to the decrease in the internal resistance values, thereby leading to an increase in the power generated.

3.3 Electrode Kinetic Study of MFCs with Different Inoculum Pre-treatments

Tafel plots, a plot between the logarithm of the current density and the overpotential i.e., the difference between the potential under the load and without a load, were drawn to study the electrocatalytic activity of the anodes in terms of exchange current density (j_0) and charge transfer resistance (R_{ct}) (Table 1). Exchange current density gives an idea about the rate of reaction taking place at the surface of electrodes by taking into account the total number of electrons exchanged during electro-oxidation and electro-reduction at equilibrium (Raman and Lan 2012). The rate of oxidation at the anode is directly proportional to the number of electrons exchanged from the electrochemically active bacteria to the anode (Raghavulu et al. 2013). The Tafel equation is as shown below:

$$j = j_0 \left[e^{-\frac{\alpha n F \eta}{RT}} - e^{\frac{(1-\alpha) n F \eta}{RT}} \right] \tag{3}$$

where j_0 is exchange current density (mA m^{-2}), j is the electrode current density (mA m^{-2}), α is the electron transfer coefficient, n is the number of electrons transferred in Faradaic charge transfer process, η is the overpotential, R is the ideal gas constant (8.314 J mol^{-1} K^{-1}), T is the absolute temperature (K), and F is the Faraday constant (96,485 C $mol^{-1}e^{-1}$) (Bard and Faulkner 2001).

Although all the MFCs were fabricated using the same electrode materials, different reaction kinetics were observed at the respective anodes. This implies that the application of different pre-treatments for sludge inoculum has resulted in variation in electrogenic activity at the anodes. The exchange current density values denote that the reaction rates were in the order of MFC-AU > MFC-AH > MFC-UH > MFC-WT. The electrochemical reactions at anode of MFC-AU were fastest, which clearly indicates the effective growth of electrogenic bacterial biofilm on its anode. The exchange current density values for MFC-WT were lowest among all, indicating that pre-treatment to inoculum had a positive effect on enhancing the electrogenic activity on anodes of other MFCs which contributed to higher voltage output.

Charge transfer resistance (R_{ct}) represents the capability of system to resist the transfer of charge from electrode against the potential difference. The R_{ct} values obtained for anode of MFCs were in the order of MFC-AU < MFC-AH < MFC-UH < MFC-WT. In MFC-WT, non-electrochemically active bacteria were higher in number, thereby decreasing the kinetics of electro-oxidation taking

Table 1 Exchange current density and charge transfer resistance obtained from Tafel plot for anodes of MFCs inoculated with differently pre-treated sludge

	MFC-WT	MFC-UH	MFC-AH	MFC-AU
Exchange current density (mA m^{-2})	2.33	5.64	7.31	11.6
Charge transfer resistance (Ω m^{-2}) $\times 10^{-3}$	2.77	1.14	0.88	0.55

place at the anode which resulted in increasing activation loss and thus increasing the charge transfer resistance (Ramasamy et al. 2008). But in the case of MFCs inoculated with pre-treated sludge, electrogenesis was the dominating process which facilitated a decrease in the activation energy required for initiating the electrochemical reaction and thus low charge transfer resistance values were obtained (Clauwaert et al. 2008).

3.4 Effect of Pre-treatment of Sludge Inoculums on Substrate Degradation

The pre-treatments applied act as a shock to the bacteria residing in the inoculum, thereby inducing selective inhibition. As a result, the group of bacteria which were not capable of surviving under those altered conditions perishes. So, there is a decrease in the total population of substrate consumers immediately after the application of the pre-treatments. This is reflected by the reduced COD removal in the MFCs with pre-treated sludge as inoculum. As the shocks were applied targeting the methanogens, the surviving population is expected to be mostly electrogenic in origin. Once the performance of the MFCs was stable, COD removal efficiencies of 71.6 ± 3.84, 76.2 ± 4.65, 79.2 ± 3.63, and 84.8 ± 2.49 % were obtained for MFC-AU, MFC-AH, MFC-UH, and MFC-WT, respectively (Fig. 4).

COD removal efficiency was highest in MFC-WT as compared to MFCs inoculated with pre-treated sludge because the activity of microbial population contributing to the COD removal in case of MFC inoculated with raw sludge is not disturbed; whereas pre-treatment has resulted in decreasing the viable bacterial population.

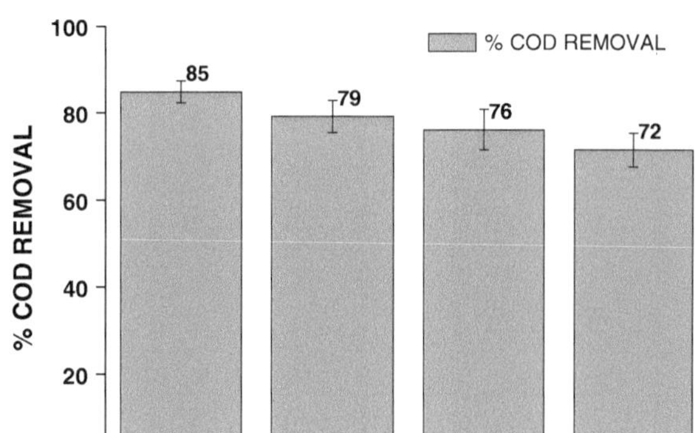

Fig. 4 COD removal efficiencies in MFCs inoculated with differently pre-treated sludge

3.5 Effect of Pre-treatment of Sludge Inoculum on Coulombic Efficiency

The oxidation of any substrate is accompanied by the removal of electrons, which need to be recovered as much as possible in the form of current in order to make MFC a feasible option for power generation. The percentage of electrons recovered to the overall amount stored in the substrate is explained by the term 'Coulombic efficiency.' More is the share of substrate consumption by the electrogenic bacteria more will be the Coulombic efficiency of the system. Once the COD removal was stable, CE values were found to be 6.23 ± 0.28, 5.43 ± 0.26, 3.20 ± 0.01, and 2.26 ± 0.09 % for MFC-AU, MFC-AH, MFC-UH, and MFC-WT, respectively (Fig. 5). CE values are higher for the MFCs inoculated with pre-treated sludge as inoculum which clearly indicates that the pre-treatments applied were effective in suppressing the nonelectrogenic population. Hence, a higher percentage of the substrate is consumed by the electrogenic bacteria and converted to current.

3.6 Methane Production Potential of Sludge with Different Pre-treatments

Methane production was measured for sludge receiving different pre-treatments in anaerobic batch reactor fed with synthetic wastewater of 3000 mg COD l^{-1} per day having acetate as a carbon source for 4 days. The effect of pre-treatments on the methane production is a function of the methanogenic population present in the inoculum. It was found that sludge pre-treated with acid and ultrasonication recorded the lowest methane production (Table 2) and the raw sludge produced highest methane production in the reactor. This clearly denotes that all other

Fig. 5 Coulombic efficiency of MFCs inoculated with sludge after different pre-treatments

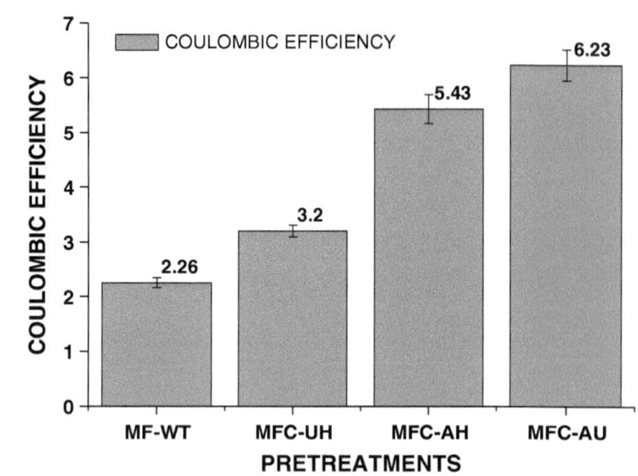

Table 2 Specific methanogenic activity of sludge after different pre-treatments

Pre-treatment	g VSS used	g CH_4-COD/g VSS.day
Without pre-treatment	0.298	0.322
Acid-ultrasonication	0.312	0.061
Acid-heat	0.280	0.070
Ultrasonication-heat	0.270	0.075

pre-treatments were effective in inhibiting methanogens. The specific methanogenic activity of acid-ultrasonication pre-treated sludge was fivefold lower than untreated sludge, indicating effective suppression of methanogens by this pre-treatment method.

4 Conclusion

Avoiding methanogenesis in MFCs poses a challenge as it reduces the power generation and Coulombic efficiency obtained from MFCs. The combined pre-treatment studies performed on anaerobic mixed inoculum demonstrated the feasibility of using combined pre-treatment for enrichment of electrochemically active bacteria in MFCs. Combined pre-treatments such as acid-ultrasonication pre-treatment, acid-heat pre-treatment, and ultrasonication-heat pre-treatment used for selective enrichment of electrochemically active anaerobic consortia showed considerable influence on the power production and substrate removal efficiency. All pre-treatment methods showed positive influence on the overall power production and Coulombic efficiency as compared to the control experiments. Individually, acid-ultrasonication pre-treatment demonstrated highest power yield.

Acknowledgments The grant received from Department of Science and Technology, Govt. of India (File No. DST/TSG/NTS/2010/61) to undertake this work is duly acknowledged.

References

Angenent LT, Sung S, Raskin L (2002) Methanogenic population dynamics during startup of a full-scale anaerobic sequencing batch reactor treating swine waste. Water Res 36:4648–4654

APHA, AWWA, WPCF (1998) Standard methods for examination of water and wastewater. 20th ed., American Public Health Association, Washington, DC.

Bard AJ, Faulkner LR (2001) Electrochemical methods. Fundamentals and applications, 2nd edn, pp 92, 100–106. John Wiley & Sons, Inc., New York, Chichester, Weinheim, Brisbane, Singapore, Toronto

Behera M, Jana PS, More TT, Ghangrekar MM (2010) Rice mill wastewater treatment in microbial fuel cells fabricated using proton exchange membrane and earthen pot at different pH. Bioelectrochemistry 79:228–233

Bhunia P, Ghangrekar MM (2007) Required minimum granule size in UASB reactor and characteristics variation with size. Bioresour Technol 98:994–999

Caccavo F Jr, Lonergan DJ, Lovley DR, Davis M, Stolz JF, McInerney MJ (1994) Geobacter sulfurreducens sp. nov., a hydrogen and acetate-oxidizing dissimilatory capable of coupling the complete oxidation of organic compounds to the reduction of iron and other metals. Arch Microbiol 159(4):336–344

Chae KJ, Choi MJ, Kim KY, Ajayi FF, Park W, Kim CW, Kim IS (2010) Methanogenesis control by employing various environmental stress conditions in two-chambered microbial fuel cells. Bioresour Technol 101:5350–5357

Clauwaert P, Aelterman P, Pham TH, Schamphelaire L De, Carballa M, Rabaey K, Verstraete W (2008) Minimizing losses in bio-electrochemical systems: the road to applications. Appl Microbiol Biotechnol 79:901–913

Debabov VG (2008) Electricity from microorganisms. Microbiology 77:123–131

Dollhopf SL, Hashsham SA, Dazzo FB, Hickey RF, Criddle CS, Tiedje JM (2001) The impact of fermentative organisms on carbon flow in methanogenic systems under constant low substrate conditions. Appl Microbiol Biotechnol 56:531–538

Du Z, Li H, Gu T (2007) A state of the art review on microbial fuel cells: a promising technology for wastewater treatment and bioenergy. Biotechnol Adv 25:464–482.

Ghangrekar MM, Shinde VB (2007) Performance of membrane-less microbial fuel cell treating wastewater and effect of electrode distance and area on electricity production. Bioresour Technol 98(15):2879–2885

van Haandel AC, Lettinga G (1994) Anaerobic sewage treatment: a practical guide for regions with a hot climate. John Wiley & Sons, Chichester, UK

Jadhav GS, Ghangrekar MM (2009) Performance of microbial fuel cell subjected to variation in pH, temperature, external load and substrate concentration. Bioresour Technol 100(2):717–723

Kim JR, Min B, Logan BE (2005) Evaluation of procedures to acclimate a microbial fuel cell for electricity production. Appl Microbiol Biotechnol 68:23–30

Liu H, Cheng S, Logan BE (2005) Power generation in fed-batch microbial fuel cells as a function of ionic strength, temperature, and reactor configuration. Environ Sci Technol 39:5488–5493

Liu H, Logan BE (2004) Electricity generation using an air-cathode single chamber microbial fuel cell in the presence and absence of a proton exchange membrane. Environ Sci Technol 38:4040–4046

Logan BE, Call D, Cheng S, Hamelers HVM, Sleutels THJA, Jeremiasse AW, Rozendal RA (2008) Microbial electrolysis cells for high yield hydrogen gas production from organic matter. Environ Sci Technol 42(23):8630–8640

Logan BE, Hamelers B, Rozendal R, Schrorder U, Keller J, Freguia S, Aelterman P, Verstraete W, Rabaey K (2006) Microbial fuel cells: methodology and technology. Environ Sci Technol 40 (17):5181–5192

Logan BE (2009) Exoelectrogenic bacteria that power microbial fuel cells. Nat Rev Microbiol 7:375–381

Logan BE, Regan JM (2006) Microbial fuel cells—challenges and applications. Environ Sci Technol 40:5172–5180

Lovley DR, Giovannoni SJ, White DC, Champine JE, Phillips EJP, Gorby YA, Goodwin S (1993) Geobacter metallireducens gen. nov. sp. nov., a microorganism capable of coupling the complete oxidation of organic compounds to the reduction of iron and other metals. Arch Microbiol 159(4):336–344

Malvankar NS, Tuominen MT, Lovley DR (2012) Biofilm conductivity is a decisive variable for high-current-density *Geobacter sulfurreducens* microbial fuel cells. Energy Environ Sci 5:5790-5797

More TT, Ghangrekar MM (2010) Improving performance of microbial fuel cell with ultrasonication pre-treatment of mixed anaerobic inoculum sludge. Bioresour Technol 101 (2):562–567

Rabaey K, Angenent L, Schroder U, Keller J (2010) Bio-electrochemical systems: from extracellular electron transfer to biotechnological application. IWA Publishing

Raghavulu SV, Modestra JA, Amulya K, Reddy CN, Venkata Mohan S (2013) Relative effect of bioaugmentation with electrochemically active and non-active bacteria on bioelectrogenesis in microbial fuel cell. Bioresour Technol 146:696–703

Raman K, Lan JCW (2012) Performance and kinetic study of photo microbial fuel cells with different electrode distances. Appl Energy 100:100–105

Ramasamy RP, Ren Z, Mench MM, Regan JM (2008) Impact of initial biofilm growth on the anode impedance of microbial fuel cells. Biotechnol Bioeng 101:101–108

Ren H, Pyo S, Lee J, Park T, Gittleson FS, Leung FCC, Kim J, Taylor AD, Lee H, Chae J (2015) A high power density miniaturized microbial fuel cell having carbon nanotube anodes. J Power Sources 273:823–830

Tice RC, Kim Y (2014) Methanogenesis control by electrolytic oxygen production in microbial electrolysis cells. Int J Hydrogen Energy 39:3079–3086

Venkata Mohan S, Mohanakrishna G, Srikanth S, Sarma PN (2008) Harnessing of bioelectricity in microbial fuel cell (MFC) employing aerated cathode through anaerobic treatment of chemical wastewater using selectively enriched hydrogen producing mixed consortia. Fuel 87:2667–2676

Wheatley A (1990) Anaerobic digestion: a waste treatment technology. In: Burkin AR (ed) Critical reports on applied chemistry, vol 31. Elsevier, London

Zhu H, Beland M (2006) Evaluation of alternative methods of preparing hydrogen producing seeds from digested wastewater sludge. Int J Hydrogen Energy 31:1980–1988

Zhuang L, Chen Q, Zhou S, Yuan Y, Yuan H (2012) Methanogenesis control using 2-Bromoethanesulfonate for enhanced power recovery from sewage sludge in air-cathode microbial fuel cells. Int J Electrochem Sci 7:6512–6523

Performance Analysis of Separators in Dual-Chambered Microbial Fuel Cell and Treatment of Combined Industrial Effluent of South Gujarat

Purvi Zaveri, Tanvi Modi, Lipi Parekh, Aditi Patel, Sameer Kureshi and Nasreen Munshi

Abstract India being the developing country faces lots of issues related to energy crisis. Increased population rate leads to generation of various types of wastes and associated treatment problems. Microbial Fuel Cells (MFCs) provide new opportunities for the sustainable production of energy from biodegradable, reduced compounds present in wastewater. MFC is a bio-electro-chemical process leading to generation of voltage difference between two electrodes by capable microbes, carrying out oxidation of substrate. In this study, combined wastewater produced by various industries of South Gujarat was investigated as an alternative to pure substrates for generation of electricity. The bacterial culture, isolated from wastewater collected from Common Effluent Treatment Plant (CETPs) of South Gujarat region, identified as *Escherichia coli,* was used for oxidation of substrates in MFC run under laboratory conditions. This study evaluated and compared the performance of MFC with two different separators, salt bridge and proton exchange membranes in terms of voltage production in dual-chambered system. The use of proton exchange membrane as separator was found to be superior as compared to salt bridge in terms of voltage generation but they were less cost effective. The use of anionic and cationic exchange membrane resulted in production of power density up to 119.4 and 200.6 mW/m^2, respectively, in MFC with synthetic wastewater, while the MFC connected with salt bridge produced 148.76 mW/m^2 of consistent power density. The organism could reduce COD by 35.72 % from effluent of CETP, Pandesara, South Gujarat and could produce open-circuit voltage of 0.72 V

Purvi Zaveri · Aditi Patel · Sameer Kureshi · Nasreen Munshi (✉)
Nirma University, Sarkhej Gandhinagar Highway, Ahmedabad 382481, India
e-mail: nasreen.munshi@nirmauni.ac.in; nasreenhaque@hotmail.com

Tanvi Modi
Nootan Science and Commerce College, Visnagar, India

Lipi Parekh
Cadila Research Center, Ahmedabad, India

S. Kumar et al. (eds.), *Proceedings of the First International Conference on Recent Advances in Bioenergy Research*, Springer Proceedings in Energy,
DOI 10.1007/978-81-322-2773-1_24

and power density of 200.6 mW/m^2 where salt bridge was used as separator. The study indicates the potential of such MFC for treatment of CETP wastewater after process optimization.

Keywords Microbial fuel cell · Wastewater · Proton exchange membrane · Salt bridge

1 Introduction

Importance of alternative sources of energy has gradually increased at global scale. Developing countries like India needs to develop sources of clean energy for ever increasing population and to combat the foreseen energy crisis. Discovery of electro-active organisms like bacteria and yeast has been reported before long back, and this gave a novel source of energy production using prokaryotic biomass (Potter 1911). Microbial Electrochemical Technologies (METs) utilize microbial metabolism to catalyze reactions such as desalination and substrate oxidation and has diverse applications (Logan and Rabaey 2012). Microbial fuel cells are electrochemical devices governed by biological activity, which comprises two basic elements, anode and cathode. Organisms in anode chamber oxidize substrate and lead to release of electron which is captured by electrode directly in case of mediator-less MFC or via a mediator in mediator requiring MFC. The captured electron is then transferred to cathode via external circuit. Internally, two compartments may be connected via salt bridge or proton exchange membrane which allows passage of proton (Jadhav and Ghangrekar 2009). Fundamentally, microbial fuel cell functions on ability of organism to transport electrons out of cell membrane; however, substrate concentration, pH, configuration of MFC, etc. affect performance of the cell. Microbial fuel cell has shown promising results in generation of electricity from various sources including wastewater (Mathuriya and Sharma 2010), salt water (Qu et al. 2013), etc., indicating its applicability to function with complex substrates.

In addition to structural variations, with emergence of advances in field of material science and electronics, many novel approaches have been developed for enhancing the functioning of MFC including the use of nano-material and incorporation of DC converters for higher voltage production (Wua et al. 2011). With all available knowledge, certain challenges like low electron recovery and unstable system performance still need to be addressed. Among many components affecting power density produced by microbial fuel cell, the separator plays major role in proton migration between two electrodes. Available separators belong to three major classes, ion exchange membrane (Anion exchange membrane (AEM), cation exchange membrane (CEM), bipolar membrane (BPM)), size selective separators (microfiltration membrane (MFM), ultrafiltration membrane (UFM), glass fibers, porous fibers), and salt bridge (Wei et al. 2011). Among the ion exchange

Table 1 List of substrates reported in MFC run

Type of substrate	Concentration of organic substrate (mg l^{-1})	Maximum power (mW/m^2)	COD removal efficiency (%)	Reference
Acetate	1000	2400	N.D.	Logan et al. (2007)
Glucose	480 (TOC)	2160	93	Catal et al. (2008)
Distillery wastewater	2100–6100 (COD)	17.76 28.15	64 61	Hampannava et al. (2011)
Food wastewater	4900 ± 350	~556	N.D.	Jia et al. (2013)
Full-strength cyanide laden wastewater (cassava mill wastewater)	16000 (COD), 86 mg l^{-1} cyanide	1800	28	Kaewkannetra et al. (2011)
Brewery wastewater	1250 ± 100	830 mW/m^3	91.7–95.7	Wen et al. (2010)
Steroid drug industrial effluent	N.D.	2230	82	Liu et al. (2012)
Swine wastewater	8320 (COD)	261	27	Min et al. (2005)
Domestic wastewater	54,000 (COD)	422	25.8	Ahn et al. (2010)
Starch processing wastewater (diluted to 1/2)	4852 (COD)	239.4	98	Lu et al. (2009)

Note N.D. Not determined

membranes, Nafion has been widely used in MFC setups and has been proven successful (Rahimnejad et al. 2014).

Use of separator has been debated a lot by researchers working on microbial fuel cells. The absence of separator leads to higher oxygen in the chamber and lower coulombic efficiency which leads to lower activity of anodic organism, whereas use of separator at the same time leads to delay in proton transfer which generates pH gradient between anode and cathode which is undesirable. If in system proton exchange membranes are being used it also adds to the cost of the functioning. Thus, selection of right type of separator for required functioning and its optimum utilization is very important (Li et al. 2011).

As mentioned in Table 1 other than pure substrates, researchers have also used wastewaters from various industries to run MFC and thus showed decrease in organic load of wastewater (Wen et al. 2010). Wastewater contains multitude of organic matter that can serve as electron source for MFC. However, there is a difference between the composition of municipal and industrial waste; both can serve the purpose. In South Gujarat, industrial wastewaters generated by various industries are taken to the combined effluent treatment plants for treatment followed by discharging them to natural water bodies. Efficiency of wastewater treatment is generally measured in terms of reduction of COD and BOD. MFC as indicated has capacity to use such incoming mix effluents as source of electricity generation (Table 1). In several cases it has been reported to reduce as high as 82–91 % of COD from synthetic wastewater after MFC run using hypochlorite as terminal electron acceptor (Jadhav et al. 2014).

Use of combined industrial effluent remains unexplored till date due to its complex nature. The golden corridor of Gujarat comprises most active industrial area of the country. Industries involved in manufacturing of varieties of products lead to generation of effluent which we hypothesized to use as anodic substrate. Complexity of effluent throws challenge for its treatment, where MFC may serve purpose of electricity generation and organic load reduction. The objective of the present study was to compare the efficiency of various separators like salt bridge and proton exchange membrane in MFC using synthetic wastewater, in terms of electricity generation. The efficiency of dual-chambered MFC was also investigated for treatment of combined industrial effluent from South Gujarat region.

2 Materials and Methods

2.1 Wastewater Source

Pandesara effluent treatment plant is one of the five CETPs situated in district Surat, South Gujarat. Incoming effluent in this CETP is combined wastewater discarded by nearly 122 different industries (Zaveri et al. 2015). Wastewater sample was collected from inlet site of the treatment plant by following APHA guidelines (Eaton 2005). Onsite sample analysis was accomplished using PCSTester 35 (Eutech Instruments, India) multimeter with respect to parameters such as pH, TDS, salinity, etc. Samples were stored properly during transportation.

2.2 MFC Configuration and Operation

The construction of microbial fuel cell prepared at laboratory scale is presented in Fig. 1. Dual chamber "H"-type system was constructed using plastic containers. Graphite electrode of dimension 14.9 cm × 3.1 cm × 0.5 cm was used as anode as well as cathode (Graphicarb Industries Ltd., Ahmedabad, Gujarat). The working volume of MFC system was 1200 ml. The setup was run in batch mode. Multimeter was used to measure the voltage difference. Synthetic wastewater (Ghangrekar and Shinde 2009) containing 1 g % sucrose was used for preparation of anolyte. But the synthetic wastewater was diluted by 20 %, hence termed as '*modified synthetic wastewater.*' For MFC with industrial wastewater as substrate Pandesara CETP inlet was used and reduction in organic load was investigated in terms of BOD and COD parallel with electricity generation. 0.5 g % cysteine was used as an oxygen scavenger to provide reducing atmosphere in the anodic chamber (Logan 2008), whereas methylene blue (1 g %) was used as mediator in anode. PBS buffer provided uniform concentration of ions in the cathodic chamber. Oxygen was used as terminal electron acceptor in cathode without sparging.

2.3 Separator

Two different types of separators were evaluated, viz., salt bridge and ion exchange membrane. The digested slurry of NaCl with agar was used to prepare salt bridge as described by Khan et al. (2012). The salt bridge was allowed to solidify at room temperature. Anionic exchange and cationic exchange membranes (Membranes International Inc., New Jersey) were procured and used in different MFCs as separators between anodic and cathodic chambers (Logan et al. 2006). Membranes were preconditioned by emersion in 5 % NaCl solution at 40 °C for membrane hydration and expansion as per manufacturer's instructions.

2.4 Electrogenic Organism and Circuit Assembly

Gram-negative coccobacillus, which was found to have electrogenic properties in lab, was used in MFC. The organism had partial sequence similarity with *E.coli* when analyzed by 16S rDNA sequencing and the sequence was submitted to GenBank (Accession no: KR149337). The pure culture of organism was maintained by periodic transfer on nutrient agar slant and was preserved at low temperatures. For activation of organism, the culture was inoculated in nutrient broth to provide 20 % inoculum for MFC system and was incubated at 30 ± 2 °C for 14–16 h at 80 rpm. Activated culture was then harvested at 7500 rpm for 10 min at room temperature. Culture was washed twice with sterile normal saline to assure least contamination with complex nutrients present in broth, resuspended in equal volume of modified synthetic wastewater for addition to anode of MFC. The final inoculum size was 20 %. Two chambers were internally connected by salt bridge, and externally the circuit was completed using multimeter where low resistance copper wires were used for connection. Power density of the system was calculated by the following formula:

$$\text{Power Density} = \frac{E^2(\text{cell})}{\text{Area}\,(\text{anode}) \times \text{Resistance}\,(\text{external})} \quad \text{(Logan et al. 2006)}$$

Data of voltage output was recorded at regular time intervals of 30 min. After initiating the setup, samples were withdrawn for BOD and COD analyses (Eaton 2005). The setup was monitored till voltage drop was observed in the system.

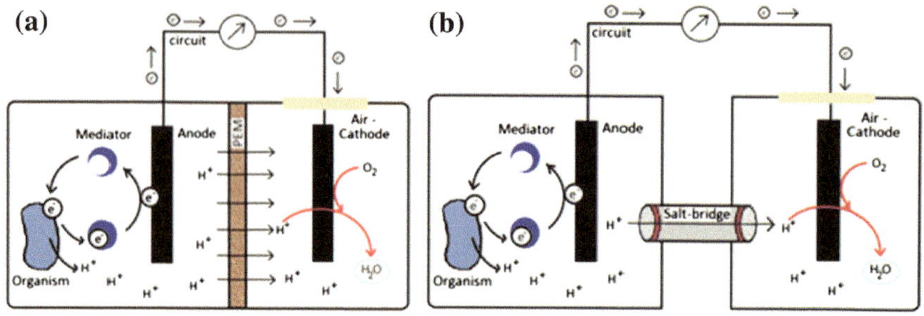

Fig. 1 Schematic representation of double-chambered microbial fuel cell construction. **a** Using proton exchange membrane as separator. **b** Using salt bridge as separator

3 Results and Discussion

3.1 MFC Performance Using Synthetic Wastewater as Anolyte

Substrate used in anode not only affects the power density and voltage production but also the microbial community composition. Synthetic or chemically defined wastewater has been used by researchers in MFC (Pant 2010). It gives an advantage of consistency in components and thus facilitates ease of optimization for other physicochemical parameters. For purpose of optimization, primarily the system preparation was carried out using standard synthetic wastewater as described by Ghangrekar and Shinde (2003). Since this MFC system displayed higher voltage only after ≈25 days with high fluctuations (data not shown), the medium was modified. Hence, system was set up using modified synthetic wastewater which was synthetic wastewater diluted by 20 %, and salt bridge was used as separator. Figure 2 depicts the open-circuit voltage (OCV) over a period of time using modified synthetic wastewater as an anolyte.

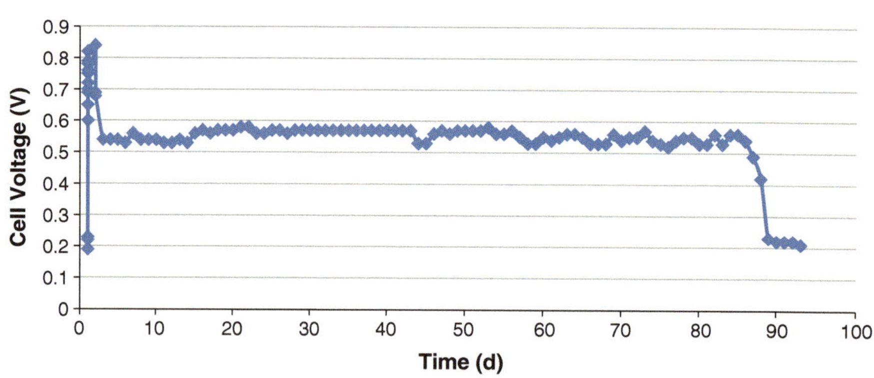

Fig. 2 Voltage production using modified synthetic wastewater as anolyte

Use of *modified synthetic wastewater* resulted in 0.82 V output for 2 days which was comparatively higher than the use of the *standard synthetic wastewater* described by Ghanghrekar and Shinde (2003). Voltage output was nearly 0.6 V for almost 3 months (90 days) and was consistent. Several media components used for growth of organism contain substantial amount of redox material which may lead to generation of potential difference initially. The initial jump of cell may be contributed by traces of such material adhered to cell even after washing steps, which was then stabilized. The system could produce 224.24 mW/m^2 power density. Using glucose as substrate, Hu (2008) could produce 161 mW/m^2 power density while adding anaerobic sludge as inoculum. In our study, all further systems of MFC were conducted using modified synthetic wastewater as anolyte as it produced consistent OCV unless otherwise mentioned.

3.2 *Influence of Different Types of Separators on MFC Performance*

Anion exchange (AEM), cation exchange (CEM), proton exchange (PEM), and bipolar membranes have been previously used in MFCs, to separate electrode chambers and decrease oxygen diffusion into the anode chamber (Kim et al. 2007). When used as separator, anion exchange membrane containing system (AEM–MFC) produced the density up to 119.4 mW/m^2 while CEM–MFC produced 200.6 mW/m^2 (Fig. 3).

After stabilization of the voltage, CEM–MFC could display slightly higher voltage as compared to AEM–MFC. The anionic exchange membranes exhibit positive charge on the surface due to quaternary ammonium groups attached. This may lead to leakage of intermediates such as acetate, butyrate, etc., and may contribute to less potential difference (Pandit et al. 2012). Although both the systems produced consistent voltage output till 27 days of incubation, MFC with AEM

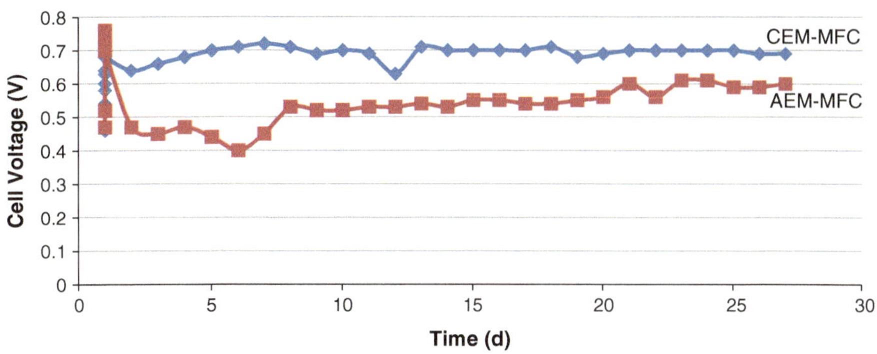

Fig. 3 Influence of different types of proton exchange membranes on voltage output in MFC

could produce least voltage among all three separators investigated. CEM–MFC displayed higher voltage output and ease of operation to the researcher but may not prove as cost effective as salt bridge.

3.3 *Performance of MFC Using Combined Industrial Effluent as Anolyte*

When combined wastewater of inlet collected from Pandesara CETP, South Gujarat was used in MFC as substrate, MFC system was found to be efficient with production of 0.72 V till 60 days of incubation (Fig. 4). Hence, proposed designed MFC was found to effectively work on actual combined industrial wastewater.

The power density obtained was 200.6 mW/m^2. Hence, the efficiency of the electrogenic bacteria in MFC system was found to be comparative whether synthetic wastewater or actual combined industrial wastewaters were used in MFC as substrates. As high as 0.7 V output were obtained consistently in CETP wastewater containing MFC. Subsequent adaptation of the organism may lead to higher voltage production in upcoming runs of MFCs. Researchers have used wastewater collected from individual industries to run MFC and treat the waste. Brewery wastewater was used by Wen et al. (2010) to generate electricity using sequential arrangement of anode and cathode. With external resistance of 300 Ω and brewery waste as anolyte, it was possible to achieve open-circuit voltage of only 0.434 V. Similarly, Liu et al. (2011) used steroidal drug industrial waste for electricity generation using bacteria. With no available reports on combined industrial effluents being used as anolyte, the present study throws opportunity of such complex effluent to be used in MFC, where salt bridge successfully can be used as separator.

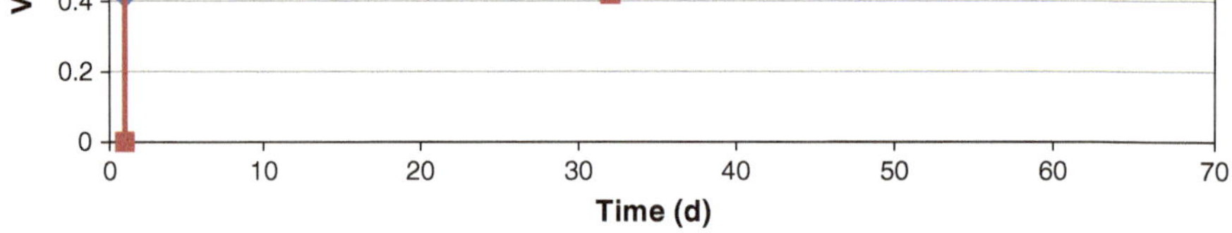

Fig. 4 Electricity generation in MFC using inlet wastewater collected from Pandesara combined effluent treatment plant of South Gujarat. *Note*: "MSW" indicates modified synthetic wastewater and "PIW" indicates Pandesara inlet wastewater

3.4 Reduction of Organic Load

As one of the important parameters of organic load reduction, the samples from MFC systems were analyzed for reduction of BOD and COD values. Table 2 shows reduction of organic load in systems using various separators with modified synthetic wastewater and actual combined wastewater as anolyte connected with salt bridge.

In MFC with modified synthetic wastewater with 1 % sucrose, the BOD reduction achieved varied between 43 and 53 %, while COD reduction values varied widely resulting in the highest 34 % COD reduction in CEM–MFC. With external resistance of 300 Ω and brewery waste as anolyte, it was possible to achieve 91.7–95.7 % of COD reduction (Wen et al. 2010). This may be because the brewery wastewater is expected to contain simple and easily assimilable compounds. When comparing the values of salt bridge and CEM containing MFCs, the BOD reduction was almost similar, while COD reduction was 30 % higher in case of CEM containing MFC. Moreover, MFC containing CEM as separator was - nearly 10 % more effective than AEM containing MFC for COD reduction. In case of inlet wastewater of Pandesara CETP as anolyte, even with use of salt bridge as separator, BOD and COD reductions achieved were significant; however, intense experimental design in future may help in achieving better system, where use of ion exchange membranes may be evaluated and compared for feasibility.

To conclude, the use of CEM in MFC was found to be the most effective in high voltage production and reduction of BOD, COD, and with ease of operation, but it adds to cost of processing. Salt bridge was found to be favorable in case of laboratory-scale experiments but needs optimization with respect to design if it is to be used at large scale; however, AEM–MFC could not produce significantly high power density as compared to systems with CEM and salt bridge. "H"-type system

Table 2 Reduction in BOD and COD values after MFC run using various connectors and combined industrial effluent

Type of separator	BOD reduction			COD reduction		
	Initial BOD mgl^{-1}	Final BOD mgl^{-1}	% Reduction	Initial COD mgl^{-1}	Final COD mgl^{-1}	% Reduction
Salt bridge-MSW	7061 ± 504	3600	49.01	16,750 ± 500	16000	04.47
CEM	7061 ± 504	3350	52.55	16,751 ± 500	11000	34.00
AEM	7061 ± 504	4050	42.64	16,752 ± 500	13000	22.38
Salt bridge-CETP	3100 ± 141.4	1350	56.45	12447	8000	35.72

Note: (1) "Salt bridge-MSW" refers to MFC with salt bridge as connector and modified synthetic wastewater as anolyte, (2) "CEM" refers to MFC with Cationic exchange membrane, (3)"AEM" refers to MFC with anionic exchange membrane (4) "Salt bridge-CETP" refers to MFC with salt bridge as connector and Pandesara CETP inlet wastewater as anolyte. In the first three MFC systems, modified synthetic wastewater was used as anolyte using synthetic wastewater resulted in 0.8 V of OCV and 0.72 when CETP wastewater was used and could sustain for quite long duration with more than 50 % of BOD reduction. Further modifications in MFC system design and optimization to achieve similar efficient decrease in COD for wastewater in short duration are still required. Adaptation of culture to the wastewater in fed batch mode of operation may help in improvement of performance parameters of MFC system.

Acknowledgments The authors express their sincere thanks to Mr. Abdul Basit, Graphicarb Industries Ltd., Ahmedabad for their kind gift of carbon felt plates required for MFC setup preparation. Authors would also like to thank In-charge, Pandesara effluent treatment plant, Surat, South Gujarat for support and help for sample collection. We would like to thank Gujarat Pollution Control Board for their support in the present study.

References

Ahn Y and Logan BE (2010) Effectiveness of domestic wastewater treatment using microbial fuel cells at ambient and mesophilic temperatures. Bioresource Technol. 101:469–475

Catal T, Li K, Bermek H and Liu H (2008) Electricity production from twelve monosaccharides using microbial fuel cells. J. Power Sources 175:196–200

Eaton (2005) Standard method for the examination of water and wastewater. APHA-AWWA-WEF, 21st Ed. Washington, D.C

Ghangrekar MM and Sindhe VB (2007) Performance of membraneless microbial fuel cell treating wastewater and effect of electrode distance and area on electricity production. Bioresource Technol. 98:2879–2885

Hampannavar US, Anupama, Pradeep NV (2011) Treatment of distillery wastewater using single chamber and double chambered MFC. Int. J. Environ. Sci. 2:114–123

Hu Z (2008) Electricity generation by a baffle-chamber membraneless microbial fuel cell. J Power Sources 179:27–33

Jadhav GS, Ghangrekar MM (2009) Performance of microbial fuel cell subjected to variation in pH, temperature, external load and substrate concentration. Bioresource Technol 100(2009):717–723

Jadhav DA, Ghade AN, Mondal D, Ghangrekar MM (2014) Comparison of oxygen and hypochlorite as cathodic electron acceptor in microbial fuel cell. Bioresource Technol 154(2014):330–335

Jia J, Tang Y, Liu B, Wu D, Ren N and Xing D (2013) Electricity generation from food waste and microbial community structure in microbial fuel cells. Bioresource Technol. 144:94–99

Kaewkannetra P, Chiwes W and Chiu TY (2011) Treatment of cassava mill wastewater and production of electricity through microbial fuel cell technology. Fuel 90:2746–2750

Khan MR, Bhattacharjee R, Amin MSA (2012) Performance of the salt bridge based microbial fuel cell. Int J Eng Technol 1(2):115–123

Kim BH, Chang IS, GaddG M (2007) Challenges in microbial fuel cell development and operation. Appl Microbiol Biotechnol 76(3):485–494

Li WW, Sheng GP, Liu XW, Yu HQ (2011) Recent advances in the separators for microbial fuel cells. Bioresource Technol 102(1):244–252

Liu G, Yates MD, Cheng S, Call DF, Sun D, Logan BE (2011) Examination of microbial fuel cell start-up times with domestic wastewater and additional amendments. Bioresource Technol 102 (15):7301–7306

Liu R, Gao E, Zhao YG, Wang A, Lu S, Wang M, Maqbool F and Huang Q (2012) Biological treatment of steroid drug industrial effluent and electricity generation in microbial fuel cell. Bioresource Technol. 123:86–91

Logan BE, Logan B, Hamelers B, Rozendal R, Schroder U, Keller J, Freguia S, Aelterman P, Verstraete W, Rabaey K (2006) Microbial Fuel Cells: Methodology and Technology. Environ. Sci.Technol. 40(7):5181–5192

Logan BE, Chen S, Watson V and Estadt G (2007) Graphite fiber brush anode for increased power production in air cathode microbial fuel cells. Env. Sci. Tech. 41:3341–3346

Logan B (2008) Ch 2. Exoelectrogens in Microbial fuel cells. Wiley Publisher, NewJersy pp:12–29

Logan BE and Rabaey K. (2012) Conversion of wastes into bioelectricity and chemicals using microbial electrochemical technologies. Science, 337:686–690

Lu N, Zho SG, Zhuang L, Zhang JT and Ni JR (2009) Electricity generation from starch processing wastewater using microbial fuel cell technology. Biochem. Eng. J. 43:246–251

Mathuriya AS, Sharma VN (2010) Bioelectricity production from various wastewaters through microbial fuel cell technology. J. Biochem. Technol. 2:133–137

Min B, Kim J, Oh S, Regan J and Logan BE (2005) Electricity generation from swine wastewater using microbial fuel cells. Water Res. 39:4961–4968

Pant D, Van Bogaert G, Diels L, Vanbroekhoven K (2010) A review of the substrates used in microbial fuel cells (MFCs) for sustainable energy production. Bioresource Technol 101(6):1533–1543

Pandit S, Ghosh S, Ghangrekar MM, Das D (2012) Performance of an anoin exchange membrane in association with cathodic parameters in dual chamber microbial fuel cell. Int. Jour. Hydrogen. Energy. 37(2012):9383–9392

Potter MC (1911) Electrical effects accompanying the decomposition of organic compounds. Royal Society (Formerly Proceedings of the Royal Society) B, 84:260–276

Qu Y, Feng Y, Liu J, He W, Shi X, Yang Q, Lv J and Logan BE (2013) Salt removal using multiple microbial desalination cells under continuous flow conditions. Desalination 317: 17–22

Rahimnejad M, Bakeri G, Najafpour G, Ghasemi M, Oh S (2014) A review on the effect of proton exchange membranes in microbial fuel cells. Biofuel Research Journal 1:7–15

Wei J, Liang P, Huang X (2011) Recent progress in electrodes for microbial fuel cells. Bioresource Technol 102(20):9335–9344

Wen Q, Wu Y, Zhao L, Sun Q, Kong F (2010) Electricity generation and brewery wastewater treatment from sequential anode-cathode microbial fuel cell. J. Zhejiang Univ. Sc. B 11(2):87–93

Wua PK, Biffingerb JC, Fitzgeraldb LA, Ringeisenb BR (2011) A low power DC/DC booster circuit designed for microbial fuel cells. Process Biochem. doi:10.1016/j.procbio.2011.06.003

Zaveri P, Munshi N, Vaidya A, Jha S, Kumar GN (2015) Functional microbial diversity dynamics in common effluent treatment plants of South Gujarat and hydrocarbon degradation. Can J Microbiol 61:389–397

Part VI
Integrated Systems

Role of Rhamnolipid: A Biosurfactant in Methane Gas Hydrate Formation Kinetics

Amit Arora, Swaranjit Singh Cameotra, Rajnish Kumar, Anil Kumar Singh, Pushpendra Kumar, Chandrajit Balomajumder and Sukumar Laik

Abstract Naturally occurring methane gas hydrate is a vast source of methane gas which is trapped in crystalline ice-like structure present in permafrost regions and under the sea in outer continental margins. It is purposed that total amount of carbon in the form of methane hydrates is almost twice the carbon content in all the fossil fuel reserves put together, and hence these are supposed to be the future potential energy resource. This paper investigates the laboratory investigations on effect of a biosurfactant rhamnolipid on methane hydrate formation kinetics. Rhamnolipid was produced by *Pseudomonas aeruginosa* strain A11. The presence of *P. aeruginosa* has been reported in Gulf of Mexico gas hydrate samples. Biosurfactant reduced the surface tension of water from 72 to 36 mN/m with CMC of 70 mg/L. The biosurfactant dose is studied at two different concentrations in the solution at 100 and 1000 ppm. Kinetic of hydrate formation and growth is compared at 0, 100, and 1000 ppm of rhamnolipid showing that rhamnolipid acts as a

Amit Arora (✉) · Chandrajit Balomajumder
Department of Chemical Engineering, Indian Institute of Technology Roorkee, Roorkee 247667, India
e-mail: aroraamitlse@yahoo.com

S.S. Cameotra · A.K. Singh
Institute of Microbial Technology, Chandigarh, India

Rajnish Kumar
Chemical Engineering and Process Development Division, National Chemical Laboratory, Pune, India

Pushpendra Kumar
Keshav Dev Malviya Institute of Petroleum Exploration, Oil and Natural Gas Corporation, Dehradun, India

Sukumar Laik
Department of Petroleum Engineering, Indian School of Mines, Dhanbad, India

S. Kumar et al. (eds.), *Proceedings of the First International Conference on Recent Advances in Bioenergy Research*, Springer Proceedings in Energy,
DOI 10.1007/978-81-322-2773-1_25

hydrate promoter at these concentrations. Thus, small dosages of rhamnolipids produced by *P. aeruginosa* strain A11 must clearly affect the gas hydrate formation kinetics in natural sites (as in Gulf of Mexico).

Keywords Methane hydrate · Potential energy resource · Rhamnolipid · Induction time · Pseudomonas aeruginosa

1 Introduction

Methane hydrates are inclusion compounds formed by water and methane at high pressure and low temperature (Sloan and Koh 2007). They are present in sedimentary deposits in permafrost regions and beneath the sea in outer continental margins (Kvenvolden 1993; Paul and Dillon 2000; Hester and Brewer 2009). One volume of gas hydrates releases about 160 volumes of methane at standard temperature and pressure (STP) (Kvenvolden 1993). It is estimated that carbon bound in methane hydrates is twice the carbon content in fossil fuel reserve, and hence these are supposed to be the future potential energy resource. The worldwide assessment of (Kvenvolden 1998a) methane hydrate is equivalent to 250 trillion cubic meters of methane gas. The amount of hydrate bound methane gas is approximately 100 times as large as that of conventional natural gas resource. As per the USA department of energy, if only very small amount of the methane stored in these hydrates could be exploited, it would be almost twice the current domestic supply of natural gas (Haq 1998; Holder et al. 1987). They are also seen as future technology for storage and transportation of gas.

1.1 Energy Potential of Gas Hydrates

The amount of methane caught in gas hydrate is not certain. Various groups have given various amounts of methane trapped in global gas hydrate deposit ranging from $\sim 10^{17}$ ft^3 or 10^5 trillion cubic feet (TCF) (Mciver 1981) to 10^8 TCF (Trofimuk et al. 1977; Kvenvolden 1988, 1998b; Gornitz and Fung 1994; Harvey and Huang 1995). Recently in review on the concerned subject by (Boswell and Collett 2011), the amount of methane caught in gas hydrates has been estimated around 10^5 TCF (Boswell and Collett 2011). As per the idea of USA department of energy, even negligible amount of methane stored in hydrates can be exploited. It is more than the current supply of natural gas in the country (Holder et al. 1987; Haq and Boulder 1998). Methane hydrates can be assumed upcoming source of hydrocarbon energy and will be a future fuel (Paul and Dillon 2000; Kvenvolden and McMenamin 1980).

2 Present Study

This paper investigates the effect of a biosurfactant rhamnolipid on methane hydrate formation kinetics in laboratory. Rhamnolipid was obtained from *P. aeruginosa* strain A11. The presence of *P. aeruginosa* has been reported in Gulf of Mexico gas hydrate samples (Lanoil et al. 2001) which helped in generating rhamnolipids at the site. Methane hydrate formation experiments were performed with 90 % saturation of distilled water (and water–surfactant solution) in porous C-type silica gel (pore volume 0.90 $cm^3\ g^{-1}$). The study is carried out in high-pressure vessel maintained at low temperature.

3 Experimental Setup, Materials, Procedure

3.1 Apparatus

Video hydrate cell was used to study the gas hydrate formation which is a mercury-free cell. It can measure the induction time for formation of hydrates and monitor the pressure drop as a function of time during hydrate formation. The system consists of constant volume hydrate cell having capacity of 250 cm^3 and pressure rating up to 3000 psi. The temperature is controlled by the thermostatic bath. A computer is attached for data acquisition of temperature and pressure versus time. The diagram of the gas hydrate cell is shown in Fig. 1.

3.2 Materials

Pure methane with 99.99 % pure methane (Chemtron Science Laboratory, Navi Mumbai, India), reverse osmosis water (Millipore SA, 67,120 Molshein, France), C-type silica gel (Merk Merck Specialities Pvt. Ltd., Mumbai, India), and rhamnolipid synthesized in laboratory from strain A11 were used for the experiment.

3.3 Procedure

First, 90 % saturated C-type silica gel was used for experiment; then 90 % saturated C-type silica gel with 100 ppm rhamnolipid aqueous solution and then 1000 ppm rhamnolipid aqueous solution was used for the experiment. The test sample was placed in hydrate cell and constant temperature was maintained immersing the cell into a temperature-controlled bath. A mixture of water and ethylene glycol (25 %) was used as a liquid for bath. A vacuum pump was used to remove the air from the

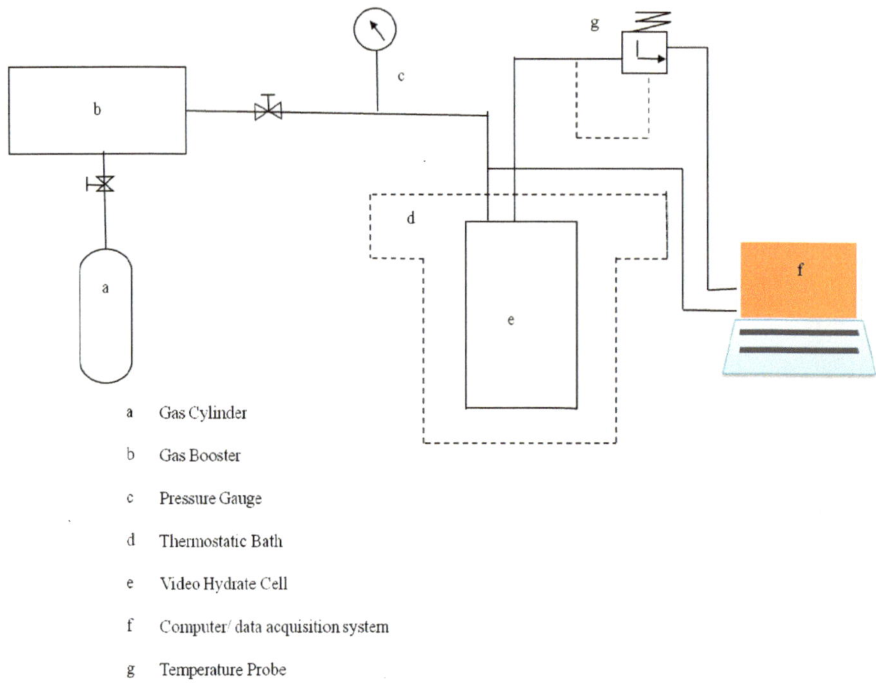

Fig. 1 Experimental setup

cell before pressurizing methane gas in the cell. The cell was pressurized up to the desired value with methane gas. Sudden pressure decline was observed for the hydrate formation and online video picture was also seen for identifying hydrate formation. When noticeable pressure drop is not observed, it signifies the completion of process of formation of hydrates. The induction time, moles of gas consumed, and rate of hydrate formation were calculated from the obtained pressure temperature date.

4 Results and Discussion

The results obtained in the present study are discussed as follows.

4.1 Induction Time

The induction time, t_i, is the passed from the beginning of the experiment at t_s to the onset of hydrate formation t_o and it is one of the kinetic parameters of hydrate

Table 1 Methane hydrate formation parameters for various types of samples

Test sample	Initial cell pressure P (Mpa)	Nucleation temperature T (K)	Nucleation pressure P (Mpa)	Induction time (min)
C-type silica gel without presence of rhamnolipids	11.62	277.16	10.46	44.24
C-type silica gel containing 100 ppm rhamnolipid	11.62	277.37	10.42	36.91

formation. The induction time for various experimental conditions is given in Table 1, and the moles of methane consumed are shown in Fig. 1.

It can be concluded from Table 1 that the induction time in the presence of 100 ppm rhamnolipid has been reduced which states that rhamnolipid is acting as promoter for the formation of methane gas hydrates.

4.2 *Moles of Gas Consumed*

While the hydrate is formed, the drop of pressure was observed, which is due to the consumption of gas of methane. The methane takes over the unoccupied cavities of water while hydrate formation. The amount of gas consumed during hydrate formation can be calculated by real gas equation:

$$n = n_{\mathrm{i}} - n_{\mathrm{f}} = \frac{V}{R}\left(\frac{P_{\mathrm{i}}}{z_{\mathrm{i}}T_{\mathrm{i}}} - \frac{P_{\mathrm{f}}}{z_{\mathrm{f}}T_{\mathrm{f}}}\right) \quad (1)$$

where n is the amount of gas consumed when hydrates form, V is the gas volume, P_{i}, T_{i} and P_{f}, T_{f} are the pressure and temperature at initial and final conditions, and R is the universal gas constant. Compressibility factors (z_{i}, z_{f}) are calculated at the corresponding pressure and temperature, respectively. The pattern of gas consumed during formation of hydrate is shown in Fig. 2.

As it is evident from Fig. 2 that maximum numbers of moles are consumed for 1000 ppm rhamnolipid, at lower dose, i.e., at 100 ppm, the number of moles of methane gas consumed is lesser than without rhamnolipid but as the concentration of rhamnolipid is increased than amount of methane gas consumed is more than that of without rhamnolipid.

Figure 3 shows the consumption of gas and pressure drops as a function of time. The growth and gas consumption regions show the two symmetric plots that could be explained as conservation of mass during the hydrate growth. Pressure drop shows the number of methane molecules leaving the gas phase to hydrate phase drop, whereas gas consumption plot shows the number of methane molecules entering into the hydrate cages. Mass remains conserved in an isochoric system,

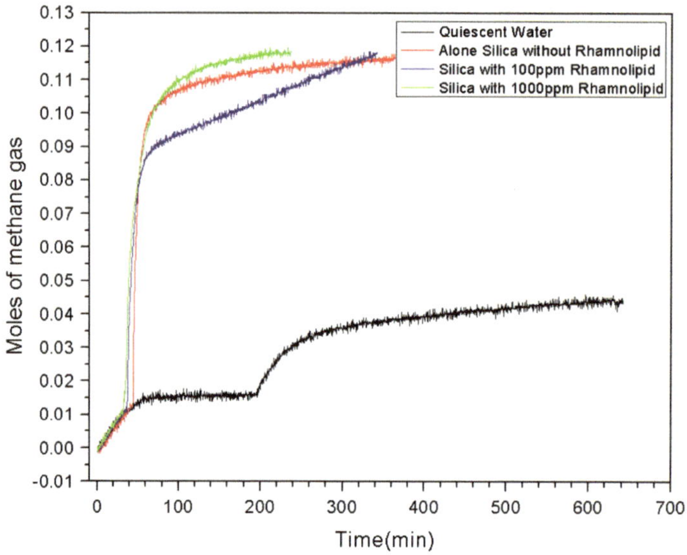

Fig. 2 Moles of methane gas consumed

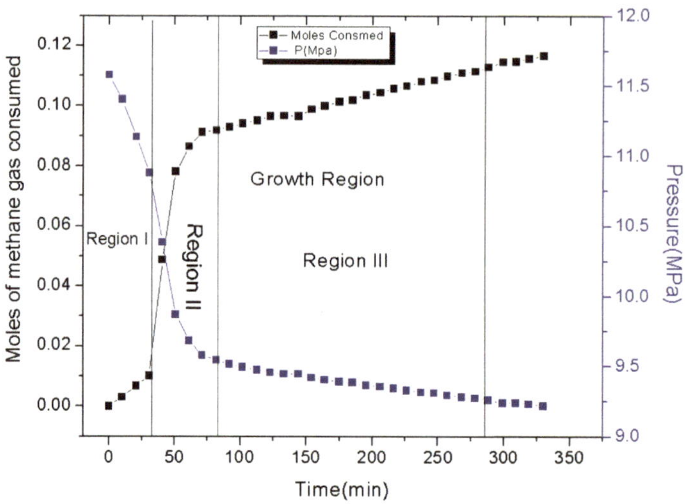

Fig. 3 Growth curve of methane hydrate formation in C-type silica gel in the presence of 100 ppm rhamnolipid

which is shown by the symmetry of the two plots about an axis passing through the intersection of the two plots and parallel to the time axis. As shown in Fig. 3, the process is splitted into three regions. First region starts at time zero where the consumption of the gas is less in the system and pressure slightly decreases because of dissolution of gas. Second region is after the induction time where gas consumption is increased; this happens because of the hydrate growth. In third region as the hydrate is formed, gas consumption is as shown in Fig. 3.

4.3 The Rate of Methane Hydrate Formation

As seen from Fig. 4, the slope of the curves after nucleation indicates an exponential behavior from which we can assume a first-order reaction. Equations 2 and 3 can be used for a first-order reaction which is defined as follows:

$$N = N_o e^{-kt} \quad (2)$$

$$\ln \frac{N}{N_o} = -kt \quad (3)$$

where N is the total number of moles at time t, N_0 is the initial number of moles, k is the rate constant (min^{-1}), and t is the time in minute. The rate constant (k) of hydrate formation can be calculated from the slope of the curve of ln (N/N_0) versus time. The hydrate formation from region 2 onward Fig. 3 was divided into five zones of time as shown in Fig. 5 (0–20, 20–40, 40–80, 80–160, 160–260 min) after nucleation. The data of these five zones are used to determine the rate constant of hydrate formation

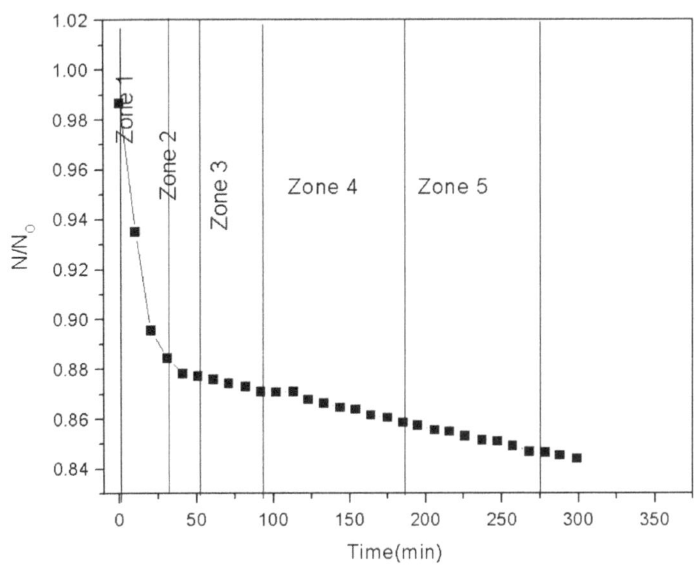

Fig. 4 *N*/*N*0 versus time plot for methane hydrate formation in the presence of C-type silica gel containing 100 ppm rhamnolipid

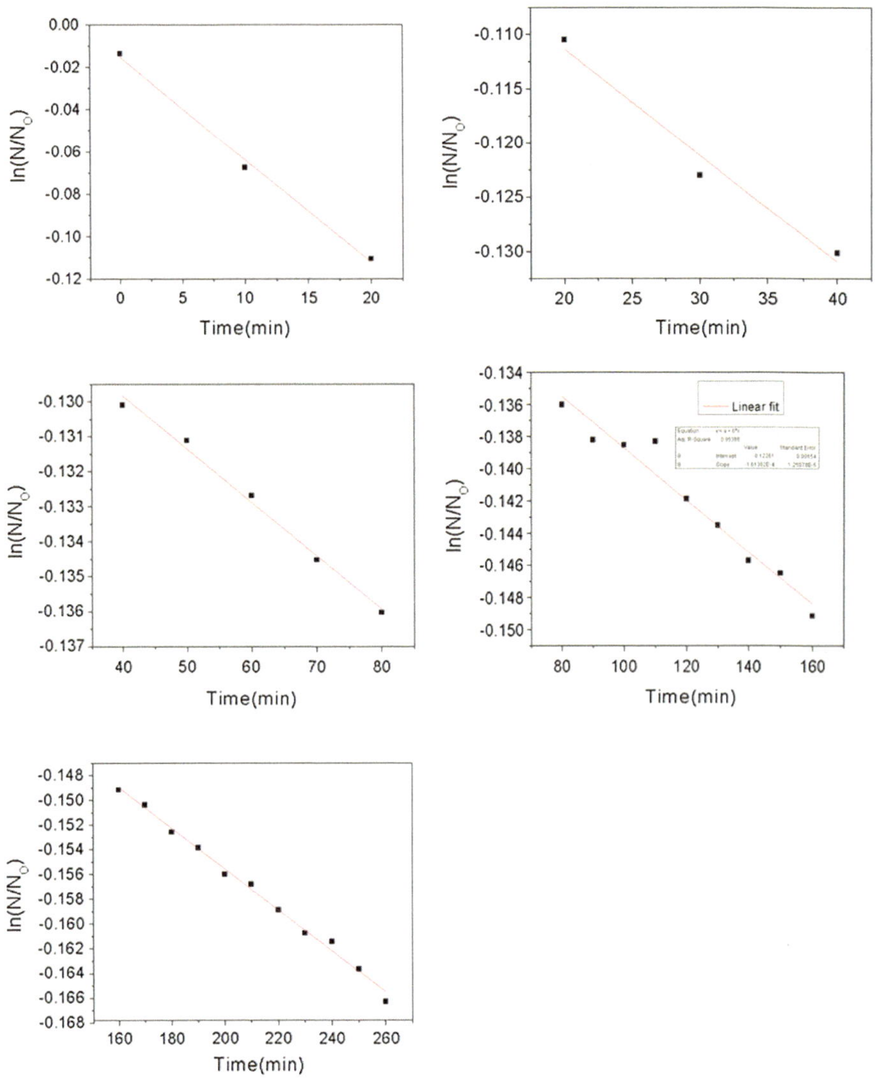

Fig. 5 Semi-logarithmic plot of change of moles of gas while methane hydrate formation in the presence of C-type silica gel containing 100 ppm rhamnolipid

in each zone. Hydrate formation rate is determined by putting the values of slopes, i.e., rate constant k calculated from Fig. 4 in the following equation:

$$\frac{dN}{dt} = -N_o k e^{-kt} \tag{4}$$

Table 2 The rate of methane gas hydrate formation for 100 ppm rhamnolipid

Type of sample	Time zone (min)	Rate constant K (min^{-1})	Formation rate (mole/min)	% over all increased rate with respect to Silica
C-type silica gel containing 100 PPM Rhamnolipid	0–20	0.00484	0.003241392	57.08272969
	20–40	9.82E−04	0.000645076	
	40–80	1.53E−04	9.9753E−05	
	80–160	1.61E−04	0.000103977	
	160–260	1.65E−04	0.0001047	

The values of rate constant and concerned rates obtained in present study in the presence of 100 ppm rhamnolipid are shown in Table 2.

The above results indicate that rhamnolipid acts as promoter for the methane hydrate formation. It is also observed that the induction time of hydrate formation is reduced in the presence of rhamnolipid (Table 1). The rate of hydrate formation is found to increase many time (Table 2).

The zeta potential of rhamnolipid synthesized in the present study from strain A11 was found to be negative; hence, it is anionic surfactant and much of work in literature on anionic surfactants such as SDS, etc. on gas hydrate formation has been reported. These results are in agreement with various studies reported on synthetic anionic surfactants like SDBS (Dai et al. 2014) ans LABS (Ganji 2007; Fazlali et al. 2013; Kumar et al. 2014).

The rhamnolipid acts as a promoter as the biosurfactant spherical micelles are formed by long-chain carbon alkyl groups which solubilize hydrocarbon gases (MacKerell 1995).

The above results indicate that rhamnolipids act as promoter which can help in synthesizing gas hydrates. Industries looking upon gas hydrates not because of their energy potentials rather it is becoming a technology these days because of their other applications such as transportation and storage of gas, desalination of water, etc. The present study clarifies the role of rhamnolipid a biosurfactant as promoter in natural sites and giving a green biodegradable promoter as much of work as promoter of gas hydrates is reported on synthetic surfactants; however, the study has given a replacement of synthetic surfactant by a green biosurfactant.

5 Conclusion

The above study clearly indicates the role of rhamnolipid in the formation of methane gas hydrate formation as promoter; thus small dosages of rhamnolipids produced by *P. aeruginosa* strain A11 must clearly enhance the gas hydrate formation kinetics in natural sites (as in Gulf of Mexico). Rhamnolipid is a green biodegradable promoter for gas hydrates formation, and is a green bio-surfactant which can substitute the synthetic surfactant which is used presently as promoter for the formation of gas hydrates.

Acknowledgments The supports from institutes namely National Chemical Laboratory (CSIR Laboratory), Maharashtra, Pune, India, Institute of Microbial Technology (CSIR Laboratory), UT, Chandigarh, India, Keshav Dev Malviya Institute of Petroleum Exploration, Oil and Natural Gas Corporation (ONGC), Uttrakhand, Dehradun, India. Indian School of Mines, Jharkhand, Dhanbad, India are highly acknowledged.

References

Boswell R, Collett TS (2011) Current perspectives on gas hydrate resources. Energy Environ Sci 4:1206–1215

Dai Y, Zhong X, Jiang X, Wang S (2014) Experiment of new additives effect on gas hydrate formation. Energy Power Eng 6:133–141

Fazlali A, Kazemi S-A, Moraveji MK, Mohammadi AH (2013) Impact of different surfactants and their mixtures on methane-hydrate formation. Energy Technol 1:471–477

Ganji H, Manteghian M, Sadaghiani ZK, Omidkhaha MR, Mofrad RH (2007) Effect of different surfactant on methane hydrate formation rate, stability and storage capacity. Fuel 86:434–441

Gornitz V, Fung I (1994) Potential distribution of methane hydrates in the world's oceans. Glob. Biogeochem Cycles 8:335–347

Haq BU (1998) Gas hydrates; greenhouse nightmare? Energy panacea or pipe dream. Geol Soc Am (GSA) Today 8(11):1–6

Harvey LDD, Haung Z (1995) Evaluation of the potential impact of methane clathrate destabilization on future global warming. J Geophys Res 100(D2):2905–2926

Hester K, Brewer PG (2009) Clathrate hydrates in nature. Ann Rev Mar Sci 1:303–327

Holder GD, Malone RD, Lawson WF (1987) Effects of gas composition and geothermal properties on the thickness and depth of natural-gas-hydrate zones. J Petrol Technol 39(9):1147–1152

Kumar SV, Kumar MG, Udayabhanu G, Mandal A, Laik S (2014) Kinetics of methane hydrate formation and its dissociation in presence of non-ionic surfactant Tergitol. J Unconventional Oil Gas Resour 54–59

Kvenvolden KA (1988) Methane hydrate—a major reservoir of carbon in the shallow geosphere. Chem Geol 71:41–51

Kvenvolden KA (1993) Gas hydrates—geological perspective and global change. Rev Geophys 31:173–187

Kvenvolden KA (1998a) A primer on the geological occurrence of gas hydrate. Geol Soc Lond Spec Pub. 137:9–30

Kvenvolden KA (1998b) Estimates of the methane content of worldwide gas-hydrate deposits, methane hydrates: resources in the near future. Paper presented at JNOC-TRC, Japan. 20–22 October

Kvenvolden KA, McMenamin MA (1980) Hydrates of natural gas: a review of their geologic occurrence. US Geol Surv Circ 825:1–11

Lanoil BD, Sassen R, La Duc MT, Sweet ST, Nealson KH (2001) Bacteria and Archaea physically associated with Gulf of Mexico gas hydrates. Appl Environ Microbiol 67(11):5143–5153

MacKerell AD (1995) Molecular dynamics simulation analysis of a sodium dodecyl sulfate micelle in aqueous solution: decreased fluidity of the micelle hydrocarbon interior. J Phys Chem 99:1846

Mciver RD (1981) Gas hydrates In: Meyer RF, Olson JC, Long-term energy resources, Pitman, Boston, pp 713–726

Paul CK, Dillon WP (2000) Natural gas hydrates occurrence, distribution and detection. Am Geophys Union Geophys Monogr 124:315

Sloan ED, Koh C (2007) Clathrate hydrates of natural gases, 3rd edn. CRC Press, Boca Raton

Trofimuk AA, Cherskii NV, Tsaryov VP (1977) The role of continental glaciation and hydrate formation on petroleum occurrence In: Meyer RF, The future supply of nature-made petroleum and gas. Pergamon Press, New York, pp 919–926

Economic Analysis of Solar PV/ Wind/Diesel Generator/Battery Connected Integrated Renewable Energy Systems for Residential Applications

Suresh Muthusamy and Meenakumari Ramachandran

Abstract Nowadays, energy becomes the basic need of each and every human being. Most of the energy demands are met by the conventional energy sources like coal, oil, natural gas, etc., which are going to exhaust in a day and cause severe energy demand in the future. Renewable energy sources become the alternative for meeting the energy demand, among which solar and wind energy resources are most commonly used. Because of the intermittent nature of solar and wind energy sources, hybrid/integrated renewable energy systems are evolved in the recent years. This paper discusses about the economic analysis and the optimal selection of solar PV, wind, diesel generator, and battery connected hybrid energy systems for residential applications. For the analysis purpose, a small middle class family in Perundurai, Erode District, Tamil Nadu has been considered for the study. Also, a suitable hybrid optimized model has been developed and the results have been discussed based on their net present cost (NPC) and cost of energy (COE), green house gas (GHG) emissions, etc.

Keywords Cost of energy · Net present cost · Integrated renewable energy systems

1 Introduction

Energy demand is growing day by day due to the increase in population, urbanization, and rapid industrialization. Most of the peoples rely on the conventional energy-based electricity production for meeting their daily needs. These conventional energy sources emit greenhouse gasses, when the fuels are being burnt. Since the oil crisis in

---

Suresh Muthusamy (✉) · Meenakumari Ramachandran
Department of Electrical and Electronics Engineering, School of Electrical Sciences, Kongu Engineering College, Perundurai, Erode 638052, Tamil Nadu, India
e-mail: infostosuresh@gmail.com

Meenakumari Ramachandran
e-mail: oremkay@gmail.com

S. Kumar et al. (eds.), *Proceedings of the First International Conference on Recent Advances in Bioenergy Research*, Springer Proceedings in Energy,
DOI 10.1007/978-81-322-2773-1_26

late 1970s, renewable energy has been considered as an alternative way of generating electrical power for meeting their needs.

Various industrial sectors and even general public and government organizations are also turning their attention toward renewable energy-based energy production for meeting their energy demands. The Government of India is also trying to maintain the energy balance between the energy demand and the source availability. Nowadays, hybrid energy systems are being mostly preferred rather than depending on single source for supplying the electric power in urban, rural, and remote areas due to the intermittent nature of solar and wind resources. Due to the integration of renewable energy-based systems, it results in high cost, complexity and several factors have been considered for integration.

As on March 31, 2014 (Renewable Energy Status Report 2014), the installed capacity of renewable energy has reached 32,269.6 MW (12.95 %) of the total potential available (249,199 MW) in India. The Ministry of New and Renewable Energy (MNRE) has set down a target of 41,400 MW, as the total installed capacity by the end of 2017. Even though there is a considerable growth in power generation, the demand for the electric power has been growing at a rapid rate, whereas the opportunities for electricity production are very high. The Government of India is trying to bridge the gap between energy demand and power production by providing the funds to the research projects and development of Ultra Mega Power Projects (UMPPs) by making use of renewable energy systems.

The remaining section of the paper is organized as follows:

Section 2 deals with IRES concepts.
Section 3 deals with the scope of the proposed work and the existing works in IRES-based systems.
Section 4 explains about the system modeling and the availability of resources.
Section 5 presents the simulation results and discussion.
Section 6 concludes with the findings of the proposed work.

2 Integrated Renewable Energy Systems

Usually, renewable energy sources are of single energy-based systems and multiple energy-based systems or hybrid/integrated energy systems (Kanase Patil et al. 2011). In case of single energy-based systems, it uses only one source along with energy storing devices and power electronic systems, whereas in case of multiple energy-based systems, it utilizes two or more renewable energy sources along with DG sets, storage devices, and power electronic devices. Several hybrid energy system combinations can be used for generating power depending upon their requirements. Hybrid energy systems are more advantageous compared to single energy-based systems because they are highly reliable and more efficient. But, a proper design and better solution has been considered, in order to find the economic and most feasible solution, i.e., it should be neither oversized nor undersized.

This paper clearly discusses about the economic analysis of solar PV/wind/DG/battery connected integrated renewable energy systems for residential load applications. For the study purpose, a small middle class family and their energy demand have been considered in Perundurai, Erode District, Tamil Nadu. A Hybrid Optimization Model has been developed and a better optimal solution with proper design has been found out from the several possible combinations based on the energy calculations like net present cost (NPC), cost of energy (COE), excess energy (EE), etc.

3 Scope of the Proposed Work

The concept of integrated renewable energy systems was first introduced by R.G. Rama kumar in the 1980s, where he worked on evaluating the design scenarios of IRES. Later, he formulated the designs for integrating the renewable energy sources from the existing, freely available sources in which a remote village had been chosen and the versatility of IRES–KB has been discussed (Ramakumar 1977; Ramakumar and Hughes 1981). Later, in the late 1980s, a linear programming approach had been developed for integrating the renewable with the aim of minimizing/reducing the total annual cost using the concept of loss of power supply probability, in terms of power and energy calculations. The main aim of the IRES model proposed by Rama kumar in linear programming approach was to minimize the total annual cost per year subject to the energy produced.

Many computer-based simulation models are available in order to design and evaluate the techno-economic analysis of the renewable energy-based systems (Akella et al. 2007; Sinha and Chandel 2014). This hybrid optimization model is one of the powerful tool which is mostly used by the researchers for designing and analyzing the hybrid power systems, which contain a mix of conventional generators, combined heat, and power, wind turbines, solar photovoltaic panels, etc.

4 System Components and Modelling

4.1 Study Area

Based on the village electrification report by MNRE released during May 2014, only few of the villages were fully electrified. Most of the villages remain unelectrified/partially electrified even though they are ready to pay for the electricity

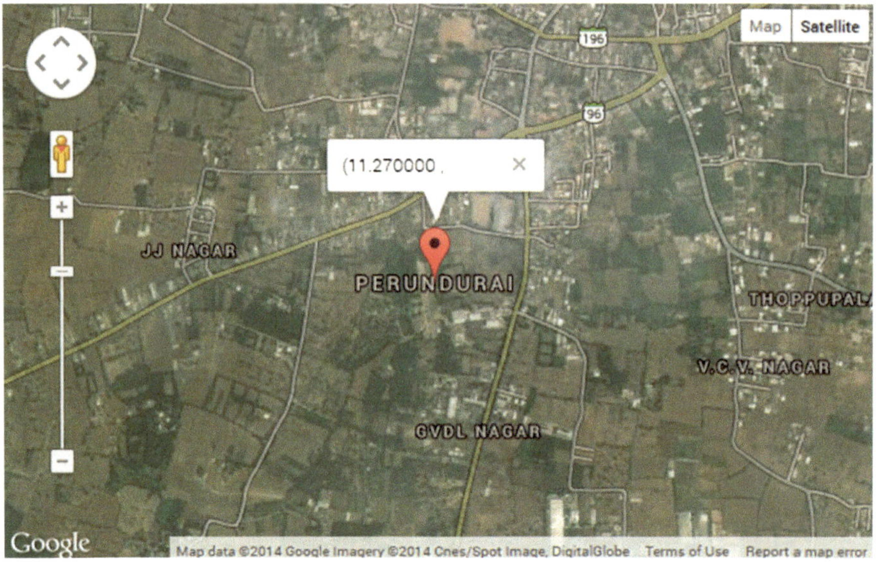

Fig. 1 Geographical location of the study area

and having a plenty of energy availability. Among the various states, some of the states like Meghalaya, Odisha, and Arunachal Pradesh are facing these problems. Even in some other states, a few of the villages are located at a remote site, where the electrical power transmission is highly complex and involves huge investments. In those cases, IRES-based systems may provide the better solution for rural electrification.

Figure 1 shows the geographical location of the study area on the map. The study area has the latitude of 11.270000 and a longitude of 77.580000 and has the DMS latitude and longitude of 11°16″12.0000″N and 77°34″48.0000″E, respectively, as shown in Fig. 2, wherein their load demand profile is shown in Table 1.

4.2 *Modelling of the Proposed System*

A suitable system has been developed using the hybrid optimization model, which consists of solar PV, wind turbine generator, diesel generator set, and battery along with the power electronic conversion systems as shown Fig. 3.

Figure 4 shows the load profile data of the selected site location for a single day, wherein there will be a maximum load of nearly 5 kW during the evening hours. The D map shows the monthly average report of the selected site from January to December and also the annual scaled report given at the last.

Fig. 2 Study area

Table 1 Load demand profile

S. No.	Name of the appliances	Power rating in W	No. of hours used	Load demand in W
1	CFL lamp	25	10	250
2	CFL lamp	40	10	400
3	Television	250	8	2000
4	Fans	50	19	950
5	Computer	200	6	1200
6	Radio	20	3	60
7	Refrigerator	700	24	16,800
8	A.C.	3.516 k	10	35,160
9	Washing machine	500	2	1000

4.3 Availability of Solar Energy Resources

The monthly average values of the solar and wind speed data related to the case study location are shown in Fig. 1. It is observed that the solar radiation of this city reaches its minimum of 4.7 kWh/m^2/day in the month of November and it is having the maximum of 6.6 kWh/m^2/day in the month of February. Also, the average of daily radiation in the whole year is about 5.46 kWh/m^2/day. Figure 5 shows the solar and wind resources with the clearness index values along with the daily solar radiation and wind speed for the selected study location.

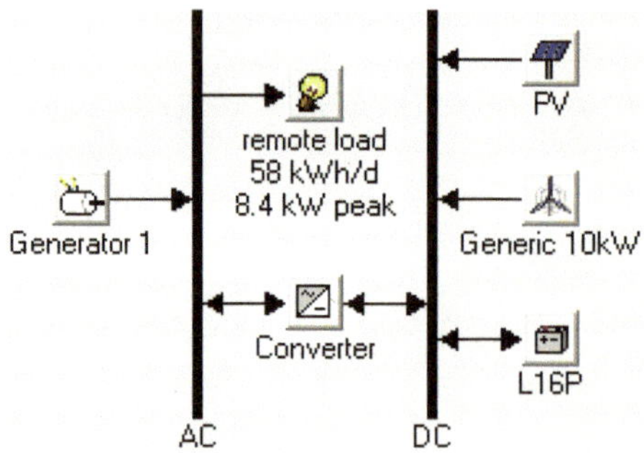

Fig. 3 Proposed system model

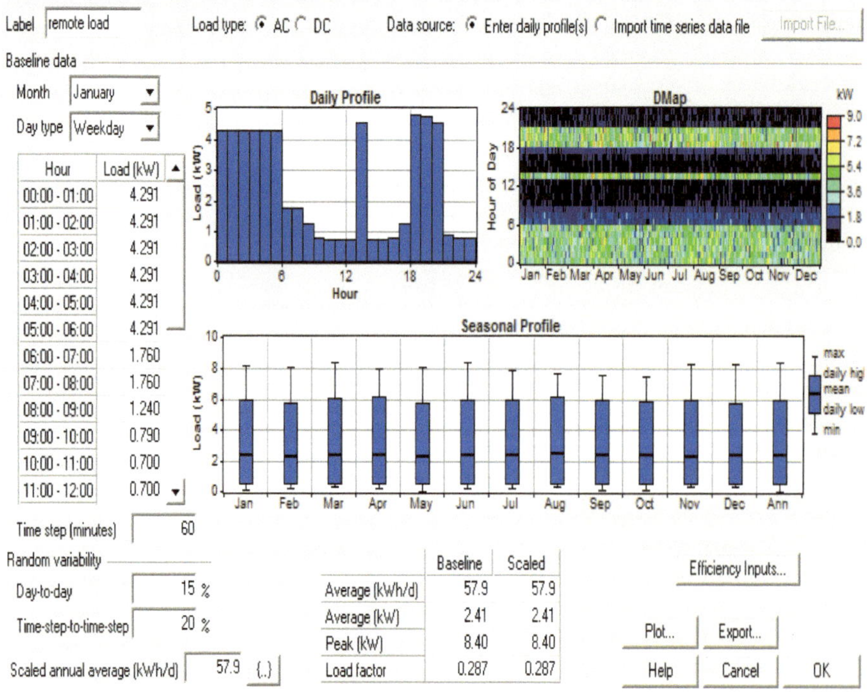

Fig. 4 Monthly load profile data

4.4 Availability of Wind Energy Resources

Although the proposed hybrid optimization model has the capability to generate the wind data, once if the user defines the four parameters which are Weibull parameter value, autocorrelation factor, diurnal pattern strength, and hour of the peak wind

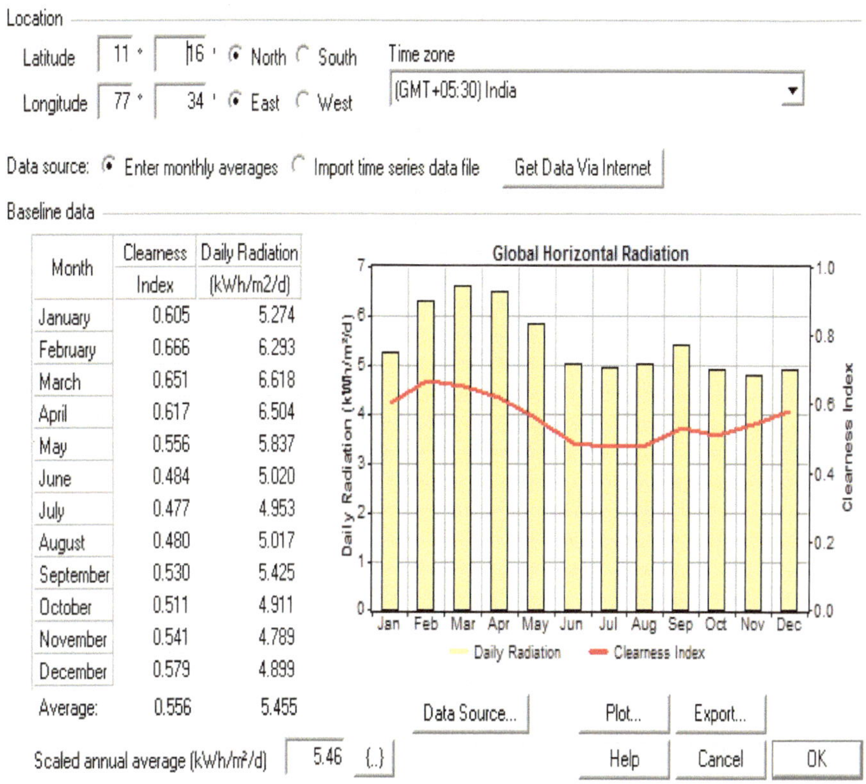

Fig. 5 Average solar radiation data over a year

speed. The average wind speed is calculated as 2.17 m/s. The availability of wind energy sources is shown in Fig. 6.

4.5 Energy Storage Devices

Lead acid batteries are included in the proposed model for storing the energy. Super capacitors and hydrogen cells may also be considered for the same purpose but, the cost of electrolysers in hydrogen cell is very high.

4.6 Power Electronic Converters

A power electronic converter is used to maintain the flow of energy between the AC and DC busses. A suitable converter design is essential in order to convert AC–DC, DC–AC, DC–DC, and AC–AC voltages.

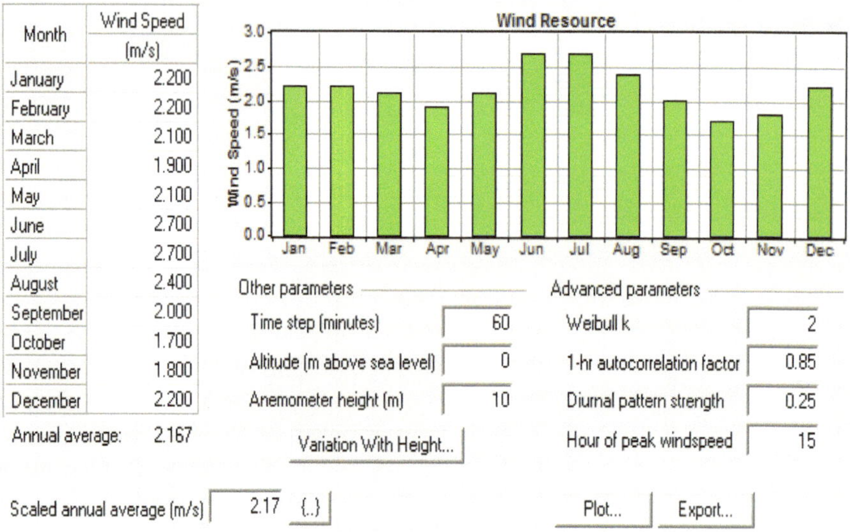

Fig. 6 Average wind speed data at a selected study location

5 Simulation Results and Discussions

For the single residential load of the selected study area, various combinations are obtained from solar PV, WTG, fuel cell, PE converter, DG, and battery by using this proposed optimization model as shown in Figs. 7 and 8. When the developed model is simulated, the possible combinations of DGs are listed out and ranked

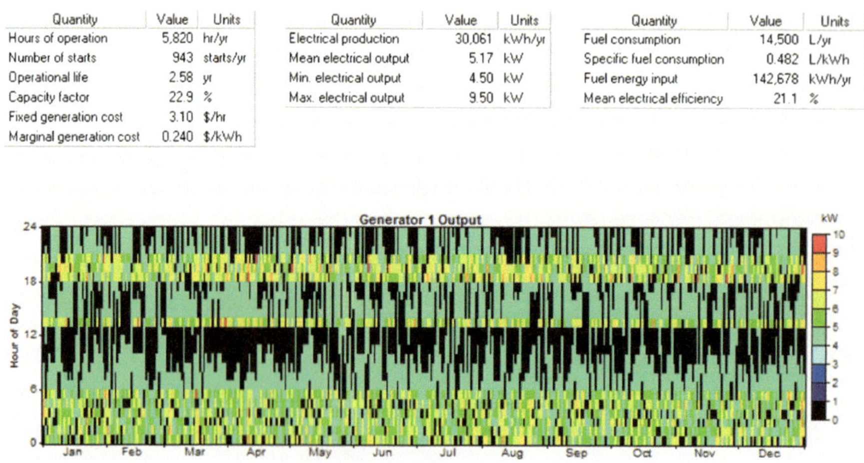

Fig. 7 Annual electricity production of diesel generator

Double click on a system below for simulation results.

PV (kW)	G10	Label (kW)	L16P	Conv. (kW)	Initial Capital	Operating Cost ($/yr)	Total NPC	COE ($/kWh)	Ren. Frac.	Diesel (L)	Label (hrs)
		15	8	6	$ 30,900	26,879	$ 374,507	1.386	0.00	14,500	5,820
	1	15	8	6	$ 60,900	27,860	$ 417,038	1.544	0.00	14,378	5,796
10		15	8	6	$ 58,900	34,654	$ 501,894	1.858	0.00	10,306	4,099
10	1	15	8	6	$ 88,900	35,580	$ 543,733	2.013	0.00	10,171	4,057
		15	8	12...	$ 12,024,900	1,545,016	$ 31,775,388	117....	0.00	14,554	5,845
	1	15	8	12...	$ 12,054,900	1,545,847	$ 31,816,016	117....	0.00	14,356	5,786
10		15	8	12...	$ 12,052,900	1,552,561	$ 31,899,838	118....	0.00	10,225	4,063
10	1	15	8	12...	$ 12,082,900	1,553,626	$ 31,943,454	118....	0.00	10,159	4,052

Fig. 8 Optimization results

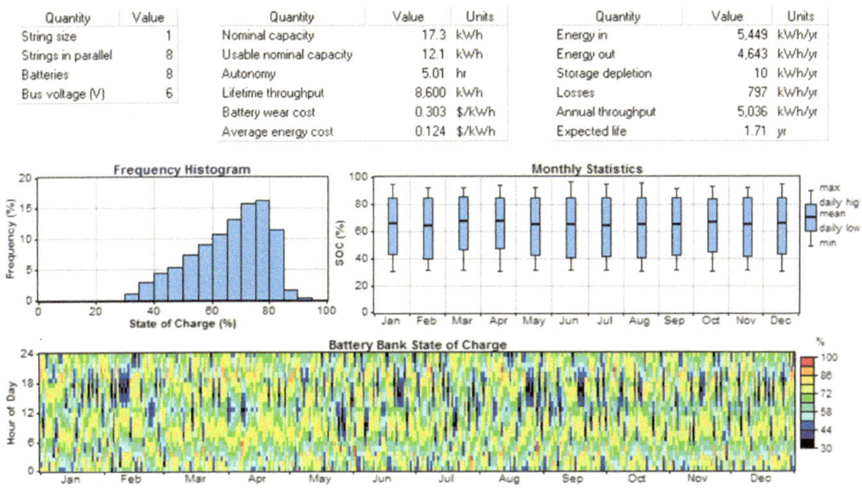

Fig. 9 State of charge of battery

according to their cost of energy (CoE) and net present cost (NPC). Sensitivity analysis can also be done by the proposed system model by eliminating the unfeasible model and ranking the feasible and economical combinations alone.

The most economic system in each category is chosen and can be used based on their ranking order. From the simulation results, it is observed that there are eight feasible combinations based on their NPC and CoE values.

From the simulation results, it is observed that the most feasible and economic solution with a minimum net present cost of about $374,507 and the cost of energy is about $1.386/kWh can be obtained by running the diesel generator.

From Fig. 9, it is found that the wind + diesel is the economically viable system next to the diesel generator of 15 kW independently, with the net present cost of $417,038 and a COE of $1.544/kWh.

From the discussion, it was inferred that for the load demand of 4300 W, the economically feasible and technically viable solution is found to supply the load either by the stand-alone diesel generator or the mix of wind + diesel generator, whose rectifier outputs and cash flow summary are shown in Figs. 10 and 11.

Quantity	Inverter	Rectifier	Units
Capacity	6.00	6.00	kW
Mean output	0.48	0.62	kW
Minimum output	0.00	0.00	kW
Maximum output	5.44	0.94	kW
Capacity factor	7.9	10.4	%

Quantity	Inverter	Rectifier	Units
Hours of operation	2,940	5,820	hrs/yr
Energy in	4,643	6,411	kWh/yr
Energy out	4,178	5,449	kWh/yr
Losses	464	962	kWh/yr

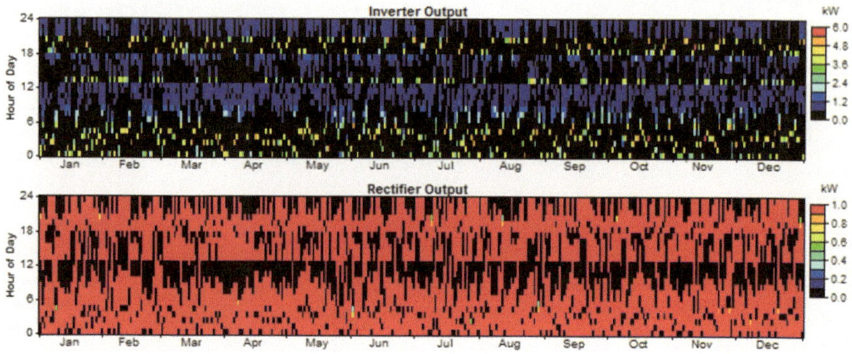

Fig. 10 Rectifier and inverter outputs

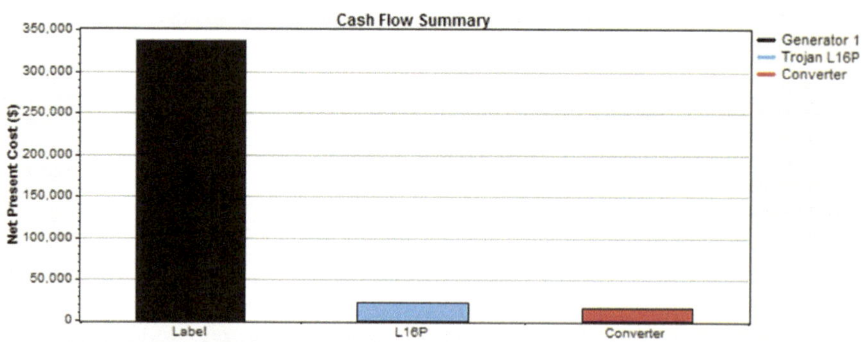

Component	Capital ($)	Replacement ($)	O&M ($)	Fuel ($)	Salvage ($)	Total ($)
Generator 1	22,500	82,332	55,799	177,941	-1,258	337,315
Trojan L16P	2,400	17,242	2,045	0	-202	21,485
Converter	6,000	2,504	7,670	0	-466	15,708
System	30,900	102,078	65,515	177,941	-1,926	374,508

Fig. 11 Cash flow summary

6 Conclusion

An optimization model consisting of solar, wind, battery, and a conventional diesel generator has been developed and simulated for the selected study area. The model developed has been found to be very useful for the Rural Electrification of Perundurai in Erode District, as it offers ‘n’ number of possible combinations for the study area. This proposed model optimizes the different renewable energy systems based on their net present cost, renewable fraction, cost of energy, fuel emissions, etc. In this work, a diesel generator has been considered in order to keep the constant output in spite of the fluctuating power outputs from solar and wind energy resources. Simulation results present the economic analysis of the integrated renewable energy systems for the selected study area. It implies that the IRES-based system has the advantages of long term availability and flexibility.

References

Akella AK, Sharma MP, Saini RP (2007) Optimum utilization of renewable energy sources in a remote area. J Renew Sustain Energy Rev 11:894–908

Kanase Patil AB, Saini RP, Sharma MP (2011) Sizing of integrated renewable energy system based on load profiles and reliability index for the state of Uttarakhand in India. J Renew Energy 36:2809–2821

Ramakumar R (1977) Technical and socio-economic aspects of solar energy and rural development in developing countries. J Solar Energy 19:643–649

Ramakumar R, Hughes WL (1981) Renewable energy sources and rural development in developing countries. IEEE Trans Educ 24:242–251

Sinha S, Chandel SS (2014) Review of software tools for hybrid renewable energy systems. J Renew Sustain Energy Rev 32:192–205

Author Index

A
Abhinav Trivedi, 15
Abhishek Sharma, 169
Aditi Patel, 319
Amit Arora, 333
Anju Arora, 23
Anuchaya Devi, 131

B
Balamurugan, P., 121
Benni, Sangeeta D., 111

C
Cameotra, Swaranjit Singh, 333
Chandrajit Balomajumder, 333

D
Dhanapati Deka, 131

G
Ghangrekar, M.M., 285, 295, 305
Gollakota, Anjani R.K., 197

H
Harmanjot Kaur, 43
Himadri Sahu, 97
Himansh Kumar, 179

J
James, Jisha P., 3

K
Karthickeyan, V., 121
Katagi, Kariyappa S., 111
Kaustubha Mohanty, 97
Khandetod, Y.P., 63
Kulkarni, Sneha S., 111
Kumar, Sachin, 43, 81

L
Lata, 23
Lipi Parekh, 319

M
Madhuri Narra, 3
Meenakumari Ramachandran, 345
Mitra, A., 285
Mohd. Zeeshan, 149
Mohit Vasudeva, 149
Mohod, A.G., 63
Mukherjee, C.K., 285
Munnolli, Ravindra S., 111
Murugan, S., 169

N
Nanda Kishore, 197
Nasreen Munshi, 319
Noori, Md.T., 285

P
Parihar, Amit Kumar Singh, 213
Prasenjit Mondal, 257
Purvi Zaveri, 319
Pushpendra Kumar, 333

R
Rai, Monika Prakash, 159
Rajendra Prasad, 245
Rajesh, P.P., 295
Rajnish Kumar, 333
Raman Rao, 43
Ram Chandra, 15
Rangan Banerjee, 213
Ratnesh Tiwari, 245
Renuka Barman, 131
Richa Arora, 81
Rupali Mahajan, 43

S. Kumar et al. (eds.), *Proceedings of the First International Conference on Recent Advances in Bioenergy Research*, Springer Proceedings in Energy,
DOI 10.1007/978-81-322-2773-1

S
Sai Gu, 197
Samanta, Sujoy Kumar, 227
Sameer Kureshi, 319
Sarma, A.K., 149, 179
Senthil, R., 121
Shalley Sharma, 23
Sharma, Nilesh Kumar, 81
Shivani Gupta, 159
Shrirame, H.Y., 63
Shuvashish Behera, 81
Singh, Anil Kumar, 333
Sonia Sharma, 23
Subramanyam, Malladi D., 197
Sukumar Laik, 333
Surender Singh, 23
Suresh Muthusamy, 345

T
Tanvi Modi, 319
Tiwari, B.R., 305

V
Varma, Anil Kumar, 257
Velmurugan Balasubramanian, 3
Verma, Amit Ranjan, 245
Verma, Priyanshu, 227
Vijay, Virendra Kumar, 15, 245
Virendra Sethi, 213

Y
Yadav, Y.K., 179, 267
Yadav, Yogender Singh, 267